Plant Invasions in Forest Ecosystems: Understanding Arrival, Expansion and Management

Plant Invasions in Forest Ecosystems: Understanding Arrival, Expansion and Management

Guest Editors

Jun Liu

Zacchaeus Compson

MDPI

Basel • Beijing • Wuhan • Barcelona • Belgrade • Novi Sad • Cluj • Manchester

Guest Editors
Jun Liu
College of Forestry
Jiangxi Agricultural University
Nanchang
China

Zacchaeus Compson
Department of Biological Sciences
University of North Texas
Denton, TX
USA

Editorial Office
MDPI AG
Grosspeteranlage 5
4052 Basel, Switzerland

This is a reprint of the Special Issue, published open access by the journal *Forests* (ISSN 1999-4907), freely accessible at: https://www.mdpi.com/journal/forests/special_issues/LFB11623YA.

For citation purposes, cite each article independently as indicated on the article page online and as indicated below:

Lastname, A.A.; Lastname, B.B. Article Title. *Journal Name* **Year**, *Volume Number*, Page Range.

ISBN 978-3-7258-7110-0 (Hbk)
ISBN 978-3-7258-7111-7 (PDF)
https://doi.org/10.3390/books978-3-7258-7111-7

Contents

forests

MDPI

Editorial

Invasion Fronts and Forest Futures: Toward Rapid-Response Actionable Ecology

Zacchaeus G. Compson [1,*] and Jun Liu [2]

[1] Department of Biological Sciences, Advanced Environmental Research Institute, University of North Texas, 1155 Union Circle #310559, Denton, TX 76203-5017, USA
[2] College of Forestry, Jiangxi Agricultural University, Nanchang 330045, China; liujun@jxau.edu.cn
* Correspondence: zacchaeus.compson@unt.edu

1. Introduction

Plant invasions are reshaping forests and forest-associated ecosystems across the globe, with cascading consequences for biodiversity, biogeochemical cycling, disturbance regimes, and critical ecosystem services. Recent syntheses emphasize both the scale of plant invasion impacts and the difficulty of prevention, as novel species continue to appear outside their native ranges [1–3]. Projections suggest that the pool of established alien species will continue to grow through mid-century, reinforcing the need for biosurveillance, prediction, and management tools that keep pace with accelerating change [4]. This Special Issue, "Plant Invasions in Forest Ecosystems: Understanding Arrival, Expansion and Management," brings together ten papers spanning forest understories, riparian corridors, mangroves, and bamboo–broadleaf ecotones, and that leverage approaches ranging from field surveys to remote sensing and spatially explicit modeling. Collectively, these papers highlight three key questions for plant invasion science. (1) Where is invasion risk highest, and why? (2) How do invaders spread across natural landscapes? (3) What post-invasion changes are most critical for ecosystem function and management?

2. Key Contributions from This Special Issue

2.1. Characterizing Invasion Risk: Susceptibility, Resources, and Context Dependence

Invasion risk is rarely uniform across a landscape; it is affected by resource availability, canopy structure, propagule pressure, and the legacy of disturbance. In forested riparian zones surrounding stream restoration sites, DeBerry and Hunter [5] identified species-specific drivers of invasion across gradients, while also finding a consistent set of predictors that repeatedly tip the scale to facilitate rapid invasion, including canopy cover, nitrogen availability, soil texture, and bioavailable phosphorus. The value of this work is not just explanatory: the authors explicitly connected these drivers to best practices that can be implemented early in the invasion process, shifting stream restoration away from reactive management toward proactive invasion resistance. At broader spatial scales, Park and Cheon [6] leveraged South Korea's National Ecosystem Survey to compare exotic and endemic plants, showing that exotic plants exhibit broader niche breadth via generalist strategies, while endemics occupy narrower, specialized niches—an asymmetry that creates conservation risk when habitats are altered or homogenized. Their analyses also suggest overlap between exotic and endemic plants in some water-related dimensions but separation along topographic and climatic variables, reinforcing a key point for managers, namely that invasion risk is multidimensional, and suitable habitat can differ depending on which niche axes are considered. In European slope forests, Lipińska et al. [7] showed that the invasive species, *Impatiens parviflora*, is associated with lower tree cover, fewer old-growth forest species, and understory conditions consistent with successional change and disturbance. Their work also suggested that shade-tolerant trees (e.g., *Abies alba*, *Fagus sylvatica*) act as constraints on spread. Importantly, this study demonstrated that invasion is not just about the invader; it is also about the state of the habitat and the degree to which the forest retains attributes of late-successional structure. Collectively, these studies underscore the heterogeneity of invasion dynamics and their context-dependent controls.

2.2. Assessing Invader Spread: Spatial Dynamics, Thresholds, and Monitoring at Scale

The spread of invasives is inherently spatial, and models must operate at scales that capture relevant landscape spatial variation. Models that ignore space (e.g., by focusing on a small geographic area) can miss critical invasion dynamics that matter most for early establishment. For example, Lu et al. [8] used spatially explicit, individual-based modeling to show that the invasion success of *Melaleuca quinquenervia* depends strongly on the initial spatial distribution of the species, as density, area covered, and the spatial configuration (e.g., clustering) of individuals can shift trajectories toward success or failure. This study underscores how early invasions are often governed by thresholds and clustering: establishment may require a sufficiently dense local founding population that can outcompete native vegetation. Several other studies demonstrate the power of remote sensing and landscape analysis for quantifying spread and targeting surveillance. In China, Li et al. [9] mapped the expansion of moso bamboo (*Phyllostachys edulis*) into multiple forest types from 2010 to 2020 using a Landsat time series, documenting substantial increases in bamboo area and emphasizing how expansion intersects with forest conversion and fragmentation processes. In coastal wetland forests, Dong et al. [10] assessed mangrove ecosystems invaded by *Spartina alterniflora* using a multi-year, remote-sensing approach to characterize invasion patterns and associated ecosystem changes. They demonstrated that invasions at the land–sea interface are now measurable at the relevant scales at which they unfold, which is essential for aligning management jurisdictions with ecological reality. At a finer spatial scale, Zeng et al. [11] analyzed the patchy expansion process at bamboo–broadleaf interfaces, describing a spread-and-fill dynamic that has direct implications for edge management and monitoring design. Collectively, these papers argue for monitoring strategies that treat invasion fronts as dynamic boundaries shaped by spatial heterogeneity, rather than static maps, and that assess invasions at spatial scales that adequately capture these dynamic processes.

2.3. Post-Invasion Changes: Community Reassembly and Potential Ecosystem Feedbacks

Invasion consequences can fundamentally alter biodiversity-ecosystem function relationships, creating self-reinforcing feedbacks when invaders modify ecosystem properties to favor their own persistence. In central Nepal, Paudel et al. [12] showed that weed invasion (*Lantana camara*) is associated with strong declines in understory diversity and shifts in soil chemistry that provide critical benefits to the invading plants, including higher total nitrogen, soil organic carbon, and available phosphorus in invaded plots. This work illustrates that invaders can restructure both biotic communities and the abiotic context of the environments they invade. In another example, Song et al. [13] explored biogeochemical feedbacks, finding that bamboo expansion is linked to decreased phosphorus cycling rates and increased phosphorus-use efficiency—factors that can favor further invader expansion—indicating that invasion can be coupled to stoichiometric nutrient shifts, with implications for productivity and competition dynamics. Finally, in the early stages of invasion, when ornamentals transition into wild populations, Nowińska and Czarna [14] documented naturalization patterns for *Crocus tommasinianus*, illuminating

https://doi.org/10.3390/f17020245

how horticultural pathways can seed new invasions long before they are recognized as management problems. Collectively, these studies provide strong evidence for the dynamic ecological changes that occur during the invasion process and suggest that these biogeochemical changes may lead to strong ecological feedbacks favoring invasions. The direct, indirect, and interrelated ways invaders interact with the changing soil environment in these systems to create feedbacks is precisely the sort of causal question that should motivate follow-up experiments and modeling.

3. Future Directions: Toward Rapid, Actionable Predictions and Causal Understanding

The papers in this Special Issue highlight priorities for the next phase of invasion science in forests, a science that is more mechanistic, more inclusive of uncertainty related to spatial heterogeneity, scale, and feedbacks, and more connected to vital management, providing rapid, clear information that managers can use to make actionable decisions. We see five near-term priorities to advance this field.

3.1. Mechanistic Inference and Causal Modeling Should Become Routine

Many invasion datasets are multivariate, spatially structured, and observational by necessity, but that does not preclude mechanistic inference. Piecewise structural equation modeling (SEM) provides a pragmatic, hypothesis-testing framework that can accommodate mixed models, nested designs, and non-independence, while remaining transparent about causal assumptions [15,16]. Complementing this, Bayesian hierarchical models can unify uncertainty across observation, process, and parameter levels, and are particularly well-suited for integrating disparate datasets (e.g., experimental plots, remote sensing, and environmental DNA detections) into a single inferential model [17,18]. For biosurveillance data specifically, occupancy detection frameworks provide a direct way to separate true presence from imperfect detection—critical for informing early invasion management strategies and determining decision thresholds [19].

3.2. Move from Correlative Risk Maps to Adaptive, Hybrid Forecasting That Couples Niche, Demography, Dispersal, and Human Vectors with Remote Sensing Data

The studies in this series reinforce the idea that invasive species establishment and spread are spatial processes with thresholds, lags, and contingency. Hybrid approaches—linking niche suitability to demographic performance, dispersal factors, landscape connectivity, and human-mediated pathways—should improve transferability and out-of-sample forecast performance [4,8]. Where possible, these models should be evaluated as forecasts (not just fits), with explicit error propagation and management-relevant performance metrics. Furthermore, multi-year imagery can be used to quantify invasion trajectories, detect frontier dynamics, and evaluate intervention effectiveness, especially when linked to permanent plots, field experiments, and process measurements [9,10,13,20]. The next step is operational: building repeatable workflows that translate imagery into (i) early-warning triggers, (ii) prioritized field validation, and (iii) post-treatment evaluation, thereby closing the loop between invasive species detection and effective management action.

3.3. Expand Biosurveillance Using Environmental DNA and Build Interoperable Networks with Reference Infrastructure

Modern genomics tools, like quantitative PCR, digital PCR, and DNA metabarcoding, can detect incipient invasions and cryptic taxa, but operational success depends on clear objectives, controls, and action thresholds that are aligned with management response capacity [21–23]. This Special Issue highlights the importance of scaling biosurveillance from projects to networks, including internationally interoperable efforts, such as the U.S.

https://doi.org/10.3390/f17020245

Fish and Wildlife Service's national environmental DNA (eDNA) strategy [24], the UK Environment Agency's DNA-based river monitoring work [25], Fisheries and Oceans Canada's eDNA initiatives [26], biosecurity-focused tools from the Cawthron Institute in Aotearoa, New Zealand [27], and EU programs, such as GuardIAS [28]. As adoption grows, curated reference libraries and automated screening tools must be treated as core infrastructure because biosurveillance is only as reliable as the taxonomic databases and reporting pipelines that support it [29].

3.4. Use Targeted Manipulative Experiments to Test Feedback Loops and Identify Management Leverage Points

Several papers in this issue implicate biogeochemical and community feedbacks (e.g., nutrient economy shifts, understory reassembly) that can facilitate plant invasions. Well-designed, factorial field experiments—combining invader removal or addition with resource manipulations, disturbance treatments, and restoration interventions—can isolate causal pathways and quantify whether feedbacks are reversible or self-reinforcing [30,31]. Critically, embedding these experiments within an adaptive management framework (i.e., with explicit hypotheses, pre-defined decision thresholds, and iterative learning) can make experiments directly relevant to operational constraints while improving generality across sites and regions [32].

3.5. Integrate Invasion Genomics and Community Genetic Structure into Prediction and Mitigation Efforts

A critical aspect not directly addressed in this Special Issue is the treatment of plant invasions as eco-evolutionary processes, where propagule pressure and dispersal interact with standing genetic variation, admixture, and rapid adaptation [33–35]. For invaders, population genomic assignment can identify sources, pathways, and bridgehead populations, as well as equip managers with genetic lineages that differ in spread potential and trait syndromes [34]. Equally important, the genetic structure of the recipient community can shape invasion resistance and post-invasion trajectories; for example, native genotypic diversity can suppress invader performance via plant–soil feedbacks [36], suggesting that mitigation and restoration should prioritize genetic provenance and diversity, not only species identity [37]. Finally, eDNA is increasingly capable of recovering within-species genetic variation and population structure, enabling genomic biosurveillance that tracks not just presence, but also lineage turnover through time [38,39]. As environmental genomic tools become more cost-effective and ubiquitous, integrating genetic information into invasive species research and management programs will enable these efforts to become more predictive, prescriptive, and effective at mitigating economic and environmental damage.

4. Conclusions

Plant invasions will remain a defining challenge for forest science and management in the coming decades [1,2]. The papers in this series show meaningful progress on three fronts: identifying context-specific drivers, quantifying spread with models and remote sensing, and linking invasions to ecosystem consequences. What is most exciting, however, is how quickly the field is converging on integrated approaches: hybrid forecasting that couples niche, demography, and dispersal; monitoring pipelines that connect satellites to plots; and biosurveillance networks that use eDNA to shift detection earlier in the invasion curve and inform the genetic contexts of invasions. This Special Issue not only accelerates that convergence but also charts a course for management that is both scientifically grounded and operationally actionable. The path forward requires sustained collaboration among researchers, managers, and technologists to iteratively close the loop of prediction, detection, and intervention.

https://doi.org/10.3390/f17020245

Author Contributions: Conceptualization, Z.G.C. and J.L.; writing—original draft preparation, Z.G.C.; writing—review and editing, Z.G.C. and J.L. All authors have read and agreed to the published version of the manuscript.

Acknowledgments: We thank all contributors to this Special Issue and the many peer reviewers whose feedback strengthened the manuscripts and greatly improved this series.

Conflicts of Interest: The authors declare no conflicts of interest.

References

1. Roy, H.E.; Pauchard, A.; Stoett, P.; Renard Truong, T. *IPBES Invasive Alien Species Assessment: Full Report*; Zenodo: Geneva, Switzerland, 2024.
2. Pyšek, P.; Hulme, P.E.; Simberloff, D.; Bacher, S.; Blackburn, T.M.; Carlton, J.T.; Dawson, W.; Essl, F.; Foxcroft, L.C.; Genovesi, P.; et al. Scientists' Warning on Invasive Alien Species. *Biol. Rev.* **2020**, *95*, 1511–1534. [CrossRef]
3. Seebens, H.; Blackburn, T.M.; Dyer, E.E.; Genovesi, P.; Hulme, P.E.; Jeschke, J.M.; Pagad, S.; Pyšek, P.; van Kleunen, M.; Winter, M.; et al. Global Rise in Emerging Alien Species Results from Increased Accessibility of New Source Pools. *Proc. Natl. Acad. Sci. USA* **2018**, *115*, E2264–E2273. [CrossRef] [PubMed]
4. Seebens, H.; Bacher, S.; Blackburn, T.M.; Capinha, C.; Dawson, W.; Dullinger, S.; Genovesi, P.; Hulme, P.E.; van Kleunen, M.; Kühn, I.; et al. Projecting the Continental Accumulation of Alien Species through to 2050. *Glob. Change Biol.* **2021**, *27*, 970–982. [CrossRef]
5. DeBerry, D.A.; Hunter, D.M. Drivers of Plant Invasion in Stream Restoration. *Forests* **2024**, *15*, 964. [CrossRef]
6. Park, B.-J.; Cheon, K. Species Composition and Ecological Niche Overlap of Alien and Endemic Plants in South Korea: Insights from the National Ecosystem Survey. *Forests* **2025**, *16*, 1485. [CrossRef]
7. Lipińska, K.; Cieśla, A.; Hrynyk, O.; Sokołowski, K.; Gawryś, R. Habitat Condition of Tilio–Acerion Forest Facilitates Successful Invasion of *Impatiens parviflora* DC. *Forests* **2025**, *16*, 1475. [CrossRef]
8. Lu, Y.; Xia, J.; Holt, R.D.; DeAngelis, D.L. Modeling the Effects of Spatial Distribution on Dynamics of an Invading *Melaleuca quinquenervia* (Cav.) Blake Population. *Forests* **2024**, *15*, 1308. [CrossRef]
9. Li, D.; Wei, J.; Wu, J.; Zhong, Y.; Chen, Z.; He, J.; Zhang, S.; Yu, L. The Expansion of Moso Bamboo (*Phyllostachys edulis*) Forests into Diverse Types of Forests in China from 2010 to 2020. *Forests* **2024**, *15*, 1418. [CrossRef]
10. Dong, D.; Gao, Q.; Huang, H. Mangroves Invaded by Spartina Alterniflora Loisel: A Remote Sensing-Based Comparison for Two Protected Areas in China. *Forests* **2024**, *15*, 1788. [CrossRef]
11. Zeng, X.; Luo, H.; Lu, J.; Zhu, X.; He, Y.; Gong, C.; Ren, Z.; Huang, D.; Song, Q.; Yang, Q. The Process of Patchy Expansion for Bamboo (*Phyllostachys edulis*) at the Bamboo–Broadleaf Forest Interface: Spreading and Filling in Order. *Forests* **2024**, *15*, 438. [CrossRef]
12. Paudel, C.K.; Tiwari, A.; Baniya, C.B.; Shrestha, B.B.; Jha, P.K. High Impacts of Invasive Weed Lantana Camara on Plant Community and Soil Physico-Chemical Properties across Habitat Types in Central Nepal. *Forests* **2024**, *15*, 1427. [CrossRef]
13. Song, S.; Wang, L.; Compson, Z.G.; Xie, T.; Liao, C.; Huang, D.; Liu, J.; Yang, Q.; Song, Q. Decreased P Cycling Rate and Increased P-Use Efficiency after *Phyllostachys edulis* (Carrière) J. Houz. Expansion into Adjacent Secondary Evergreen Broadleaved Forest. *Forests* **2024**, *15*, 1518. [CrossRef]
14. Nowińska, R.; Czarna, A. Naturalization of the Ornamental Plant *Crocus tommasinianus* Herb. (Iridaceae) in Forest Ecosystems: A Case Study from Poland. *Forests* **2024**, *15*, 1851. [CrossRef]
15. Shipley, B. Confirmatory Path Analysis in a Generalized Multilevel Context. *Ecology* **2009**, *90*, 363–368. [CrossRef]
16. Lefcheck, J.S. piecewiseSEM: Piecewise Structural Equation Modelling in R for Ecology, Evolution, and Systematics. *Methods Ecol. Evol.* **2016**, *7*, 573–579. [CrossRef]
17. Gelman, A.; Carlin, J.B.; Stern, H.S.; Rubin, D.B. (Eds.) *Bayesian Data Analysis*, 3rd ed.; Chapman & Hall: Boca Raton, FL, USA, 2013; ISBN 978-1-4398-4095-5.
18. Kery, M.; Royle, J.A. *Applied Hierarchical Modeling in Ecology: Analysis of Distribution, Abundance and Species Richness in R and BUGS: Volume 1: Prelude and Static Models*; Academic Press: New York, NY, USA, 2016; ISBN 978-0-12-801486-8.
19. MacKenzie, D.I.; Nichols, J.D.; Lachman, G.B.; Droege, S.; Andrew Royle, J.; Langtimm, C.A. Estimating Site Occupancy Rates When Detection Probabilities Are Less Than One. *Ecology* **2002**, *83*, 2248–2255. [CrossRef]
20. Fassnacht, F.E.; White, J.C.; Wulder, M.A.; Næsset, E. Remote Sensing in Forestry: Current Challenges, Considerations and Directions. *For. Int. J. For. Res.* **2024**, *97*, 11–37. [CrossRef]
21. Deiner, K.; Bik, H.M.; Mächler, E.; Seymour, M.; Lacoursière-Roussel, A.; Altermatt, F.; Creer, S.; Bista, I.; Lodge, D.M.; de Vere, N.; et al. Environmental DNA Metabarcoding: Transforming How We Survey Animal and Plant Communities. *Mol. Ecol.* **2017**, *26*, 5872–5895. [CrossRef]

 https://doi.org/10.3390/f17020245

22. Morisette, J.; Burgiel, S.; Brantley, K.; Daniel, W.M.; Darling, J.; Davis, J.; Franklin, T.; Gaddis, K.; Hunter, M.; Lance, R.; et al. Strategic Considerations for Invasive Species Managers in the Utilization of Environmental DNA (eDNA): Steps for Incorporating This Powerful Surveillance Tool. *Manag. Biol. Invasions Int. J. Appl. Res. Biol. Invasions* **2021**, *12*, 747–775. [CrossRef]
23. Burgiel, S.; Stepien, C.A. Environmental DNA as a Tool for Invasive Species Detection and Management. Available online: https://repository.library.noaa.gov (accessed on 25 January 2026).
24. Richards, H. Skin, Scales, and Scat. U.S. Fish & Wildlife Service. Available online: https://www.fws.gov/story/2024-07/skin-scales-and-scat-national-edna-strategy (accessed on 24 January 2026).
25. Exploring the Potential of DNA Techniques to Monitor Macroinvertebrate Communities and Invasive Species in Rivers. Available online: https://www.gov.uk/government/publications/exploring-the-potential-of-dna-techniques-to-monitor-macroinvertebrate-communities-and-invasive-species-in-rivers (accessed on 24 January 2026).
26. Government of Canada; Fisheries and Oceans Canada (DFO). Environmental DNA (eDNA). Available online: https://www.dfo-mpo.gc.ca/science/biotech-genom/dna-adn/index-eng.html (accessed on 24 January 2026).
27. eDNA Tools for Conservation in Aotearoa New Zealand. Available online: https://edna.cawthron.org.nz/tools/ (accessed on 24 January 2026).
28. Katsanevakis, S.; Zaiko, A.; Olenin, S.; Costello, M.J.; Gallardo, B.; Tricarico, E.; Adriaens, T.; Jeschke, J.; Sini, M.; Burke, N.; et al. GuardIAS—Guarding European Waters from Invasive Alien Species. *Manag. Biol. Invasions* **2024**, *15*, 701–730. [CrossRef]
29. Zaiko, A.; Scheel, M.; Schattschneider, J.; von Ammon, U.; Scriver, M.; Pochon, X.; Pearman, J.K. Pest Alert Tool—A Web-Based Application for Flagging Species of Concern in Metabarcoding Datasets. *Nucleic Acids Res.* **2023**, *51*, W438–W442. [CrossRef]
30. Ehrenfeld, J.G. Ecosystem Consequences of Biological Invasions. *Annu. Rev. Ecol. Evol. Syst.* **2010**, *41*, 59–80. [CrossRef]
31. Kulmatiski, A.; Beard, K.H.; Stevens, J.R.; Cobbold, S.M. Plant–Soil Feedbacks: A Meta-Analytical Review. *Ecol. Lett.* **2008**, *11*, 980–992. [CrossRef]
32. Walters, C.J. *Adaptive Management of Renewable Resources*; The Blackburn Press: Caldwell, NJ, USA, 2002; ISBN 978-0-02-947970-4.
33. Sherpa, S.; Després, L. The Evolutionary Dynamics of Biological Invasions: A Multi-Approach Perspective. *Evol. Appl.* **2021**, *14*, 1463–1484. [CrossRef]
34. McGaughran, A.; Dhami, M.K.; Parvizi, E.; Vaughan, A.L.; Gleeson, D.M.; Hodgins, K.A.; Rollins, L.A.; Tepolt, C.K.; Turner, K.G.; Atsawawaranunt, K.; et al. Genomic Tools in Biological Invasions: Current State and Future Frontiers. *Genome Biol. Evol.* **2024**, *16*, evad230. [CrossRef]
35. Mounger, J.; Ainouche, M.L.; Bossdorf, O.; Cavé-Radet, A.; Li, B.; Parepa, M.; Salmon, A.; Yang, J.; Richards, C.L. Epigenetics and the Success of Invasive Plants. *Philos. Trans. R. Soc. B Biol. Sci.* **2021**, *376*, 20200117. [CrossRef] [PubMed]
36. Cheng, C.; Liu, Z.; Zhang, Q.; Tian, X.; Ju, R.; Li, B.; van Kleunen, M.; Chase, J.M.; Wu, J. Genotype Diversity Enhances Invasion Resistance of Native Plants via Soil Biotic Feedbacks. *Ecol. Lett.* **2024**, *27*, e14384. [CrossRef] [PubMed]
37. Heuertz, M.; Carvalho, S.B.; Galindo, J.; Rinkevich, B.; Robakowski, P.; Aavik, T.; Altinok, I.; Barth, J.M.I.; Cotrim, H.; Goessen, R.; et al. The Application Gap: Genomics for Biodiversity and Ecosystem Service Management. *Biol. Conserv.* **2023**, *278*, 109883. [CrossRef]
38. Johnson, M.D.; Freeland, J.R.; Parducci, L.; Evans, D.M.; Meyer, R.S.; Molano-Flores, B.; Davis, M.A. Environmental DNA as an Emerging Tool in Botanical Research. *Am. J. Bot.* **2023**, *110*, e16120. [CrossRef] [PubMed]
39. Andres, K.J.; Lodge, D.M.; Andrés, J. Environmental DNA Reveals the Genetic Diversity and Population Structure of an Invasive Species in the Laurentian Great Lakes. *Proc. Natl. Acad. Sci. USA* **2023**, *120*, e2307345120. [CrossRef]

 https://doi.org/10.3390/f17020245

forests

MDPI

Article

Species Composition and Ecological Niche Overlap of Alien and Endemic Plants in South Korea: Insights from the National Ecosystem Survey

Byeong-Joo Park [1] and Kwangil Cheon [2,*]

[1] Baekdudaegan National Arboretum, Bonghwa 36209, Republic of Korea; bzpark@koagi.or.kr
[2] Ecosystem Service Team, National Institute of Ecology, Seocheon 33657, Republic of Korea
* Correspondence: ndz1000@nie.re.kr; Tel.: +82-41-950-5463

Abstract

Biodiversity conservation in South Korea faces increasing challenges from alien plant invasions. These invasions threaten endemic species uniquely adapted to specialized habitats, making it crucial to understand their ecological interactions. This study quantitatively compared the species composition, ecological niches, and species turnover patterns of alien and endemic plants in South Korea using data from the National Ecosystem Survey. Non-metric multidimensional scaling (NMDS) and multi-response permutation procedure (MRPP) analyses revealed significant compositional heterogeneity between groups. Kernel density estimation (KDE) revealed niche overlap in water-related factors (precipitation, water yield), but clear separation in topographic and climatic variables (altitude, slope, temperature). Alien plants exhibited broader niche breadths, confirming their ecological generalist traits, whereas endemic species displayed narrower niches confined to specialized habitats. Zeta diversity analysis indicated slower species turnover in alien species, suggesting niche assimilation and habitat homogenization. Both groups fit a power-law model, emphasizing deterministic environmental filtering. These findings highlight the ecological risks posed by alien species to stability of endemic plant communities and underscore the importance of targeted, science-based management strategies.

Keywords: zeta diversity; niche breadth overlap; alien plant; plant invasion

1. Introduction

The recently adopted Kunming-Montreal Global Biodiversity Framework identifies the management of invasive alien species and biodiversity conservation (Targets 6 and 23) as key challenges. It emphasizes the importance of establishing a management system for alien species, controlling their spread, and analyzing their impacts on ecosystems [1].

The vascular flora of South Korea comprises 3975 taxa (3307 species and 668 infraspecific taxa, including subspecies, varieties, forms, and hybrids) classified into 951 genera and 183 families. This includes 31 lycophytes, 333 ferns and allies, 30 gymnosperms, and 3581 angiosperms [2]. Of these, 3878 taxa are confirmed to occur in Korea, while 97 remain uncertain, and three are considered extinct or possibly extinct. The alien vascular flora comprises 619 taxa 6 (96 families, 353 genera, 595 species, 6 subspecies, 11 varieties, 1 forma, and 6 hybrids) [3]. A revised checklist identifies 373 endemic taxa (64 families, 179 genera, 304 species, 6 subspecies, 49 varieties, and 14 nothospecies) [4]. The flora of the Korean Peninsula shares many biogeographic similarities with those of neighboring regions such as northeastern China, Japan, and the Russian Far East. For example, a recent study of Northeast Asia's vascular plants found that Korea ranks among the top in species richness and shares many dominant genera and families with other temperate East Asian regions [5].

The colonization and spread of alien plant species have consistently been regarded as primary threats to ecosystem function and biodiversity stability, and the international community recognizes them as critical drivers of biodiversity loss. The spread of alien plants threatens the structure and function of native plant communities by causing resource competition, ecological niche overlap, and ecosystem disruption, ultimately leading to biodiversity instability and reduced ecosystem services [6,7].

The invasion of alien plants adds species to specific habitats and alters the ecological niches and resource utilization patterns of native and endemic plants within existing ecosystems. This process intensifies competition within native plant communities, especially among endemic species adapted to specialized habitats. In this context, comparative studies of the ecological interactions between alien and endemic plants are essential for understanding their status, temporal dynamics, and implications for ecosystem structure and health [8].

Several domestic and international studies have examined the relationships between alien plants and species in specialized habitats, including resource competition, spatial arrangements between alien and endemic plants, and predictions of invasiveness and resource overlap based on niche analyses [9–11]. Furthermore, previous studies have examined the physiological characteristics of alien plants, focusing on their environmental adaptation, such as assessing their range within vegetation types based on soil moisture conditions [12–15].

Research on alien plants in South Korea has mainly focused on their relationships with environmental factors, invasion risks, and the spatial distribution of hotspot areas using GIS approaches [16–22]. However, studies comparing alien plants with other plant communities, such as endemic plants or native plants, and analyzing their ecological niches, are lacking. One study suggested the potential for examining success in becoming established depending upon various environmental factors between native and alien plants within ecological restoration sites [23]; however, this was limited to a specific habitat category (e.g., disturbed areas). In this context, it is necessary to secure basic data to establish and develop management strategies through supplementary analyses of the ecological niches of alien plants at the national level. In summary, clarifying the management and geographical distribution of alien plants in South Korea is crucial. Additionally, the interpretation that alien plants cause biodiversity loss presupposes their impact on the survival of native plants within their habitats. However, quantitative comparative studies on species interactions remain scarce. Therefore, while studies on environmental factors within single populations of alien plants are important for their management, quantitative analyses of their correlations with other species, ecological niche overlap, and environmental suitability are also necessary [24,25].

In particular, endemic plants, owing to their narrow distributions and environmental specificity, are more vulnerable to climate change and habitat disruption than alien plants. They also possess unique phytogeographical characteristics and value [26–28]. Thus, more precise habitat-based analyses of alien plants are required to establish and develop conservation and response strategies.

The coexistence patterns of alien and endemic plants can be comprehensively analyzed through quantitative comparisons of ecological niche structures and resource utilization patterns [29]. Recent studies have attempted to analyze their competitive structures and ecological isolation using various statistical techniques, including ecological niche overlap analysis, estimation of niche breadth for environmental variables, spatial species composition, and zeta diversity analysis [9,30]. To address the lack of comprehensive data on these parameters, this study aimed to compare species composition, ecological niches, and spatial species turnover patterns of alien and endemic plants using regional ecological data collected from the National Ecosystem Survey of South Korea. The specific objective was to identify ecological differences between the two groups and their underlying factors. Specifically, this study incorporated ecosystem service indicators as environmental factors to assess the potential for alien plant dispersal and its ecological implications within the broader framework of human–nature interactions. In summary, research on ecological niche theory will provide a strong foundation for habitat-based species conservation and restoration, as well as for developing and implementing effective policies for ecosystem and biodiversity conservation [31].

This study was designed based on the following three hypotheses:

Hypotheses 1. *The species compositions of alien and endemic plants are heterogeneous.*

Hypotheses 2. *The ecological niches of alien and endemic plants overlap in specific environmental factors, resulting in differences in niche breadth.*

Hypotheses 3. *The differences in the ecological niches of the two groups originate from a deterministic process.*

Based on these hypotheses, we examine and interpret the ecological assimilation process of alien plants and the spatial separation structure of endemic plants, providing fundamental data for effective biodiversity conservation and alien plant management strategies.

2. Materials and Methods

2.1. Biotic Data and Selection of Target Sites

In this study, we tested the proposed hypotheses by analyzing extracted data with various statistical tools (Figure 1).

The 800 survey grids were established using the 1:25,000 digital topographic map system, the standard unit of the National Ecosystem Survey conducted by the Korean Ministry of Environment [32]. Each grid reflects the spatial extent of the corresponding map and was structured to facilitate the use of national vegetation survey data in ecological analysis. This national survey system facilitates integrated analyses of plant species composition and environmental factors based on quantified ecological data. In this study, we considered each grid a single plot and assessed the presence of specific plant species.

Spatial distribution data on flora were obtained from 800 plots using the Spatial Join tool in ArcGIS (ver. 10.8; Esri, Redlands, CA, USA). Plant species were extracted from each grid, and species richness was then calculated. Species richness was defined as the number of plant species recorded in each plot [33]. The survey period for plant species ranged from 2006 to 2023. Alien plants, the target group of this study, identified based on the Korean Invasive Alien Species Database [34], while endemic plants were identified using the species list of Chung et al. and the Ministry of Environment [4,35].

We selected 80 plots with the highest species richness (top 10%) for both alien and endemic plants, resulting in 160 plots in total, and compared their ecological characteristics and species composition. Plots with high species richness represent the core niche space within the ecological niche of the flora and serve as a sample group enabling a clear comparison of preferences for environmental resources [36]. Furthermore, plots with a high probability of species occurrence were selected to analyze specific habitat niches, a sampling method that enhances the reliability of environmental and species distribution analyses [37,38].

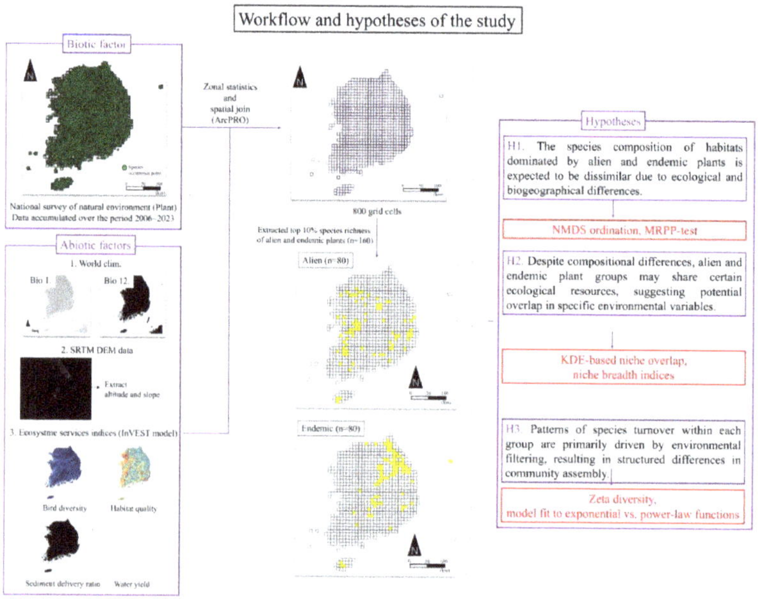

Figure 1. Workflow chart of research directions, hypothesis setting, and analysis (yellow highlight: the grids with the top 10% species richness of alien and endemic plants among the 800 survey units).

2.2. *Abiotic and Ecosystem Service Variables*

We identified a total of eight factors using ArcGIS to determine the environmental conditions of the 160 plots. The environmental factors assessed were temperature, precipitation, altitude, slope, habitat quality (HQ), bird diversity, water yield (WY), and sediment delivery ratio (SDR). These variables were chosen because they represent key climatic, topographic, and ecosystem service factors known to influence the distribution and ecological characteristics of plant species, particularly alien and endemic species, as supported by previous studies [39–53]. Temperature and precipitation layers were derived from WorldClim Bioclim data (v.2.1; 30 arc-second resolution, representing 1970–2000 climate averages), and altitude and slope were calculated using NASA's Shuttle Radar Topography Mission (SRTM) digital elevation model (30 m resolution).

Data on ecosystem service indicators, such as HQ, bird diversity, WY, and SDR, were computed using the InVEST program (v3.14.1; Natural Capital Project, Stanford University, CA, USA) based on land cover maps and spatial environmental data provided by the Korean National Institute of Ecology [54].

HQ was calculated based on habitat type, habitat sensitivity, and threat levels. Habitat types were derived from the Korean Land Use and Land Cover (LULC) dataset, and the calculation formula is as follows [54]:

$$HQ = H \times (1 - (\frac{D^2}{D^2 + K^2})) \tag{1}$$

HQ: Habitat quality
H: The habitat suitability of land use type
D: Threat level by habitat type
K: Half-saturation constant

Bird diversity was defined as species richness per grid cell based on data from the National Ecosystem Survey [54]. For WY, values for each pixel were derived by subtracting the annual actual evapotranspiration (*AET*) from the annual precipitation (*PRE*). The formula is as follows [54]:

$$\mathrm{WY} = \left(1 - \frac{AET}{PRE}\right) \times P \tag{2}$$

WY: Water yield
AET: The annual actual evapotranspiration
PRE: The annual precipitation

SDR was calculated based on rainfall- and erosion-related factors as well as slope steepness. The calculation formula is as follows [54]:

$$\mathrm{SDR} = \mathrm{R} \times \mathrm{Ks} \times \mathrm{LS} \times \mathrm{C} \times \mathrm{P} \tag{3}$$

SDR: The sediment delivery ratio
R: Rainfall erosivity factor
Ks: Soil erodibility factor
LS: Slope length and steepness factor
C: Cover management factor
P: Support practice factor

All spatial data values were extracted for each plot using ArcGIS and normalized to a 0–1 scale for cross-variable comparisons. The distribution patterns of alien and endemic plants differ depending on these indicators [55–57]. Ecosystem service indicators were calculated using the InVEST program.

2.3. Analytical Methods

We analyzed species composition differences between alien and endemic plant habitats using non-metric multidimensional scaling (NMDS), selecting two axes with the highest R^2 values [58–60]. The multi-response permutation procedure (MRPP) was applied to test group differences in species composition, with T and A statistics used to interpret heterogeneity and within-group agreement [58,61].

To analyze ecological niche overlap and breadth, all environmental variables were normalized (min–max scaling) to a 0–1 range for comparability across units. Ecological niche overlap for each environmental factor was visualized using kernel density estimation (KDE) and quantified with Schoener's D index [62,63]. The KDE-based probability density function integrates to 1 over the entire domain, which is a fundamental property of probability density functions. Schoener's D index ranges from 0 to 1 and quantifies the area of overlap between the two groups, indicating the degree to which the groups actually share ecological niche space for each environmental factor [62,63].

The KDE formula is presented in Equation (4).

$$f_h(x) = \frac{1}{nh} \times \sum_{i=1}^{n} Kf \times \left(\frac{x - x_i}{h}\right) \tag{4}$$

$f_h(x)$: Estimated probability density value at point x
n: Number of samples
h: Bandwidth (smoothing parameter)

Kf: Kernel function (e.g., Gaussian normal distribution kernel)

x_i: Observed values

The smoothing parameter (h, bandwidth) determines the shape of the KDE curve, was calculated using Silverman's rule of thumb (Equation (5)) [64].

$$h = 1.06 \times \sigma \times n^{-1/5} \quad (5)$$

σ: Sample standard deviation

n: Number of data points

The Gaussian normal distribution kernel used in this study was:

$$K(u) = \frac{1}{\sqrt{2\pi}} \times e^{-\frac{1}{2}u^2} \quad (6)$$

K(u): Kernel function

u: The normalized value of the distance between the data point and the center of the kernel.

To compare the ecological niche breadth between groups for each environmental factor, we employed KDE bandwidth, interquartile range (IQR), and Shannon entropy [33,64–66]. Each indicator was applied using 1000 bootstrap resampling iterations for the two groups to derive the mean and standard error [33,67]. Based on these results, the statistical significance between the two groups was analyzed with an independent samples *t*-test. By presenting both ecological niche overlap area and niche breadth indicators, we aimed to accurately interpret the functions and interaction patterns of each habitat based on ecological niche structure.

We analyzed zeta diversity to quantify multi-site species turnover and shared species patterns across plots [30,68]. Zeta ratio analysis was also applied to assess species persistence and regional similarity [30,68,69]. The zeta ratio represents the probability of a species occurring in the nth plot given its occurrence in the $n - 1$ plot. The equation used is presented below.

$$\zeta - ratio = \frac{\zeta_n}{\zeta_{n-1}} \quad (7)$$

ζ_n: The number of species shared between n areas

ζ_{n-1}: The number of species shared between (n−1) areas.

We compared power-law and exponential models to infer ecological assembly processes. The power-law model indicates deterministic, niche- and environment-based filtering, whereas the exponential model aligns with neutral theory and stochastic turnover [30,68,70]. Model fit was evaluated using Akaike's information criterion (AIC), with lower AIC values indicating better fit [71].

As for the analysis program, biotic and abiotic factors were extracted using ArcGIS and PC-ORD (ver. 7.0; MjM Software Design, Gleneden Beach, OR, USA) was employed for NMDS and MRPP tests. R Studio (ver.4.5.1; R Foundation for Statistical Computing, Vienna, Austria) with the packages 'ggplot2', 'vegan', 'boot,' and 'zetadiv' was employed for KDE modeling, niche overlap analysis, and zeta diversity analysis.

3. Results

3.1. Ordination and MRPP-Test

We performed NMDS ordination to analyze the species composition heterogeneity of alien and endemic plant habitats (Figure 2). The stress value in the two-dimensional ordination was 17.581, below the acceptable threshold of 20 in ecological analyses

(p = 0.004), indicating reliable results [58]. The R^2 values for the first and second axes were 0.609 and 0.211, respectively, with an overall R^2 value of 0.820. When comparing the species composition of alien and endemic plant habitats, the two groups partially overlapped, but their centroids were separated in two-dimensional space, indicating heterogeneous composition. The environmental variables influencing the arrangement of species composition were included gradient, HQ, SDR, precipitation, temperature, and bird diversity.

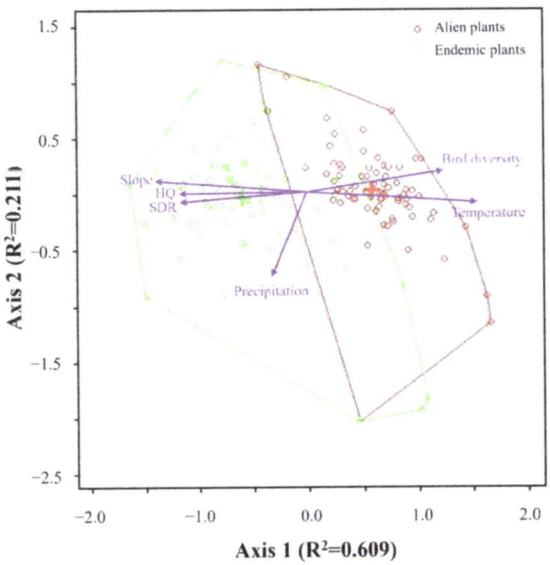

Figure 2. NMDS ordination (cut-off R^2 = 0.3) of species composition for each group. Purple arrows represent environmental vectors fitted using a joint plot in PC-ORD 7.0. The length and direction of the arrows indicate the strength (R^2 > 0.3) and direction of correlation with the ordination axes. The cross symbol indicates the centroid of species composition for each group.

The MRPP results (Table 1) showed T = −73.8937 and A = 0.2814, indicating significant differences between groups, while within-group similarity remained relatively high (p < 0.001).

Table 1. Summary of MRPP results comparing species composition across different groups. The T statistic indicates the separation between groups, while A represents the chance-corrected within-group agreement. Significant results (p < 0.05) suggest that community composition differs among groups.

Compared Plant Groups	T	A	*p*-Value
Alien vs. Endemic	−73.894	0.281	<0.001

3.2. Ecological Niche Comparisons of Habitats of Alien and Endemic Plants

We performed KDE analysis to determine the degree of sharing of available resources between each group based on the overlap area of the ecological niches of alien and endemic plant habitats. The degree of shared resource use reflects the level of competition between the two groups for growth and distribution [72,73]. To compare the degree of overlap between the ecological niches of alien and endemic plants, KDE-based overlap (Schoener's

D) was calculated for eight environmental variables (Figure 3, Table 2). Schoener's D values quantify the area of overlap between two distributions on a scale from 0 to 1; the higher the value, the greater the resource sharing (or competition) between the two groups.

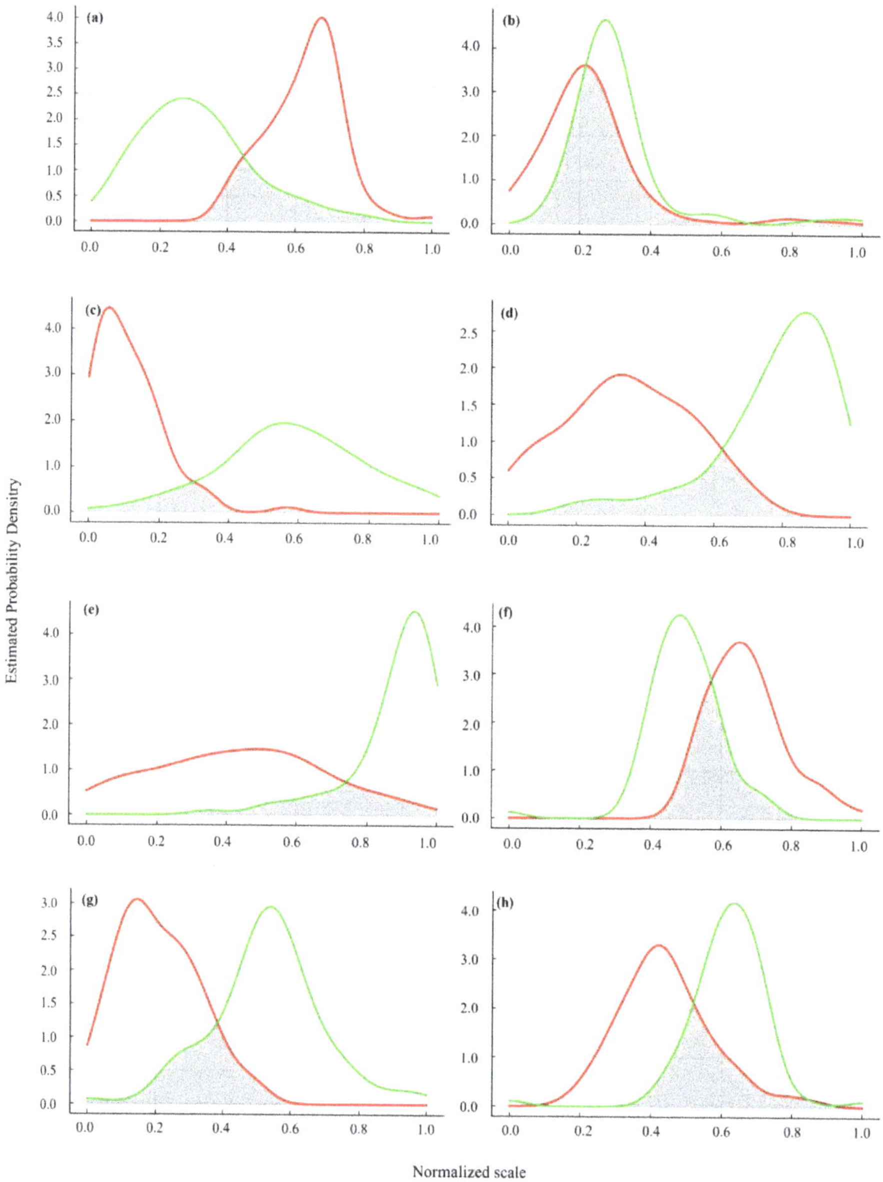

Figure 3. Kernel density estimation (KDE) curves showing ecological niche distributions of Alien (red) and Endemic (green) species groups across eight normalized environmental gradients. Gray shading indicates the overlapping region between the two distributions. The x-axis represents the normalized scale of each environmental variable, and the y-axis represents the estimated probability density. Environmental variables are as follows: (**a**) temperature, (**b**) precipitation, (**c**) altitude, (**d**) slope, (**e**) habitat quality, (**f**) bird diversity, (**g**) sediment delivery ratio, (**h**) water yield.

Table 2. Overlap metrics for 8 Environmental variables (Schoener's D).

Variable	Schoener's D
Temperature	0.276
Precipitation	0.719
Altitude	0.157
Slope	0.261
Habitat quality	0.255
Bird diversity	0.419
Water yield	0.397
Sediment delivery ratio	0.277

Precipitation showed the highest overlap (0.719), indicating similar water-resource niches for the two communities. Bird diversity (0.419) and water yield (0.397) followed, and these ecosystem service indicators are interpreted as factors related to habitat connectivity and moisture conditions within habitats, respectively.

In contrast, altitude showed the lowest overlap at 0.157, followed by gradient (0.261) and temperature (0.276). These results imply that the two groups occupy distinct ecological niches shaped by topographic and temperature conditions. These environmental variables suggest that their distributions are influenced by different ecological characteristics.

In summary, alien plants share ecological niches with endemic plants in relation to moisture-related variables but tend to be confined to specific habitats under topographic and temperature conditions.

The interpretation of the x-axis (normalized scale) and y-axis (estimated probability density) for each environmental factor in the KDE graph reveals that alien plants were generally concentrated in the mid-range or slightly lower normalized scale, whereas endemic plants, except for temperature and bird diversity, were concentrated in the relatively higher normalized scale. Concentration in the lower normalized scale indicates that these plants are abundant even under conditions with limited environmental resources. Higher normalized scales indicate that these plants can particularly grow and develop in environments with abundant resources or require specific conditions for growth. Accordingly, this analysis demonstrated that endemic plants were more specialized in their environmental requirements for growth compared to alien plants.

We analyzed three niche breadth indicators—KDE bandwidth, IQR, and Shannon entropy—for eight environmental variables (Figure 4) to compare the ecological niches of alien and endemic plants. KDE bandwidth represents the quantitative range of ecological niche breadth, while IQR indicates the core area where 50% of the data is distributed based on the KDE curve. Shannon entropy is used to confirm the evenness of the distribution.

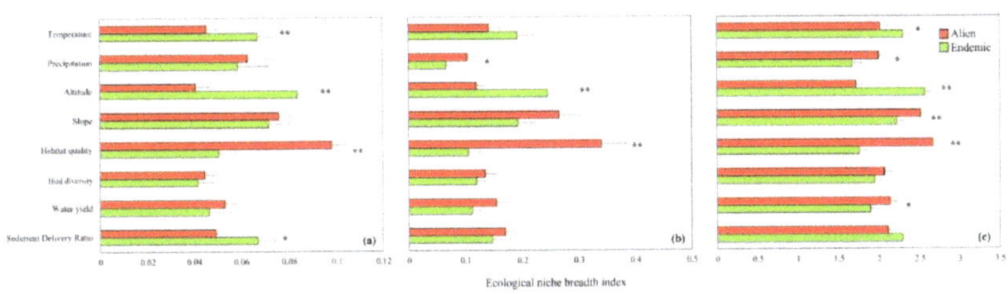

Figure 4. Ecological niche breadth of Alien (red) and Endemic (green) species groups across eight environmental variables, based on (**a**) KDE bandwidth, (**b**) interquartile range (IQR), and (**c**) Shannon entropy. Error bars represent standard errors derived from 1000 bootstrap replicates. Statistical significance between groups was assessed using two-sample *t*-tests ($p < 0.05$ is indicated by * and $p < 0.01$ by **).

3.3. Zeta Diversity Analysis

Based on the above ecological niche analysis results, we can raise the following question: Do the distribution characteristics of the two groups (alien and endemic plants) according to environmental variables affect changes in species composition or species turnover patterns? To answer this question, we derived a graph of the decrease in zeta diversity and analyzed the associated ecological processes (Figure 5).

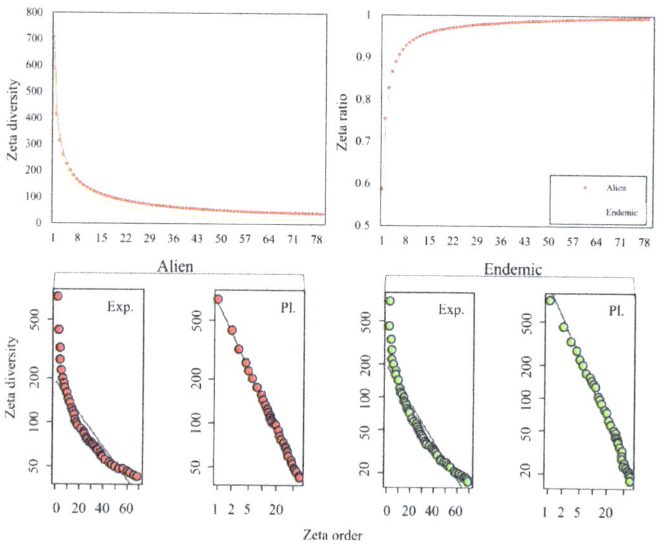

Figure 5. Comparisons of zeta diversity decline and the zeta ratio for each group. To determine the ecological processes for each habitat associated with alien and endemic species groups, AICs of exponential and power-law function regression were calculated.

Zeta diversity decreased with increasing spatial order in both alien and endemic plants, but the decline was slower in alien plants. The zeta ratio graph tended to converge to 1 as the zeta order increased in both groups. Table 3 presents the AIC values of the power-law and exponential function models based on the graph of the decrease in zeta diversity. The AIC values of the power-law function and exponential function models for alien plants were −530.3144, and −109.6002, respectively, and the values for endemic plants were −302.4994 and −85.1348, respectively. These results indicate that the power-law function model better explained the ecological processes in both groups.

Table 3. Comparison of AIC values for exponential and power-law models fitted to zeta diversity decay in alien and endemic plant groups.

Index	Power Law	Exponential
Alien	−530.3144	−109.6002
Endemic	−302.4994	−85.1348

4. Discussion

4.1. (Hypothesis 1) The Species Compositions of Alien and Endemic Plants Are Heterogeneous

The NMDS analysis showed clear distinctions in species composition between alien and endemic plant groups, and the MRPP test confirmed statistically significant differences.

This implies that the two plant groups formed under different environmental requirements [58]. The primary distribution areas of alien plants in South Korea are ports, landfills, and areas near airports, and their distribution ranges extend to roads, construction sites, and development areas [22,74]. In South Korea, some endemic plants occur along roadsides and forest edges, but most are restricted to specialized habitats. Most species are limited to habitats with strong environmental specificity, such as high mountain areas, cliffs, and rocky areas, where human access is limited [75–79]. A similar trend is found worldwide, with native plants generally adapted to habitats limited by environmental filters such as altitude, precipitation, and soil, and showing high sensitivity to anthropogenic disturbance. Although the NMDS contour overlay revealed heterogeneous spatial patterns, some areas of overlap between alien and endemic plants were also observed. This can be interpreted as reflecting a transitional spatial structure where alien plants are gradually invading habitats where endemic plants have been established for a long time [80]. The habitats of the two groups are not completely separated, and alien plants are gradually invading the habitats of endemic plants. After disturbance, native and alien plants may coexist in specific habitats if resources are sufficient. However, as succession progresses, these habitats become more heterogeneous [23]. Alien plants exhibit distinct species composition and distribution patterns over time to avoid competition with native species and utilize newly available resources [9]. With respect to altitude and gradient, alien plants were distributed under lower environmental conditions than endemic plants. The distance between the density centroids was longer, indicating that differentiation occurred due to the adaptation of endemic species to the unique climate conditions resulting from vertical distribution and geographical isolation. Furthermore, the relationship between gradient and the ecological niche of alien plants in this study was consistent with a previous study that found alien plants frequently occurred in vegetation established at low gradients [18].

In consideration of the direction of the joint plot of environmental factor vectors, alien plants were correlated with bird diversity and temperature, while endemic plants were correlated with altitude, HQ, SDR, and precipitation [58]. This result is consistent with the highest density points on the normalized scale found in the KDE-based ecological niche analysis in this study. In other words, alien plants were concentrated in areas with high temperature and bird diversity on a normalized scale, while endemic plants exhibited a high distribution density pattern in environments with high altitude and gradients. These results align with the joint plot analysis results of NMDS ordination.

These findings indicate that the species composition of each group did not originate from a stochastic process, but rather from a deterministic process, i.e., the filtering effect of specific environmental factors. This study will discuss this environmental filtering mechanism in more detail in the following sections (Sections 4.2 and 4.3) to clarify its close connection with community structure, ecological niche breadth, and spatial species turnover (zeta diversity).

4.2. *(Hypothesis 2) The Ecological Niches of Alien and Endemic Plants Overlap in Specific Environmental Factors, Resulting in Differences in Niche Breadth*

We analyzed ecological niche overlap (Schoener's D) and niche breadth (KDE bandwidth, IQR, Shannon entropy) for eight environmental variables to quantitatively compare niche characteristics of alien and endemic plants [62–66,70].

For temperature, alien plant densities were concentrated in warmer ranges, reflecting their adaptation to high-temperature environments and suggesting that their distribution may expand under ongoing climate warming [7,42]. Endemic species showed higher densities in moist habitats (precipitation and WY), indicating reliance on stable water resources and vulnerability to habitat drying [55,56,79].

Regarding HQ, alien species were broadly distributed across the entire normalized scale, whereas endemic plants were concentrated in high-HQ areas, demonstrating their preference for undisturbed, biodiversity-rich habitats [26,54,75,78]. The broad distribution of alien plants suggests high adaptability to diverse disturbance regimes [6,24,80].

Alien species were more common in areas with higher bird diversity, implying that heterogeneous or recovering habitats facilitate their dispersal [52,53]. In contrast, endemic species were associated with areas of lower bird diversity, suggesting environmental uniqueness or microhabitat specialization [77,79]. For sediment SDR, alien species occurred mostly in low-SDR regions, such as managed areas and gentle slopes, whereas endemic plants were concentrated in eroded or rocky terrain, indicating strong topographic specificity [46,54–56].

KDE-based overlap analysis showed the largest overlap for water-related variables (precipitation, WY), suggesting high competition potential for water resources between groups [58,76]. Endemic KDE curves peaked in high precipitation and WY ranges, while alien KDE curves peaked in intermediate to low conditions. Previous studies similarly demonstrated that alien species, such as *Bromus tectorum* in Nevada grasslands, alter hydrological cycles by reducing deep soil moisture availability and increasing surface moisture use [8,12–15].

Conversely, topographic variables such as altitude, slope, and temperature showed lower overlap, indicating clearer niche separation between alien and endemic plants [81]. Endemic species' wider niche breadth for altitude and temperature highlights ecological adaptation to temperature gradients, consistent with NMDS ordination showing strong compositional diversity along elevation gradients [82,83].

Across all factors, alien plants exhibited a wider ecological niche breadth than endemic species in five or six environmental variables, supporting their classification as ecological generalists capable of thriving under diverse conditions. Endemic species had significantly broader niche breadth only in temperature and altitude, consistent with their adaptation to specialized environments and narrower ecological tolerance.

HQ-related metrics reinforced these findings: alien plants had significantly wider niche breadth ($p < 0.01$) with even KDE distributions across HQ values, whereas endemic species were concentrated at high HQ levels (0.8–1.0), highlighting their vulnerability to anthropogenic disturbance.

Together, these metrics—kernel bandwidth (habitat dispersion), IQR (variation around the median), and Shannon entropy (environmental resource diversity)—provide quantitative insight into ecological connectivity and resource utilization [9,29]. Our findings highlight alien plants as ecological generalists occupying broad environmental ranges, while endemic species are specialists restricted to narrow conditions [84]. This supports our hypothesis that alien and endemic species exhibit overlapping niches for moisture-related variables but contrasting niche breadth patterns overall.

Our findings are consistent with recent predictive modeling studies that emphasize the importance of mechanistic niche models in projecting invasive species potential under climate change scenarios [85]. Further, analyses of niche shifts and regional risk in recent literature parallel our observations of niche breadth differences between alien and endemic plants [86].

4.3. (Hypothesis 3) The Differences in the Ecological Niches of the Two Groups Originate from a Deterministic Process

Zeta diversity analysis revealed a slower decline in species turnover among alien plants compared to endemic plants. The zeta ratio graphs for both groups showed an upward trend, converging toward 1; however, the convergence occurred at a faster zeta order in alien plants. Furthermore, the decline in zeta diversity for both groups supported a power-law function model, suggesting that species turnover and changes in species composition are driven by a deterministic process [70].

Contrary to our prediction that alien plants would generally show large variations in species composition depending on local environmental conditions and regional heterogeneity, the findings of this study indicate that alien plants adapt swiftly to diverse habitats, and habitats became stabilized with the repeated appearance of specific species [87]. This is consistent with the slower decline observed for alien plants in the graph [30,68].

This pattern was also observed in the zeta ratio, with alien plants showing a tendency to converge to 1 more rapidly than endemic plants, particularly between the 4th and 25th orders. This indicates that alien plants undergo rapid niche assimilation, swiftly adapting to diverse habitats, occupying ecological niches, and homogenizing species composition [88]. This process may induce competition between the two groups as the overlap of available resources increases [89].

In contrast, endemic plants showed relatively large changes in species composition between habitats, and the convergence of the zeta ratio appeared slower even at higher orders, indicating a species composition structure in which regional specificity was relatively maintained compared to alien plants. These results suggest that endemic plants, like rare species, follow distribution strategies strongly linked to environmental gradients or isolated habitat characteristics [27,43,90,91].

Additionally, the decrease in zeta diversity based on the power-law function suggests changes in species composition in line with selective distribution and environmental suitability rather than random distribution. Combined with the previously previous reports on ecological niche breadth and overlap area, this strongly supports the distinct ecological separation and species turnover structures between the two groups. Consequently, endemic and alien plants are shaped by different environmental filters, forming distinct ecological structures in both spatial patterns and species composition. The zeta diversity decline curve for alien plants was gentler compared to endemic plants, which may indicate the spatial persistence of common species and a wide environmental tolerance range [30,70].

These results support the potential for alien plants to penetrate and spread across various habitats in South Korea. In this context, the gentle slope of zeta diversity in the alien plant group, as found in this study, underpins the ecological process of stable species turnover. Simultaneously, endemic plants may be more vulnerable to climate change or habitat fragmentation due to ecological niche conservatism and local specificity [92,93].

5. Conclusions

This study quantitatively compared species composition, ecological niches, and species turnover patterns of alien and endemic plants in South Korea, using national-scale data from the National Ecosystem Survey. Alien plants were identified as ecological generalists with broad niche breadth across multiple environmental factors, whereas endemic plants exhibited specialized niches confined to specific conditions.

Zeta diversity analysis showed that alien plants maintain relatively homogeneous species composition across regions, highlighting their rapid establishment and the potential risks of habitat homogenization. Their wide ecological niche, particularly for ecosystem service indicators such as habitat quality (HQ), suggests their potential to spread into areas of high conservation value.

This study demonstrates the utility of national biodiversity datasets for multidimensional ecological assessments and provides a foundation for effective monitoring and management strategies. However, as our analysis was based on grid-level presence-absence data, future studies should incorporate species population dynamics, functional traits, and impact indices to better inform management priorities. Science-based and selective strategies, rather than indiscriminate control, are essential to balance alien plant management with the ecological roles of native and endemic communities.

Author Contributions: Conceptualization, B.-J.P. and K.C.; software, B.-J.P.; formal analysis, B.-J.P. and K.C.; investigation, B.-J.P. and K.C.; writing—original draft, B.-J.P.; writing—review and editing, K.C.; data curation, B.-J.P.; visualization, B.-J.P. All authors have read and agreed to the published version of the manuscript.

Funding: This research was funded by the National Institute of Ecology (Project No. NIE-B-2025-03) and Korea Environment Industry & Technology Institute (KEITI) through Climate Change R&D Project for New Climate Regime, funded by Korea Ministry of Environment (MOE) (RS-2022-KE002369).

Data Availability Statement: The datasets used and/or analyzed during the current study are available from the corresponding author on reasonable request.

Acknowledgments: This research was conducted as part of the National Institution of Ecology's research project, "Development of Policy Decision Support System Base on Ecosystem Services Assessment (Project No. NIE-B-2025-03)" and Korea Environment Industry & Technology Institute (KEITI) through Climate Change R&D Project for New Climate Regime, funded by Korea Ministry of Environment (MOE) (RS-2022-KE002369)." in Republic of Korea.

Conflicts of Interest: The authors declare no conflicts of interest.

References

1. Kunming–Montral Global Biodiversity Framework: 2030 Targets. Available online: https://www.cbd.int/gbf/targets/ (accessed on 2 July 2025).
2. Son, D.C.; Jang, J.E.; Park, B.K.; Lee, K.H.; Kang, E.S.; Chang, K.S.; Jung, S.Y.; Gil, H.Y.; Kim, S.J.; Choi, K. An updated checklist of the vascular flora native to Korea. *J. Asia-Pac. Biodivers.* **2025**, *18*, 651–687. [CrossRef]
3. Kang, E.S.; Lee, S.R.; Oh, S.H.; Kim, D.K.; Jung, S.Y.; Son, D.C. Comprehensive review about alien plants in Korea. *Korean J. Plant Taxon.* **2020**, *50*, 89–119. [CrossRef]
4. Chung, G.Y.; Jang, H.D.; Chan, K.S.; Choi, H.J.; Kim, Y.S.; Kim, H.J.; Son, D.C. A checklist of endemic plants on the Korean Peninsula II. *Korean J. Plant Taxon.* **2023**, *53*, 79–101. [CrossRef]
5. Wang, S.Q.; Dong, X.Y.; Ye, L.; Wang, H.F. Flora of Northeast Asia. *Plants* **2023**, *12*, 2240. [CrossRef] [PubMed]
6. Simberloff, D.; Martin, J.-L.; Genovesi, P.; Maris, V.; Wardle, D.A.; Aronson, J.; Courchamp, F.; Galil, B.; García-Berthou, E.; Pascal, M.; et al. Impacts of biological invasions: What's what and the way forward. *Trends Ecol. Evol.* **2013**, *28*, 58–66. [CrossRef] [PubMed]
7. Early, R.; Bradley, B.A.; Dukes, J.S.; Lawler, J.J.; Olden, J.D.; Blumenthal, D.M.; Gonzalez, P.; Grosholz, E.D.; Ibañez, I.; Miller, L.P.; et al. Global threats from invasive alien species in the twenty–first century and national response capacities. *Nat. Commun.* **2016**, *7*, 12485. [CrossRef] [PubMed]
8. Levine, J.M.; Vilà, M.; Antonio, C.M.D.; Dukes, J.S.; Grigulis, K.; Lav Areorel, S. Mechanisms underlying the impacts of exotic plant invasions. *Proc. R. Soc. Lond. B Biol. Sci.* **2003**, *270*, 775–781. [CrossRef]
9. Broennimann, O.; Treier, U.A.; Müller-Schärer, H.; Thuiller, W.; Peterson, A.T.; Guisan, A. Evidence of climatic niche shift during biological invasion. *Ecol. Lett.* **2007**, *10*, 701–709. [CrossRef]
10. Gallien, L.; Münkemüller, T.; Albert, C.H.; Boulangeat, I.; Thuiller, W. Predicting potential distributions of invasive species: Where to go from here? *Divers. Distrib.* **2010**, *16*, 331–342. [CrossRef]
11. Petitpierre, B.; Kueffer, C.; Broennimann, O.; Randin, C.; Daehler, C.; Guisan, A. Climatic niche shifts are rare among terrestrial plant invaders. *Science* **2012**, *335*, 1344–1348. [CrossRef]

12. Cline, J.F.; Uresk, D.W.; Rickard, W.H. Comparison of water used by a sagebrush–bunchgrass community and a cheatgrass community. *J. Range Mngmt.* **1977**, *30*, 199–201. [CrossRef]
13. Rickard, W.H.; Vaughan, B.E. Plant Community Characteristics and Responses. In *Shrub–Steppe: Balance and Change in a Semi–Arid Terrestrial Ecosystem*; Rickard, W.H., Rogers, L.E., Vaughan, B.E., Liebetrau, S.F., Eds.; Elsevier: Amsterdam, The Netherlands, 1988; pp. 109–179. ISBN 9780444988870.
14. Melgoza, G.; Nowak, R.S.; Tausch, R.J. Soil water exploitation after fire: Competition between *Bromus tectorum* (cheatgrass) and two native species. *Oecologia* **1990**, *83*, 7–13. [CrossRef]
15. Dyer, A.R.; Rice, K.J. Effects of competition on resource availability and growth of a California bunchgrass. *Ecology* **1999**, *80*, 2697–2710. [CrossRef]
16. Park, J.S.; Lee, H.; Choi, D.; Kim, Y. Spatially varying relationships between alien plant distributions and environmental factors in South Korea. *Plants* **2021**, *10*, 1377. [CrossRef] [PubMed]
17. Adhikari, P.; Lee, Y.H.; Park, Y.S.; Hong, S.H. Assessment of the spatial invasion risk of intentionally introduced alien plant species (IIAPS) under environmental change in South Korea. *Biology* **2021**, *10*, 1169. [CrossRef]
18. Lim, B.S.; Seok, J.E.; Lim, C.H.; Kim, G.S.; Shin, H.C.; Lee, C.S. Distribution, Effect, and Control of Exotic Plants in Republic of Korea. *Biology* **2023**, *12*, 826. [CrossRef] [PubMed]
19. Kil, J.H.; Shim, K.C.; Park, S.H.; Koh, K.S.; Suh, M.H.; Ku, Y.B.; Suh, S.U.; Oh, H.K.; Kong, H.Y. Distributions of Naturalized Alien Plants in South Korea1. *Weed Tech.* **2004**, *18*, 1493–1495. [CrossRef]
20. Lee, Y.M.; Park, S.H.; Jung, S.Y.; Oh, S.H.; Yang, J.C. Study on the current status of naturalized plants in South Korea. *Korean J. Plant Taxon.* **2011**, *41*, 87–101. [CrossRef]
21. Kim, C.G.; Kil, J. Alien flora of the Korean Peninsula. *Biol. Invasions* **2016**, *18*, 1843–1852. [CrossRef]
22. Choi, S.H.; Jeong, Y.H.; Park, S.M.; Lee, J.W.; Oh, H.S. Current status of invasive alien plants in Ramsar wetlands on Jeju Island, Republic of Korea. *BioInvasions Rec.* **2023**, *12*, 931–941. [CrossRef]
23. Woo, J.H.; Lee, M.K.; Lee, H.I.; Lee, C.B. Joint influence of forest strata attributes and abiotic factors on herbaceous plant abundance in South Korean forest restoration sites: Native versus alien species. *Forests* **2024**, *15*, 1924. [CrossRef]
24. Huston, M.A. Management strategies for plant invasions: Manipulating productivity, disturbance, and competition. *Divers. Distrib.* **2004**, *10*, 167–178. [CrossRef]
25. Pearson, D.E.; Ortega, Y.K.; Eren, Ö.; Hierro, J.L. Community assembly theory as a framework for biological invasions. *Trends Ecol. Evol.* **2018**, *33*, 313–325. [CrossRef] [PubMed]
26. Chung, M.Y.; Son, S.; Suh, G.U.; Herrando-Moraira, S.; Lee, C.H.; López-Pujol, J.; Chung, M.G. The Korean Baekdudaegan Mountains: A glacial refugium and a biodiversity hotspot that needs to be conserved. *Front. Genet.* **2018**, *9*, 489. [CrossRef] [PubMed]
27. Pouteau, R.; Biurrun, I.; Brunel, C.; Chytrý, M.; Dawson, W.; Essl, F.; Fristoe, T.; Haveman, R.; Hobohm, C.; Jansen, F.; et al. Potential alien ranges of European plants will shrink in the future, but less so for already naturalized than for not yet naturalized species. *Divers. Distrib.* **2021**, *27*, 2063–2076. [CrossRef]
28. Chae, H.H.; Kim, Y.C.; Son, S.W. Korean and worldwide research trends on rare plant and endemic plant in Korea. *Korean J. Environ. Ecol.* **2022**, *36*, 257–276. [CrossRef]
29. Soberón, J.; Nakamura, M. Niches and distributional areas: Concepts, methods, and assumptions. *Proc. Natl. Acad. Sci. USA* **2009**, *106*, 19644–19650. [CrossRef]
30. Hui, C.; McGeoch, M.A. Zeta diversity as a concept and metric that unifies incidence-based biodiversity patterns. *Am. Nat.* **2014**, *184*, 684–694. [CrossRef] [PubMed]
31. Koo, K.A.; Park, S.A. Review of Ecological Niche Theory from the Early 1900s to the Present. *Korean J. Environ. Ecol.* **2021**, *35*, 316–335. [CrossRef]
32. National Institute of Ecology. *Manual of the 5th National Survey of Natural Environment*; National Institute of Ecology: Seocheon, Republic of Korea, 2019; pp. 11–50. ISBN 9791189730727.
33. Magurran, A.E. *Measuring Biological Diversity*; Blackwell Publishing: Oxford, UK, 2004; pp. 72–161. ISBN 9780632056330.
34. Information of Korean Alien Species. Available online: https://kias.nie.re.kr/home/main/main.do (accessed on 8 July 2025).
35. National Institute of Biological Resources: Biodiversity of the Korean Peninsula. Available online: https://species.nibr.go.kr/index.do (accessed on 8 July 2025).
36. Zurell, D.; Thuiller, W.; Pagel, J.; Cabral, J.S.; Münkemüller, T.; Gravel, D.; Dullinger, S.; Normand, S.; Schiffers, K.H.; Moore, K.A.; et al. Benchmarking novel approaches for modelling species range dynamics. *Glob. Change Biol.* **2016**, *22*, 2651–2664. [CrossRef]
37. Guisan, A.; Thuiller, W. Predicting species distribution: Offering more than simple habitat models. *Ecol. Lett.* **2005**, *8*, 993–1009. [CrossRef]
38. Broennimann, O.; Fitzpatrick, M.C.; Pearman, P.B.; Petitpierre, B.; Pellissier, L.; Yoccoz, N.G.; Thuiller, W.; Fortin, M.J.; Randin, C.; Zimmermann, N.E.; et al. Measuring ecological niche overlap from occurrence and spatial environmental data. *Glob. Ecol. Biogeogr.* **2012**, *21*, 481–497. [CrossRef]

39. Oliver, C.D.; Larson, B.C. *Forest Stand Dynamics*; McGraw-Hill: New York, NY, USA, 1990; pp. 27–467. ISBN 9780070478299.
40. Wilson, E.P. *The Diversity of Life*; The Belknap Press of Harvard University Press: Cambridge, MA, USA, 1992; pp. 278–424. ISBN 9780080289144.
41. Kimmins, J.P. *Forest Ecology: A Foundation for Sustainable Management*; Prentice–Hall Inc.: Saddle River, NJ, USA, 1997; pp. 33–596. ISBN 9780137174219.
42. Thuiller, W.; Lavorel, S.; Araújo, M.B.; Sykes, M.T.; Prentice, I.C. Climate change threats to plant diversity in Europe. *Proc. Natl. Acad. Sci. USA* **2005**, *102*, 8245–8250. [CrossRef] [PubMed]
43. Lee, C.B.; Chun, J.H.; Kim, H.H. Plant & Forest: Elevational Patterns and Determinants of α and β Plant Diversity on the Ridge of the Baekdudaegan Mountains, South Korea. *J. Agric. Life Sci.* **2014**, *48*, 93–104.
44. Stevens, G.C. The elevational gradient in altitudinal range: An extension of Rapoport's latitudinal rule to altitude. *Am. Nat.* **1992**, *140*, 893–911. [CrossRef]
45. Körner, C. The use of 'altitude' in ecological research. *Trends Ecol. Evol.* **2007**, *22*, 569–574. [CrossRef]
46. Kong, W.S.; Kim, K.; Lee, S.; Park, H.; Cho, S.H. Distribution of High Mountain Plants and Species Vulnerability Against Climate Change. *J. Environ. Impact Assess.* **2014**, *23*, 119–136. [CrossRef]
47. Trigas, P.; Panitsa, M.; Tsiftsis, S. Elevational Gradient of Vascular Plant Species Richness and Endemism in Crete—The Effect of Post-Isolation Mountain Uplift on a Continental Island System. *PLoS ONE* **2013**, *8*, e59425. [CrossRef]
48. Isbell, F.; Calcagno, V.; Hector, A.; Connolly, J.; Harpole, W.S.; Reich, P.B.; Scherer-Lorenzen, M.; Schmid, B.; Tilman, D.; Ruijven, J.; et al. High plant diversity is needed to maintain ecosystem services. *Nature* **2011**, *477*, 199–202. [CrossRef]
49. Cardinale, B.; Duffy, J.; Gonzalez, A.; Hooper, D.U.; Perrings, C.; Venail, P.; Narwani, A.; Mace, G.M.; Tilman, D.; Wardle, D.A.; et al. Biodiversity loss and its impact on humanity. *Nature* **2012**, *486*, 59–67. [CrossRef]
50. Maestre, F.T.; Quero, J.L.; Gotelli, N.J.; Escudero, A.; Ochoa, V.; Delgado-Baquerizo, M.; García-Gómez, M.; Bowker, M.A.; Soliveres, S.; Escolar, C.; et al. Plant species richness and ecosystem multifunctionality in global drylands. *Science* **2012**, *335*, 214–218. [CrossRef]
51. Lavorel, S.; Grigulis, K.; McIntyre, S.; Williams, N.S.; Garden, D.; Dorrough, J.; Berman, S.; Quétier, F.; Thébault, A.; Bonis, A. Assessing functional diversity in the field–methodology matters! *Funct. Ecol.* **2008**, *22*, 134–147. [CrossRef]
52. Sekercioglu, C.H. Increasing awareness of avian ecological function. *Trends Ecol. Evol.* **2006**, *21*, 464–471. [CrossRef]
53. Kissling, W.D.; Rahbek, C.; Böhning-Gaese, K. Food plant diversity as broad–scale determinant of avian frugivore richness. *Proc. R. Soc. B* **2007**, *274*, 799–808. [CrossRef]
54. National Institute of Ecology. *The Manual of Assessment Map of Ecosystem Service*; Design Crepas: Seoul, Republic of Korea, 2022; pp. 54–61. ISBN 9791166981463.
55. Pimentel, D.; Harvey, C.; Resosudarmo, P.; Sinclair, K.; Kurz, D.; McNair, M.; Crist, S.; Shpritz, L.; Fitton, L.; Saffouri, R.; et al. Environmental and economic costs of soil erosion and conservation benefits. *Science* **1995**, *267*, 1117–1123. [CrossRef]
56. Montgomery, D.R. Soil erosion and agricultural sustainability. *Proc. Natl. Acad. Sci. USA* **2007**, *104*, 13268–13272. [CrossRef]
57. Lambers, H.; Oliveira, R.S. *Plant Physiological Ecology*, 5th ed.; Lambers, H., Ed.; Springer: Cham, Switzerland, 2019; pp. 187–263. ISBN 9783030231971.
58. McCune, B.; Grace, J.B. *Analysis of Ecologcial Communities*; MjM Software Design: Gleneden Beach, OR, USA, 2002; pp. 179–204. ISBN 9780972129008.
59. Anderson, A.J.B. Ordination methods in ecology. *J. Ecol.* **1971**, *59*, 713–726. [CrossRef]
60. Clarke, K.R. Non-parametric multivariate analyses of changes in community structure. *Aust. J. Ecol.* **1993**, *18*, 117–143. [CrossRef]
61. Mielke, P.W.; Berry, K.J. *Permutation Methods: A Distance Function Approach*; Springer Science & Business Media: New York, NY, USA, 2007; pp. 11–124. ISBN 9781441924162.
62. Schoener, T.W. Nonsynchronous spatial overlap of lizards in patchy habitats. *Ecology* **1970**, *51*, 408–418. [CrossRef]
63. Fieberg, J.; Kochanny, C.O. Quantifying home–range overlap: The importance of the utilization distribution. *J. Wildl. Manage.* **2005**, *69*, 1346–1359. [CrossRef]
64. Silverman, B.W. *Density Estimation for Statistics and Data Analysis*; Chapman and Hall: London, UK, 1986; pp. 1–19. ISBN 9780412246203.
65. Shannon, C.E. A mathematical theory of communication. *Bell Syst. Tech. J.* **1948**, *27*, 379–423. [CrossRef]
66. Jost, L. Entropy and diversity. *Oikos* **2006**, *113*, 363–375. [CrossRef]
67. Davison, A.C.; Hinkley, D.V. *Bootstrap Methods and Their Application*; Cambridge University Press: Cambridge, UK, 1997; pp. 212–309. ISBN 9780521574716.
68. McGeoch, M.A.; Latombe, G.; Andrew, N.R.; Nakagawa, S.; Nipperess, D.A.; Roigé, M.; Marzinelli, E.M.; Campbell, A.H.; Vergés, A.; Thomas, T.; et al. Measuring continuous compositional change using decline and decay in zeta diversity. *Ecology* **2019**, *100*, e02832. [CrossRef]

69. Park, B.J.; Lee, J.W.; Kim, J.D.; Heo, T.I.; Lee, D.H.; Lee, J.H.; Nam, K.B.; Kim, H.J.; Shin, C.H.; Byeon, J.G. Forest Stand Structure of *Abies nephrolepis* Population in Mt. Hwangbyeong. *J. Agric. Life Sci.* **2020**, *54*, 35–45. [CrossRef]
70. Latombe, G.; McGeoch, M.A.; Nipperess, D.A.; Hui, C. zetadiv: Functions to Compute Compositional Turnover Using Zeta Diversity. *Methods Ecol. Evol.* **2018**, *9*, 431–442.
71. Akaike, H. A new look at the statistical model identification. *IEEE Trans. Autom. Control* **1974**, *19*, 716–723. [CrossRef]
72. Sale, P.F. Overlap in resource use, and interspecific competition. *Oecologia* **1974**, *17*, 245–256. [CrossRef]
73. Wandrag, E.M.; Catford, J.A.; Duncan, R.P. Quantifying niche availability, niche overlap and competition for recruitment sites in plant populations without explicit knowledge of niche axes. *J. Ecol.* **2019**, *107*, 1791–1803. [CrossRef]
74. Oh, M.; Heo, Y.; Lee, E.J.; Lee, H. Major environmental factors and traits of invasive alien plants determine their spatial distribution: A case study in Korea. *J. Ecol. Environ.* **2021**, *45*, 277–286. [CrossRef]
75. Cañadas, E.M.; Fenu, G.; Peñas, J.; Lorite, J.; Mattana, E.; Bacchetta, G. Hotspots within hotspots: Endemic plant richness, environmental drivers, and implications for conservation. *Biol. Conserv.* **2014**, *170*, 282–291. [CrossRef]
76. Walas, Ł.; Taib, A. Environmental regionalization and endemic plant distribution in the Maghreb. *Environ. Monit. Assess.* **2022**, *194*, 100. [CrossRef] [PubMed]
77. Fois, M.; Fenu, G.; Cañadas, E.M.; Bacchetta, G. Disentangling the influence of environmental and anthropogenic factors on the distribution of endemic vascular plants in Sardinia. *PLoS ONE* **2017**, *12*, e0182539. [CrossRef]
78. Pouteau, R.; Brunel, C.; Dawson, W.; Essl, F.; Kreft, H.; Lenzner, B.; Meyer, C.; Pergl, J.; Pyšek, P.; Seebens, H.; et al. Environmental and socioeconomic correlates of extinction risk in endemic plants. *Divers. Distrib.* **2022**, *28*, 53–64. [CrossRef]
79. Park, B.J.; Heo, T.I.; Cheon, K.I. Vegetation Structure and Habitat Characterization: An Ecological Basis for the Conservation of the Korean Endemic Plant, Taihyun's Abelia (*Zabelia tyaihyonii* (Nakai) Hisauti & H. Hara, 1951; Caprifoliaceae). *Forests* **2025**, *16*, 1042. [CrossRef]
80. Pauchard, A.; Shea, K. Integrating the study of non–native plant invasions across spatial scales. *Biol. Invasions* **2006**, *8*, 399–413. [CrossRef]
81. Lanchier, N.; Neuhauser, C. A spatially explicit model for competition among specialists and generalists in a heterogeneous environment. *Ann. Appl. Probab.* **2006**, *16*, 1385–1410. [CrossRef]
82. Sexton, J.P.; McIntyre, P.J.; Angert, A.L.; Rice, K.J. Evolution and ecology of species range limits. *Annu. Rev. Ecol. Evol. Syst.* **2009**, *40*, 415–436. [CrossRef]
83. Freeman, B.G.; Scholer, M.N.; Ruiz-Gutierrez, V.; Fitzpatrick, J.W. Climate change causes upslope shifts and mountaintop extirpations in a tropical bird community. *Proc. Natl. Acad. Sci. USA* **2018**, *115*, 11982–11987. [CrossRef]
84. Slatyer, R.A.; Hirst, M.; Sexton, J.P. Niche breadth predicts geographical range size: A general ecological pattern. *Ecol. Lett.* **2013**, *16*, 1104–1114. [CrossRef]
85. Fenollosa, E.; Pang, S.E.H.; Briscoe, N.; Guisan, A.; Salguero-Gómez, R. Powerful yet challenging: Mechanistic niche models for predicting invasive species potential distribution under climate change. *Ecography* **2025**, e07775. [CrossRef]
86. Wang, J.G.; Wu, J.W.; Li, W.J. Ecological niche changes and risk regionalization of the invasive plant *Praxelis clematidea*. *Ecol. Evol.* **2025**, *15*, e71546. [CrossRef] [PubMed]
87. MacDougall, A.S.; Gilbert, B.; Levine, J.M. Plant invasions and the niche. *J. Ecol.* **2009**, *97*, 609–615. [CrossRef]
88. Catford, J.A.; Jansson, R.; Nilsson, C. Reducing redundancy in invasion ecology by integrating hypotheses into a single theoretical framework. *Divers. Distrib.* **2009**, *15*, 22–40. [CrossRef]
89. Davis, M.A.; Grime, J.P.; Thompson, K. Fluctuating resources in plant communities: A general theory of invasibility. *J. Ecol.* **2000**, *88*, 528–534. [CrossRef]
90. Thuiller, W.; Lavorel, S.; Araújo, M.B. Niche properties and geographical extent as predictors of species sensitivity to climate change. *Glob. Ecol. Biogeogr.* **2005**, *14*, 347–357. [CrossRef]
91. Barber, Q.E.; Nielsen, S.E.; Hamann, A. Assessing the vulnerability of rare plants using climate change velocity, habitat connectivity, and dispersal ability: A case study in Alberta, Canada. *Reg. Environ. Change* **2016**, *16*, 1433–1441. [CrossRef]
92. Brown, J.H. On the relationship between abundance and distribution of species. *Am. Nat.* **1984**, *124*, 255–279. [CrossRef]
93. D'Amen, M.; Mod, H.K.; Gotelli, N.J.; Guisan, A. Disentangling biotic interactions, environmental filters, and dispersal limitation as drivers of species co-occurrence. *Ecography* **2018**, *41*, 1233–1244. [CrossRef]

forests

MDPI

Article

Habitat Condition of *Tilio–Acerion* Forest Facilitates Successful Invasion of *Impatiens parviflora* DC

Kateryna Lipińska *, Adam Cieśla, Olena Hrynyk, Karol Sokołowski and Radosław Gawryś

Department of Forest Ecology, Forest Research Institute (PL), Braci Leśnej 3, Sękocin Stary, 05-090 Raszyn, Poland
* Correspondence: k.fyalkowska@ibles.waw.pl

Abstract

Impatiens parviflora DC. occurs in various plant communities. Its occurrence has been confirmed in Poland across 13 natural habitats protected under the Habitats Directive. The aim of our work is to determine the differences between the plots with and without *I. parviflora* in terms of the species richness and ecological conditions of the 9180* habitat-type forest. Using data from 315 plots on which a phytosociological relevés was carried out, we analyzed the geographical variability, the Shannon-Winner index and the indicator species for old forests. Flora diversity was represented using the DCA, and the IndVal index was calculated to determine the species that best characterize the differentiated groups. The highest percentage of monitoring plots with *I. parviflora* is located in the Sudetes Mountains (67.7%) and the lowest in the Bieszczady Mountains (7.5%). Plots with *I. parviflora* were characterized by significantly lower tree cover, a higher number of tree species in the stand, a lower height of both the understory and herb layer and a lower number of old forest species. *Impatiens parviflora* does not affect the total number of species in the understorey but is associated with a lower proportion of species typical of old forests. The presence of *I. parviflora* also correlates with a higher proportion of young trees in the understorey, suggesting a link with successional processes and habitat disturbance. The spread of *I. parviflora* is limited by shade-loving trees such as *Abies alba* Mill. and *Fagus sylvatica* L. The diversity of the distribution of *I. parviflora* depends on local conditions, so conservation efforts should take into account the local ecological context.

Keywords: alien species; habitat protection; Natura 2000; slope forests

1. Introduction

The Habitats Directive [1] states that natural habitat types (terrestrial or aquatic areas characterized by geographical, abiotic and biotic features and which are entirely natural or semi-natural) threatened with extinction in their natural range or whose natural range is small due to their regression or spatial limitation, or which are outstanding examples of typical features of one or more biogeographical regions, are to be conserved within the borders of the European Union. Protection refers to a series of measures that are necessary for the conservation or restoration of natural habitats and to which the Member States of the European Union have committed themselves. Of particular importance are the priority natural habitat types, i.e., habitats at risk of disappearing, which are mainly found in the EU.

One of the priority habitats protected by the Habitats Directive is the *Tilio–Acerion* forests of slopes, scree and ravines (code 9180*), which, according to the CORINE classification [2], correspond to the Mixed Ravine and Slope Forests (code 41.4). Habitat type 9180* comprises species-rich deciduous forests with a constant and often dominant proportion of *Acer pseudoplatanus* L. Habitat type 9180* occurs in all mountainous regions of Europe on steep slopes with skeleton-rich soils and active erosion processes. The slopes are usually north-facing, which creates special microclimatic conditions characterized by high humidity, low solar radiation and cool temperatures [3]. In Poland, due to these requirements, they only occur in the mountainous regions of southern Poland and in its foothills. The area of this habitat type in Poland is 20.4 km^2 [4].

Impatiens parviflora DC. is an annual plant that occurs naturally in the mountainous regions of Central Asia. In Europe, the species appeared in the 1830s—the first wild specimens were found in 1831 near the botanical garden in Geneva (Switzerland), where the species was cultivated [5]. In Western Europe, it was mainly found growing near botanical gardens until the end of the 19th century. The first mentions of its occurrence in Poland (in a natural environment) date back to 1850. The current distribution area in Poland covers the entire country [6]. *I. parviflora* has a wide range of ecological requirements for growth and development, which varies in different parts of Europe [7] and is most commonly found in forest communities [5]. It occurs in coniferous and mixed coniferous forests, in mixed forests, deciduous forests (most often in hornbeam forests and beech forests) and even in alder and riparian forests [8]. It also occurs in various non-forest plant communities—disturbed or undisturbed [9–14]. Reproduction of *I. parviflora* is related to its ability to produce a large number of seeds (2000 to 10,000 seeds from one plant) and to their rapid germination [15]. The seeds can remain alive in the soil for up to 3–5 years. The time from germination to flowering is 8–9 weeks. The seeds are actively dispersed up to 3–4 m away. They can also be transported by water currents, mammalian hairs and soil, into the root ball of garden plants [5,16]. Due to the wide geographical distribution of the species and the possibility of mass occurrence, *I. parviflora* is perceived as a threat to 13 natural habitats protected under the Habitats Directive (3220; 3240; 6430; 8210; 8220; 9110; 9130; 9160; 9170; 9180*; 9190; 91E0*; 91F0) [17]. The presence of this species also has a direct impact on the assessment of the conservation status of a particular phytocoenosis [18].

The aim of our work is to determine the differences between the plots with and without *I. parviflora* in terms of the species richness and ecological conditions of the 9180* habitat type forest. In this way, we can learn at a national level about the ecological factors that favour the occurrence of *I. parviflora* in this rare and naturally valuable forest type.

2. Materials and Methods

2.1. Data Collection

Our research is based on 315 plots on which a phytosociological relevés was carried out in 2021. These plots are used to monitor the conservation status of habitat type 9180* as part of the State Environmental Monitoring (SEM) resulting from the provisions of the Habitats Directive [19]. Their distribution is shown in Figure 1.

The phytosociological relevés were carried out according to the method developed by Braun-Blanquet [20]. The percentage cover of each vascular plant species was estimated using a seven-degree cover class scale (r—1–2 plants; +—2–5 plants; 1—species covers < 5%; 2—species cover 5%–25%; 3—species cover 25%–50%; 4—species cover 50%–75%; 5—species cover 75%–100%) [20]. The area of the phytosociological relevés was between 100 and 400 m^2. Within the area of the phytosociological relevés, we estimated cover (%) of the vegetation layers (A—tree layer: height > 5 m, B—shrubs (height 0.5–5 m), C—herbaceous vegetation and trees less than 0.5 m high) and their average height (cm).

The height above sea level was determined using a GPS device. The slope gradient (%) was also measured at each sample area. Plant communities were identified for each relevés. But because relevés are located systematically in vegetation patches, it is not always possible to identify plant communities in the same levels of organization.

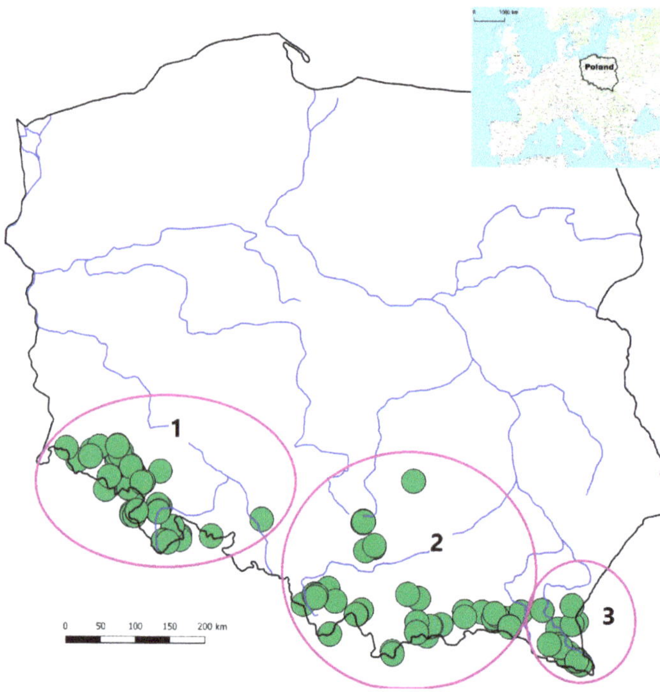

Figure 1. Map of the study area; green dots—locations of phytosociological relevés; 1—Sudetes Mountains; 2—Western Carpathians and Świętokrzyskie Mountains; 3—Bieszczady Mountains.

2.2. Data Analysis

Two plot groups were distinguished: IP0—plots without *Impatiens parviflora* DC. occurrence, IP1—plots with *I. parviflora* occurrence. Based on the floristic lists contained in the phytosociological relevés, the number of plant species in each vegetation layer was determined for each plot. To take into account the geographical variability, the sample plots were divided into groups according to their location: 1—Sudetes Mountains; 2—Western Carpathians and Świętokrzyskie Mountains; 3—Bieszczady Mountains (Figure 1). The Sudetes Mountains occupy the southern part of western Poland. It is a mountain range made of crystalline rock. This mountain range has been repeatedly destroyed and uplifted by erosion. The Carpathians, which lie to the east of the Sudetes Mountains, are a younger and higher mountain range than the Sudetes Mountains. The Carpathians can be divided into the Western Carpathians, which consist mainly of sedimentary rock, and the Eastern Carpathians (Bieszczady Mountains), which consist of sandstone and shale. The Bieszczady Mountains are a gentle mountain range characterized by a large forest cover and a low population density. The Świętokrzyskie Mountains are the lowest mountain range located north of the Western Carpathians. The landscape in the Świętokrzyskie Mountains is characterized by steep slopes, deeply carved valleys, outliers and stone run. The Western Carpathians and the Świętokrzyskie Mountains were grouped together into one group as they have similar floristic communities. Species cover from the Braun-Blanquet scale was converted into the percentage of the phytosociological releve area: r = 0.01%, + = 0.1%, 1 = 5%, 2 = 17.5%, 3 = 37.5%, 4 = 62.5%, 5 = 87.5% [20]. Only vascular plants were analyzed. The Shannon–Wienner diversity index were calculated for each plot. In addition, the average of the ecological indicator values was calculated for each plot. The ligth index (L), the trophism soil index (Tr) and the humidity soil index (W) were taken into account. The numerical values of the individual species were taken from Zarzycki [21]. Zarzycki's ecological indicators are better suited to Polish conditions. In contrast to Ellenberg's indicator numbers, the values of the indices according to Zarzycki do not range from 1 to 9, but from 1 to 5. The value of a particular ecological indicator on the sample plot was calculated as the average value of the individual species occurring on the plot, weighted according to their cover. The indicator species for old forests (plant species which show fidelity to forests which have documented long-term continuity in the area under consideration) were selected according to Dzwonko and Loster [22]. Statistical analyzes were carried out using R (version 4.3.1) [23]. The mean values of the variables between two groups were compared using a Student *t*-test—package "stats" [23]. Rarefaction curves were calculated using the iNEXT package [24,25].

The relationship between the proportion of *I. parviflora* in the surface cover and tree species surface cover was determined using GLM analyses (Poisson distribution was used)—package "stats". Only species of tree layer whose frequency, in this layer, exceeded 10% were included in the analysis. In this way, we were able to exclude species that only occurred sporadically in the habitat. The species eliminated in this way showed no significant correlation with the *I. parviflora* coverage. For the analysis, the percentage cover of *I. parviflora* was rounded to whole numbers. To achieve the best fit of the model, the function "step" from the package "stats" was used, which performed multiple comparisons of different combinations of the given variables to achieve the lowest possible AIC (Akaike Information Criterion) value of the model. The variables used in the analysis were checked for the presence of collinearities by calculating the Variance Inflation Factor using the "vif" function from the "car" package [26].

Flora diversity was represented using the DCA analysis of the "vegan" package [27]. The relationship between plot order and environmental variables was determined using the permutation test ("vegan" package, function "envfit", 999 iterations). The *p*-value was adjusted using the Bonferroni correction. In the DCA analysis, the species coverage rate was converted from the Braun-Blanquet scale [20] to the ordinal scale of van der Maarel [28]: r = 1; + = 2; 1 = 3; 2 = 5; 3 = 7; 4 = 8; 5 = 9.

To determine the species that best characterize the IP0 and IP1 groups an index called IndVal (indicator value) was calculated [29]. The index is a product of two components A (specificity, i.e., the proportion of individual species of species i occurring in class j) and B (fidelity, i.e., the proportion of plots in class j containing species i). The IndVal index was calculated using the package "indicspecies" [30]. To calculate the IndVal index, the cover-abundance values was converted from the Braun-Blanquet scale to the Van der Maarel scale [28].

We additionally performed linear regression analyses using a package "stats" [23]. Three models were fitted using (1) the presence/absence of *I. parviflora*, (2) cover of *I. parviflora* expressed in the van der Maarel scale (species cover was converted from the Braun-Blanquet scale [20] to the ordinal scale of van der Maarel [28]), and (3) percentage cover of *I. parviflora* (species cover from the Braun-Blanquet scale was converted into the percentage of the phytosociological releve area: r = 0.01%; + = 0.1%; 1 = 5%; 2 = 17.5%; 3 = 37.5%; 4 = 62.5%; 5 = 87.5%) as predictors of the Shannon-Winner index (H).

3. Results

The frequency of *Impatiens parviflora* DC. in the 315 analyzed plots was 46%, the arithmetic mean of the coverage of this species in the phytosociological relevés was 13.2 (SE = 1.3%). The analyzed phytosociological data of 9180* habitat type plots represent six plant communities (Table 1). The proportion of plots where *I. parviflora* was detected ranged from 0% to 79.3% in the individual communities. At the plots in the Bieszczady Mountains (N = 80) *I. parviflora* occurred at 7.5% of the plots, in the Western Carpathians (N = 74) at 40.5% and in the Sudetes Mountains (N = 161) at 67.7%.

Table 1. Number (N) of plots in individual plant communities and frequency (F) of identified plots with *Impatiens parviflora* DC.

Plant Community	Total		Bieszczady Mountains		Western Carpathians		Sudetes Mountains	
	N	F (%)	N	F (%)	N	F (%)	N	F (%)
Acer pseudoplatanus-Aruncus sylvestris	6	50	0	0	0	0	6	50
Aceri-Tilietum Faber 1936	82	79.3	0	0	6	16.7	76	84.2
Arunco dioici-Aceretum pseudoplatani Moor 1952	85	44.7	12	0	28	42.9	45	57.8
Phyllitido-Aceretum pseudoplatani Moor 1952	91	25.3	63	9.5	28	60.7	0	0
Sorbo aucupariae-Aceretum pseudoplatani Celiński et Wojterski (1961 n.n.) 1978	17	0	5	0	12	0	0	0
Tilio platyphylli-Acerion Klika 1955	34	47.1	0	0	0	0	34	47.1
Total	315	46	80	7.5	74	40.5	161	67.7

The analysis of the occurrence of *I. parviflora* in different plant communities and mountain ranges showed that the occurrence of the communities varies between the mountains, which also applies to the abundance of the species. The highest frequency of the species was found in the Sudetes Mountains, the lowest in the Bieszczady Mountains. In the Bieszczady Mountains, 80 plots were described (25.4% of the total number of plots). *I. parviflora* was found only in *Phyllitido–Aceretum pseudoplatani* (63 plots), where its abundance was very low (9.5%). In the other communities, *I. parviflora* was not recorded despite numerous detections (*Arunco dioici–Aceretum pseudoplatani*—12 plots; *Sorbo aucupariae–Aceretum pseudoplatani*—5 plots). In the Western Carpathians, 74 plots (23.5%) were described. The species reached the highest frequency in *Phyllitido–Aceretum pseudoplatani* (60.7%; 28 plots), and it also had a significant proportion in *Arunco dioici–Aceretum pseudoplatani* (42.9%; 28 plots). The occurrence of *I. parviflora* in *Aceri–Tilietum* was low (16.7%; 6 plots). The species was absent in *Sorbo aucupariae–Aceretum pseudoplatani* and *Tilio platyphylli–Acerion*. Most plots were described in the Sudetes Mountains (161 plots, i.e., 51.1% of the total). The highest frequency values were also found here. *I. parviflora* was most common in *Aceri–Tilietum* (84.2%; 76 plots) and very common in *Arunco dioici–Aceretum pseudoplatani* (57.8%; 45 plots). It was also very common in *Tilio platyphylli–Acerion* (47.1%; 34 plots). However, the species was not found in *Phyllitido–Aceretum pseudoplatani*, which is practically absent in the Sudetes Mountains.

Compared to IP0 plots, IP1 plots were characterized by a significantly lower tree cover, a higher number of tree species in the stand, a lower height of both the understory and herb layer and a lower number of old forest species (Table 2). These areas were also characterized by a larger area and were located at a lower altitude above sea level. Compared to the IP0 plots, lower habitat preferences of the flora present in the IP1 plots were found in terms of substrate moisture and higher light incidence on the forest floor.

Table 2. Comparison of the average values (median) of environmental variables between plots with *I. parviflora* (IP1) and without (IP0).

Variable	Abbreviation	IP0 N = 170		IP1 N = 145		*t*-Test
		Mean	SE	Mean	SE	*p*
Tree layer (%)	A	78.59	1.32	74.28	1.14	0.014
Shrub layer (%)	B	21.31	1.3	19.18	1.36	0.259
Herbal layer (%)	C	70.91	1.51	68.31	1.8	0.270
Above sea level.	npm	620.51	17.74	443.67	11.06	0.000
Shannon–Wiener index	H	2.97	0.03	2.94	0.03	0.569
Height A of tree layer (m)	H_A	26.71	0.38	26.96	0.31	0.606
Height B of shrub layer (m)	H_B	3.76	0.18	2.83	0.13	0.000
Height C of herbal layer (cm)	H_C	47.94	1.65	33.93	1.41	0.000
Number of species in A-tree layer	N_A	2.98	0.08	3.59	0.12	0.000
Number of species in B-shrub layer	N_B	3.18	0.18	3.5	0.2	0.224
Number of species in C-herbal layer	N_C	23.21	0.69	22.27	0.63	0.316
Old forests species (number of species)	GSL	13.43	0.44	11.46	0.4	0.001
Trophism soil index	Tr	3.83	0.02	3.87	0.02	0.145
Humidity soil index	W	3.35	0.01	3.28	0.01	0.001
Light index	L	2.43	0.03	2.54	0.03	0.004

The results from linear regression analyses are in Table S1 in Supplementary Materials. In all models, regression coefficients were negative and statistically non-significant ($p > 0.05$), indicating that neither the presence nor the cover of the *I. parviflora* significantly affected plant species diversity.

The analysis of the rarefaction curve showing species richness in the IP0 plots was higher than in the IP1 plots, with the same number of sample plots (Figure 2).

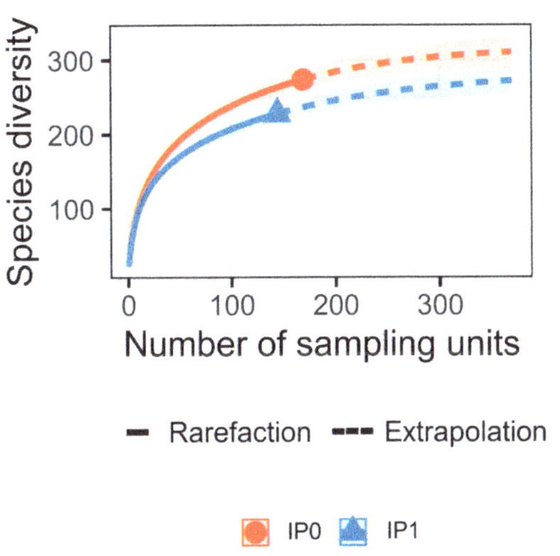

Figure 2. Rarefaction/extrapolation curves of cumulative species richness (Hill's numbers qΔ with q = 0) against the number of sampling plots with *Impatiens parviflora* DC. (blue) and without *I. parviflora* (orange). The shaded area represents 95% confidence intervals obtained using the bootstrap method based on 999 repetitions.

The analysis of the IndVal indicator species showed that the IP0 plots have 17 species from the herb layer, including two tree species, while the IP1 plots have 11 species, including eight tree species (Table 3).

Table 3. List of plant species that are indicators of plots with and without *I. parviflora* in the species composition. Only the species of herbal layer (including trees up to 0.5 m in height) are statistically significant ($p < 0.05$) based on the IndVal analysis. Sequence of the species by IndVal index.

IP0 Plots Species	IndVal	IP1 Plots Species	IndVal
Athyrium filix-femina (L.) Roth	0.650	*Acer platanoides* L.	0.667
Lunaria rediviva L.	0.578	*Dryopteris filix-mas* (L.) Schott	0.667
Oxalis acetosella L.	0.556	*Acer pseudoplatanus* L.	0.614
Impatiens noli-tangere L.	0.549	*Fraxinus excelsior* L.	0.595
Rubus hirtus Waldst. et Kit. agg.	0.525	*Poa nemoralis* L.	0.509
Symphytum cordatum Waldst. et Kit. ex Willd.	0.517	*Quercus petraea* (Matt.) Liebl.	0.415
Asplenium scolopendrium L.	0.515	*Tilia cordata* Mill.	0.414
Glechoma hirsuta Waldst. et Kit.	0.515	*Carpinus betulus* L.	0.413
Dentaria bulbifera L.	0.497	*Tilia platyphyllos* Scop.	0.410
Abies alba Mill.	0.497	*Mycelis muralis* (L.) Dumort.	0.372
Sambucus nigra L.	0.480	*Corylus avellana* L.	0.354
Rubus idaeus L.	0.470	-	-
Polystichum aculeatum (L.) Roth	0.469	-	-
Petasites albus (L.) Gaertn.	0.456	-	-
Circaea lutetiana L.	0.438	-	-
Salvia glutinosa L.	0.437	-	-
Chrysosplenium alternifolium L.	0.414	-	-

Of the 11 tree species in the stand layer whose frequency in the phytosociological relevés was above 10%, the average cover was highest in *Acer pseudoplatanus* and *Fagus sylvatica* L. (Table 4). These species also achieved the highest frequency. Of this group of trees, only *Picea abies* (L.) H.Karst., had no significant association with the occurrence of *I. parviflora*. *Fagus sylvatica*, *Abies alba*, *Carpinus betulus* L., *Ulmus glabra* Huds., *Acer pseudoplatanus* and *Acer platanoides* L. had the strongest limiting effect on *I. parviflora* (in decreasing order). *Quercus petraea* (Matt.) Liebl, *Fraxinus excelsior* L., *Tilia cordata* Mill. and *Tilia platyphyllos* Scop. had a positive association with the occurrence of *I. parviflora*. Of the species studied, the occurrence of *I. parviflora* is best explained by the growth of *Fagus sylvatica*.

The DCA analysis showed a positive correlation between the cover of *I. parviflora* and the location of the monitoring plots in the Sudetes Mountains (Figure 3). The plots at this location are characterized by high light incidence on the forest floor, the occurrence of *Aceri–Tilietum* association and low height of the individual vegetation layers. The plots in the Sudetes Mountains are also characterized by the lowest humidity of the substrate, low ground cover and a low proportion of old forest species. The DCA analysis showed a negative correlation between the cover of *I. parviflora* and the plots in the Bieszczady Mountains, which are usually represented by the *Phyllitido–Aceretum pseudoplatani* plant community, the greatest height of the undergrowth and understorey layer, a high trophism soil index and a low light index. The plots located in the Carpathian Mountains showed a negative correlation with the cover of *I. parviflora* and are characterized by the highest altitude above sea level, species richness, low cover of layer A and high cover of layer C. The cover of layer B and the number of species in this layer do not correlate with the arrangement of the sample plots.

Table 4. Descriptive statistics of surface cover in the sample plot of selected tree species (F > 10%) and the result of the GLM analysis (family = Poisson) in the form of an exponential function of the parameter explaining the cover of *I. parviflora* (all variables in one model). AIC = 4932.4. *—*p*-value < 0.05. Abbreviations: F—Species frequency on sample plots; M—Mean; SE—Standard error; exp (coefficient)—exponential function from the value of the model coefficient.

Gatunek	F (%)	M	SE	Exp (Coefficient)
Intercept				9.182 *
Abies alba	17.1	2.81	0.49	0.977 *
Acer platanoides	30.8	7.56	0.86	0.997 *
Acer pseudoplatanus	75.9	30.34	1.53	0.994 *
Carpinus betulus	21.0	3.85	0.57	0.981 *
Fagus sylvatica	56.2	15.03	1.20	0.973 *
Fraxinus excelsior	18.4	4.56	0.64	1.013 *
Picea abies	14.0	1.45	0.28	0.999
Quercus patrea	12.4	2.04	0.38	1.022 *
Tilia cordata	21.6	5.55	0.79	1.011 *
Tilia platyphyllos	26.7	6.71	0.79	1.004 *
Ulmus glabra	16.8	3.02	0.52	0.981 *

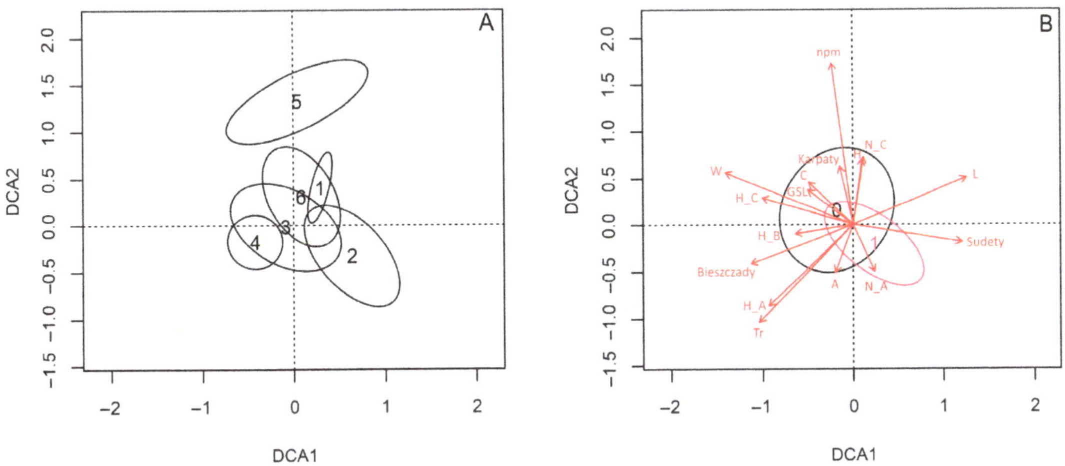

Figure 3. Result of the DCA analysis. (**A**) DCA1 = 0.361; (**B**) DCA2 = 0.294. (**A**) Centroid location (ellipse = 1SD from centroid) of plots according to plant communities belonging to natural habitat type 9180* (1—*Arunco dioici–Aceretum pseudoplatani*, 2—*Aceri–Tilietum*, 3—*Lunario–Aceretum pseudoplatani*, 4—*Phyllitido–Aceretum pseudoplatani*, 5—*Sorbo aucupariae–Aceretum pseudoplatani*, and 6—*Tilio platyphylli–Acerion*). (**B**) Interactions between the centroid locations (ellipse = 1SD from centroid) of plots according to the presence of *I. parviflora* (0—not present, 1—present) and environmental variables (arrows). The variables from Table 2 and the location of the plots were taken into account; Bieszczady—Bieszczady Mountains Karpaty—Western Carpathians, Sudety—Sudetes Mountains, other abbreviations as in Table 2 (Only those variables whose adjustment was statistically significant at $p < 0.05$ based on the permutation test were shown).

4. Discussion

Invasive species are one of the most important direct causes of biodiversity loss and changes in the functioning of forest ecosystems [31,32]. Even non-invasive species that are not native to a particular ecosystem pose such a threat [33]. It is generally believed that *I. parviflora* can pose a significant threat to deciduous forests, especially riparian (91E0*,

91F0), oak-hornbeam (9160, 9170) and beech forests (9110, 9130, 9150) that occur in humid habitats [17,34]. In our studies, the habitat plots in which *I. parviflora* occurred did not differ significantly from plots without *I. parviflora* in terms of shrub and herb layer cover, and the average number of species was similar. However, the plots with *I. parviflora* were characterized by a lower number of old growth forests species, a higher number of species in the tree layer and a lower cover of this layer. Old growth forests have a floristic composition that differs significantly from that of younger forests [35]. Old growth forest species are generally more shade-tolerant than other forest plant species and avoid dry and moist sites [36]. This group also includes species of geophytes found by Florianova and Münzbergová to be negatively affected by *I. parviflora* in oak-hornbeam forests [37]. Although there is no direct correlation between the occurrence of *I. parviflora* and the average number of herbaceous plants in the plots, our results based on the analysis of indicator species show that the occurrence of old forest species characterizes plots without *I. parviflora*. However, *I. parviflora* was found to be associated with certain species of ground cover, mainly with the regeneration of tree species up to a height of 0.5 m. It is likely that woody species will become strong competitors for *I. parviflora* with increasing height in the long term [38]. Most likely, the relationship between the occurrence of woody species up to 0.5 m in height in the herb layer on the IP1 plots and the more numerous occurrences of species in old forests on the IP0 plots is the reason why there are no differences in the average number of species between plots (with and without *I. parviflora*).

Most alien species are unable to invade the interior of the forest and are restricted to the boundary between the forest and, for example, agricultural land [39]. *I. parviflora* most commonly invades forest habitats in overexposed areas such as forest edges and under natural or artificial gaps in the canopy [40]. Studies from Slovenia [41] confirm the preference of *I. parviflora* in *Tilio–Acerion* forests for disturbed stands with low canopy closure and modified ecological conditions. An important finding is that *I. parviflora* does not grow in plots together with *Athyrium filix-femina* (L.) Roth. This native species most likely prevents the invasion of *I. parviflora* due to its size and considerable competitive ability [42]. It should be noted that areas with *I. parviflora* were characterized by a lower height of the herb layer and shrub layer. The greater height of the lower vegetation layers leads to greater shading of the forest floor [43]. As shown in studies [9,44,45], *I. parviflora* is not competitive in full shade. It is unable to compete effectively for limited resources in the absence of light. This view is supported by our research, which shows that an increasing proportion of *Abies alba* and *Fagus sylvatica* in the stand has a limiting effect on the surface cover of *I. parviflora*. Researchers from Slovakia, who examined plots in oak forests with *Quercus robur* and *Quercus petraea* with *Fagus sylvatica*, showed that the colonization of the plot by *I. parviflora* was only effective if there was no *Fagus sylvatica* in the stand [46].

All the plots (both with and without *I. parviflora*) studied had an average tree cover of over 70% (A), while the average cover value in the plots with *I. parviflora* was significantly lower. The average value of the light index is also significantly higher in the plots with *I. parviflora* and the average value of the moisture index is significantly lower. This indicates that the presence of *I. parviflora* is associated with a slightly greater light incidence on the forest floor. This may be due to the presence of gaps, the presence of trees with openwork crowns (*Fraxinus excelsior*) or the presence of evenly thinned stands [46]. Dobravolskaitė [47] also found a correlation between an increase in the amount of light and the number of *I. parviflora*. At the same time, Dobravolskaitė [47] highlights that *I. parviflora* can also occur in Lithuanian forests in places that are too dark for native species. In addition, lower light exposure leads to higher humidity of the forest floor, which can partially compensate for the lack of light [43]. The results of the monitoring of shady fir forest (habitat type 91P0—

Holy Cross fir forests (*Abietetum polonicum*)) show that *I. parviflora* occurs sporadically in this habitat, but does not form dense stands, so it does not pose a threat to the species composition of the ground cover [48]. Also, in the *Luzulo–Fagetum* beech forests (habitat type 9110), where tree density reached 90%, *I. parviflora* was only found in clumps up to 15% cover [8]. The opposite reaction of *I. parviflora* to light is described by Krokaite [45], who states that an increase in shading favours greater cover of this species. Since they also indicate a negative correlation with the number of herbaceous plant species, it is possible that the light avoidance of *I. parviflora* is due to competition from other plants occurring in the ground layer. Čuda [43] describes the low competitive capabilities of *I. parviflora* and states that *I. parviflora* owes its dispersal success not to its competitive ability, but to its ability to avoid competition through tolerance to light deficiency.

Hejda [42] considers it unlikely that *I. parviflora* affects the native species richness of colonized communities, which could be related to its limited ability to form a dense canopy and a modest root system. Obidziński and Symonides [9] state that *I. parviflora* prefers degraded habitats, and therefore floristically poor communities. The negative relationship between *I. parviflora* and species richness therefore results from its adaptive abilities and its ability to colonize areas where other plant species cannot grow [47]. In our study, the hypothesis of the occurrence of *I. parviflora* in disturbed areas can be supported by the lower number of old-growth forest species in the *I. parviflora* plots. The same applies to the results of the IndVal analysis, which indicate the simultaneous occurrence of numerous woody species in the understorey with *I. parviflora*. Both results may indicate areas in which secondary or primary forest succession is taking place [49].

We found no differences in the number of species in herbal layer in the plots. Similar results were obtained by Hejda [42] for floodplain forests and mesophilic forests with beech and oak, and by Diekmann [50] for different types of lowland communities with *I. parviflora* in Germany. However, our analysis of the rarefaction curves showed differences in species richness at the landscape level. The lower species richness at the landscape level of the *I. parviflora* plot could be primarily due to the fact that it was found in a lower number of plant communities compared to plots without *I. parviflora*. Added to this is its occurrence above sea level. Studies on the spatial variability of *I. parviflora* occurrence have shown that the average altitude above sea level in the plots with *I. parviflora* was almost 180 m lower than in the IP0 plots. In its natural range, *I. parviflora* occurs at an altitude of up to 3000 m above sea level. Studies carried out in Germany (in low mountain ranges at an altitude of 300–1200 m) showed low frost hardiness and a low survival rate of the seedlings [51]. In our research *I. parviflora* was not found in the highest elevations of 9180* habitat type representing the *Sorbo aucupariae–Aceretum pseudoplatani* community. The limiting effect of altitude above sea level on the occurrence of *I. parviflora* was also demonstrated by Lapin et al. in Austria [52]. No differences in the fertility of the substrate, expressed by the trophism soil index (Tr), were found between forests plots with and without *I. parviflora*. Invasive species generally prefer sites with higher fertility, but a necessary condition is temporal coincidence with the availability of dispersal opportunities [53,54]. This suggests that for *I. parviflora*, climate rather than soil fertility is the stronger limiting factor in mountain forests. Areas with and without *I. parviflora* differ not only in their altitude above sea level, but also in their geographical distribution. The areas of 9180* habitat type forests where *I. parviflora* occurs, are concentrated in the south-west of Poland, while the 9180* habitat type forests are equally numerous in the south-east of Poland. This is due to the fact that habitat type 9180* is phytosociologically very diverse within its distribution area throughout the country. In the Bieszczady Mountains (south-east Poland), the soils are more fertile than in the mountain ranges to the west, and the proportion plots with *I. parviflora* is only 7.5%. In the Bieszczady Mountains, *I. parviflora* occurs only sporadically and is restricted to isolated areas of *Phyllitido–Aceretum pseudoplatani*. In the Sudetes Mountains, proportion plots with *I. parviflora* is almost 70%. The results show that the Sudetes Mountains are the main distribution area of this species in *Tilio–Acerion* forests. In the Western Carpathians, the most important habitats for *I. parviflora* are sycamore communities, especially those with rich substrate and high humidity. The differences in the attendance of *I. parviflora* between the Bieszczady Mountains and the other mountain ranges could be due to the lower anthropogenic pressure in the Bieszczady Mountains and less disturbed sites favouring the distribution of *I. parviflora* [9,40,42,46]. But also to the lack of communities where *I. parviflora* finds optimal conditions, such as *Aceri–Tilietum* (Table 1), which is confirmed by the results of the DCA analysis (Figure 3A). However, the lack of differences in trophism soil index values within the country may also be due to the uneven distribution of *I. parviflora* within the range of habitat 9180*.

The results we have presented refer to the entire range of habitat type 9180* in Poland. Our research also shows that within a natural habitat—a unit that can include many different communities and geographical areas—the influence of ecological factors may depend on local conditions. Therefore, limiting the analysis to a smaller area of habitat occurrence and thus limiting the variability of the analyzed data may lead to different results. The same result can be expected if the analysis is limited to a specific plant community. However, looking at the overall variability of habitat across the country allows general conclusions to be drawn and highlights ecological factors that should be considered when planning research or conservation actions on a smaller spatial scale.

5. Conclusions

Our work is based on a comparison of species composition and ecological conditions in study plots in habitat 9180* with and without *I. parviflora*. The sample plots are located in the entire range of habitat 9180* in Poland. No differences were found in the number of species in the herb layer between the two types of plots analyzed (with and without *I. parviflora*). However, it was found that the presence of *I. parviflora* is associated with a lower number of species typical of old-growth forests. At the same time, *I. parviflora* presence is accompanied by numerous tree species in the herb layer, which could indicate a connection with stand succession processes and confirms the thesis of other authors regarding the association of *I. parviflora* with disturbed sites. *I. parviflora* cannot cope with limited light conditions, so its spread is limited by shade-loving species, especially woody species such as *Abies alba* and *Fagus sylvatica*. The frequency of occurrence of *I. parviflora* varies within the range of habitat type 9180*, most likely due to the different human pressure and the presence of communities where this species finds optimal development conditions (e.g., *Aceri–Tilietum*). We observed a negative relationship between altitude above sea level and the occurrence of *I. parviflora*, indicating climatic constraints. The impact of *I. parviflora* on habitat type 9180* is dependent on local environmental conditions. Therefore, conservation measures should be planned based on the local habitat context.

Supplementary Materials: The following supporting information can be downloaded at: https://www.mdpi.com/article/10.3390/f16091475/s1, Table S1: Results of linear regression models testing the effect of *Impatiens parviflora* (presence and cover) on Shannon-Wienner index (H).

Author Contributions: Conceptualization, K.L. and R.G.; methodology, R.G.; formal analysis, R.G.; investigation, O.H.; resources, A.C.; data curation, O.H.; writing—original draft preparation, K.L.; writing—review and editing, K.S.; supervision, A.C.; project administration, A.C.; funding acquisition, A.C. All authors have read and agreed to the published version of the manuscript.

Funding: This research and APC were conducted on behalf of The Chief Inspectorate of Environmental Protection (CIEP) and financed by the National Fund for Environmental Protection and Water Management, grant number 540-103.

Data Availability Statement: Data is unavailable due to ethical restrictions. The data presented in this study are available on request from the corresponding author.

Acknowledgments: We would like to thank Andrew Carr for his assistance with the linguistic correction. We also sincerely thank the editor and the anonymous reviewers for their valuable comments and suggestions, which have significantly improved the quality of this paper.

Conflicts of Interest: The authors declare no conflicts of interest.

References

1. Directive, H. Council Directive 92/43/EEC of 21 May 1992 on the conservation of natural habitats and of wild fauna and flora. *Off. J. Eur. Union* **1992**, *206*, 50.
2. European Commission: Directorate-General for Environment. *Corine Biotopes Manual—A Method to Identify and Describe Consistently Sites of Major Importance for Nature Conservation. Volume III, Data Specifications, Part II*; Publications Office of the European Union: Luxembourg, 1991.
3. European Commission. *Interpretation Manual of European Union Habitats—EUR28*; European Commission: Brussel, Belgium, 2013.
4. Article 17 Web Tool. Available online: https://nature-art17.eionet.europa.eu/article17/ (accessed on 12 January 2024).
5. Trepl, L. *Über Impatiens parviflora DC. als Agriophyt in Mitteleuropa*; Dissertationes Botanicae; Schweizerbart Science Publishers: Stuttgart, Germany, 1984; Volume 73; pp. 1–400.
6. Zając, A.; Zając, M. *Atlas Rozmieszczenia Roślin Naczyniowych w Polsce*; Pracownia Chorologii Komputerowej Instytutu Botaniki Uniwersytetu Jagiellońskiego: Kraków, Poland, 2001.
7. Godefroid, S.; Koedam, N. Comparative ecology and coexistence of introduced and native congeneric forest herbs: *Impatiens parviflora* and *I. noli-tangere*. *Plant Ecol. Evol.* **2010**, *143*, 119–127. [CrossRef]
8. Chmura, D.; Orczewska, A. Udział *Impatiens parviflora* DC. w zbiorowiskach leśnych Wyżyny Śląskiej i Płaskowyżu Głubczyckiego. *Arch. Ochr. Sr.* **2004**, *30*, 117–136.
9. Obidziński, T.; Symonides, E. The influence of the groundlayer structure on the invasion of small balsam (*Impatiens parviflora* DC.) to natural and degraded forest. *Acta Soc. Bot. Pol.* **2000**, *69*, 311–318. [CrossRef]
10. Chmura, D.; Sierka, E. Relation between invasive plant *Impatiens parviflora* and species richness of forest floor vegetation. *Pol. J. Ecol.* **2006**, *54*, 417–428.
11. Sołtys, A.; Barabasz-Krasny, B. Ekspansja wybranych gatunków obcego pochodzenia we florze i szacie roślinnej Ojcowskiego Parku Narodowego (Południowa Polska). [Expansion of the selected alien plant species into the flora of the Ojców National Park (Southern Poland)]. *Prądnik. Pr. I Mater. Muz. im. Prof. W. Szafera* **2010**, *20*, 333–376.
12. Towpasz, K.; Stachurska-Swakoń, A. The analysis of the forest flora of the Strzyżowskie Foothills from the perspective of presence of anthropogenic species. *Acta Univ. Lodz. Folia Biol. Et. Oecologica* **2011**, *7*, 99–110. [CrossRef]
13. Chmura, D. *Biology and Ecology of an Invasion of Impatiens Parviflora DC. in Natural and Semi-Natural Habitats*; Wydawnictwo ATH: Bielsko-Biała, Poland, 2014. Available online: https://www.researchgate.net/publication/265250389 (accessed on 14 September 2025).
14. Barabasz-Krasny, B.; Możdżeń, K.; Sołtys-Lelek, A.; Stachurska-Swakoń, A. Biological traits of *Impatiens parviflora* DC. under different habitat conditions. *Not. Bot. Horti Agrobo.* **2018**, *46*, 277–285. [CrossRef]
15. Perrins, J.; Fitter, A.; Williamson, M. Population biology and rates of invasion of three introduced *Impatiens* species in the British Isles. *J. Biogeogr.* **1993**, *20*, 33–44. [CrossRef]
16. Coombe, D.E. Biological Flora of the British Isles, *Impatiens parviflora* DC. *J. Ecol.* **1956**, *44*, 701–713. [CrossRef]
17. Tokarska-Guzik, B.; Dajdok, Z.; Zając, M.; Zając, A.; Urbisz, A.; Danielewicz, W.; Hołdyński, C. *Rośliny Obcego Pochodzenia w Polsce ze Szczególnym Uwzględnieniem Gatunków Inwazyjnych*; Generalna Dyrekcja Ochrony Środowiska: Warszawa, Poland, 2012; pp. 1–197.
18. Świerkosz, K.; Bodziarczyk, J. 9180* Sycamore and maple-lime forests of slopes, screes and ravines Tilio platyphyllis-Acerion pseudoplatani. In *Monitoring of Natural Habitats. Methodological Guide for 5 Natural Habitats: 3110, 6210, 6520, 7230 and 9180**; Mróz, W., Ed.; GIOŚ: Warszawa, Poland, 2013; pp. 99–108.
19. Mróz, W.; Kornatowska, B. Monitoring of natural habitats. In *Methodological Guide for 5 Habitats (3110, 6210, 6520, 7230, 9180*)*; Library of Environmental Monitoring: Warszawa, Poland, 2017.
20. Braun-Blanquet, J. *Pflanzensoziologie. Grundzüge der Vegetationskunde, 3:865*; Springer Verlag: Wien, Austria, 1964.

21. Zarzycki, K.; Trzcińska-Tacik, H.; Różański, W.; Szeląg, Z.; Wołek, J.; Korzeniak, U. Ekologiczne liczby wskaźnikowe roślin naczyniowych Polski. In *Różnorodność Biologiczna Polski, 2*; Instytut Botaniki im. W. Szafera PAN: Kraków, Poland, 2002.
22. Dzwonko, Z.; Loster, S. Wskaźnikowe gatunki roślin starych lasów i ich znaczenie dla ochrony przyrody i kartografii roślinności. Typologia zbiorowisk i kartografia roślinności w Polsce. IGiPZ PAN. *Pr. Geogr.* **2001**, *178*, 120–132.
23. R Core Team. *R: A Language and Environment for Statistical Computing*; R Foundation for Statistical Computing: Vienna, Austria, 2023. Available online: https://www.R-project.org/ (accessed on 14 September 2025).
24. Chao, A.; Gotelli, N.J.; Hsieh, T.C.; Sander, E.L.; Ma, K.H.; Colwell, R.K.; Ellison, A.M. Rarefaction and extrapolation with Hill numbers: A framework for sampling and estimation in species diversity studies. *Ecol. Monogr.* **2014**, *84*, 45–67. [CrossRef]
25. Hsieh, T.C.; Ma, K.H.; Chao, A. iNEXT: An R package for interpolation and extrapolation of species diversity (Hill numbers). *Methods Ecol. Evol.* **2016**, *7*, 1451–1456. [CrossRef]
26. Fox, J.; Weisberg, S. *An {R} Companion to Applied Regression*, 3rd ed.; Sage Publications: Thousand Oaks, CA, USA, 2019. Available online: https://www.john-fox.ca/Companion/ (accessed on 14 September 2025).
27. Oksanen, J.; Simpson, G.; Blanchet, F.; Kindt, R.; Legendre, P.; Minchin, P.; O'Hara, R.; Solymos, P.; Stevens, M.; Szoecs, E.; et al. Vegan: Community Ecology Package. R package Version 2.6-4. 2022. Available online: https://CRAN.R-project.org/package=vegan (accessed on 14 September 2025).
28. Van der Maarel, E. Tranformation of cover-abundance values in phytosociology and its effect on community similarity. *Vegetation* **1979**, *39*, 97–104. [CrossRef]
29. Dufrene, M.; Legendre, P. Species assemblages and indicator species: The need for a flexible asymmetrical approach. *Ecol. Monogr.* **1997**, *67*, 345–366. [CrossRef]
30. De Cáceres, M.; Legendre, P. Associations between species and groups of sites: Indices and statistical inference. *Ecology* **2009**, *90*, 3566–3574. [CrossRef]
31. Convention on Biological Diversity. 1992. Available online: https://www.cbd.int (accessed on 15 October 2023).
32. Zav'âlova, L. *Pro Sučasnì Pìdhodi do Vyvčennâ Fìtoìnvazìjna Teritotìï Ob'êktìv PZF Ukraìni. [About Modern Approaches to the Investigation of Plant Invasions in Protected Areas of Ukraine. Dynamics of Biological and Landscape Diversity of Protected Areas] Dinamìka Bìologìčnogo ta Landšaftnogo Rìznomanìttâ Zapovìdnih Teritorìj*; Drukarnâ Ruta: Kam'yanets-Podilskyi, Ukraine, 2016; pp. 46–49.
33. Myroslav, S.V.; Vira, P.V.; Raisa, B.I.; Liudmyla, Z.V.; Oksana, K.O.; Olga, K.M. Historical overview of the studies of alien flora of Ukraine. *Acta Horti Bot. Bucurestiensis* **2018**, *45*, 5–32.
34. Adamowski, W.; Chmura, D.; Hołdyński, C. *Opracowanie Zasad Kontroli i Zwalczania Inwazyjnych Gatunków Obcych Wraz z Przeprowadzeniem Pilotażowych Działań i Edukacją Społeczną ze Środków Unii Europejskiej w Ramach Programu Infrastruktura i Środowisko 2014–2020*; Projekt nr POIS.02.04.00-00-0100/16pn. Karta informacyjna gatunku; Government of Poland: Warsaw, Poland, 2020; 10p.
35. Schmidt, M.; Mölder, A.; Schönfelder, E.; Engel, F.; Schmiedel, I.; Culmsee, H. Determining ancient woodland indicator plants for practical use: A new approach developed in northwest Germany. *For. Ecol. Manag.* **2014**, *330*, 228–239. [CrossRef]
36. Hermy, M.; Honnay, O.; Firbank, L.; Grashof-Bokdam, C.; Lawesson, J.E. An ecological comparison between ancient and other forest plant species of Europe, and the implications for forest conservation. *Biol. Conserv.* **1999**, *91*, 9–22. [CrossRef]
37. Florianova, A.; Münzbergová, Z. Invasive *Impatiens parviflora* has negative impact on native vegetation in oak-hornbeam forests. *Flora* **2017**, *226*, 10–16. [CrossRef]
38. Łysik, M. Ten years of change in ground-layer vegetation of European beech forest in the protected area (Ojcow national park, south Poland). *Pol. J. Ecol.* **2008**, *56*, 17–31.
39. Honnay, O.; Verheyen, K.; Butaye, J.; Jacquemyn, H.; Bossuyt, B.; Hermy, M. Possible effects of habitat fragmentation and climate change on the range of forest plant species. *Ecol. Lett.* **2002**, *5*, 525–530. [CrossRef]
40. Jarcuška, B.; Slezák, M.; Hrivnák, R.; Senko, D. Invasibility of alien *Impatiens parviflora* in temperate forest understories. *Flora* **2016**, *224*, 14–23. [CrossRef]
41. Kermavnar, J.; Kozamernik, E.; Kutnar, L. Assessing the Heterogeneity and Conservation Status of the Natura 2000 Priority Forest Habitat Type Tilio-Acerion (9180*) Based on Field Mapping. *Forests* **2023**, *14*, 232. [CrossRef]
42. Hejda, M. What Is the Impact of Impatiens parviflora on Diversity and Composition of Herbal Layer Communities of Temperate Forests? *PLoS ONE* **2012**, *7*, e39571. [CrossRef] [PubMed]
43. Depauw, L.; Perring, M.P.; Landuyt, D.; Maes, S.L.; Blondeel, H.; De Lombaerde, E.; Brūmelis, G.; Brunet, J.; Closset-Kopp, D.; Czerepko, J.; et al. Light availability and land-use history drive biodiversity and functional changes in forest herb layer communities. *J. Ecol.* **2020**, *108*, 1411–1425. [CrossRef]
44. Čuda, J.; Skálová, H.; Janovský, Z.; Pyšek, P. Competition among native and invasive *Impatiens* species: The roles of environmental factors, population density and life stage. *AoB Plants* **2015**, *7*, plv033. [CrossRef]

45. Krokaite, E.; Janulionien, R.; Jociene, L.; Rekašius, T.; Rajackait, G.; Paulauskas, A.; Marozas, V.; Kupcinskien, E. Relating Invasibility and Invasiveness: Case Study of *Impatiens parviflora*. *Front. Ecol. Evol.* **2022**, *10*, 845947. [CrossRef]
46. Kermavnar, J.; Kutnar, L. Habitat Degradation Facilitates the Invasion of Neophytes: A Resurvey Study Based on Permanent Vegetation Plots in Oak Forests in Slovenia (Europe). *Plants* **2024**, *13*, 962. [CrossRef]
47. Dobravolskaitė, R. Alien species *Impatiens parviflora* invasion into forest communities of Lithuania. *Bot. Lith.* **2012**, *18*, 3–12. [CrossRef]
48. Cieśla, A.; Mionskowski, M.; Müller, I.; Perzanowska, J.; Korzeniak, J.; Gawryś, R.; Kolada, A.; Barańska, A.; Bielczyńska, A.; Bociąg, K.; et al. Stan ochrony siedlisk przyrodniczych w Polsce w latach 2013–2018. *Biul. Monit. Przyr.* **2021**, *24/4*, 7–121.
49. Seidl, R.; Potterf, M.; Müller, J.; Turner, M.G.; Rammer, W. Patterns of early post-disturbance reorganization in Central European forests. *Proc. R. Soc. B* **2024**, *291*, 20240625. [CrossRef]
50. Diekmann, M.; Effertz, H.; Baranowski, M.; Dupré, C. Weak effects on plant diversity of two invasive *Impatiens* species. *Plant Ecol.* **2016**, *217*, 1503–1514. [CrossRef]
51. Laube, J.; Sparks, T.; Bässler, C.; Menzel, A. Small differences in seasonal and thermal niches influence elevational limits of native and invasive balsams. *Biol. Conserv.* **2015**, *191*, 682–691. [CrossRef]
52. Lapin, K.; Oettel, J.; Steiner, H.; Langmaier, M.; Sustic, D.; Starlinger, F.; Kindermann, G.; Frank, G. Invasive Alien Plant Species in Unmanaged Forest Reserves, Austria. *NeoBiota* **2019**, *48*, 71–96. [CrossRef]
53. Davis, M.A.; Grime, J.P.; Thompson, K. Fluctuating resources in plant communities: A general theory of invasibility. *J. Ecol.* **2000**, *88*, 528–534. [CrossRef]
54. Blumenthal, D.; Mitchell, C.E.; Pysek, P.; Jarosík, V. Synergy between pathogen release and resource availability in plant invasion. *Proc. Natl. Acad. Sci. USA* **2009**, *106*, 7899–7904. [CrossRef]

forests

MDPI

Article

Mangroves Invaded by *Spartina alterniflora* Loisel: A Remote Sensing-Based Comparison for Two Protected Areas in China

Di Dong [1,2,3,†], Qing Gao [1,†] and Huamei Huang [1,2,*]

[1] South China Sea Institute of Planning and Environment Research, State Oceanic Administration, Guangzhou 510300, China; dongdide90@163.com (D.D.); gaoqing182020@163.com (Q.G.)
[2] Technology Innovation Center for South China Sea Remote Sensing, Surveying and Mapping Collaborative Application, Ministry of Natural Resources, Guangzhou 510300, China
[3] Key Laboratory of Marine Environmental Survey Technology and Application, Guangzhou 510300, China
* Correspondence: scsmeei@163.com
† These authors contributed equally to this work.

Abstract: Mangroves are one of the world's most productive and ecologically important ecosystems, and they are threatened by the widespread invasion of *Spartina alterniflora* Loisel in China. As few studies have examined the spatial pattern differences of *S. alterniflora* invasion and the nearby mangroves in different latitudes, we chose the Zhangjiang Estuary and the Dandou Sea, two representative mangrove–salt marsh ecotones in the north and south of the Tropic of Cancer, as the study areas for comparison. The object-based image analysis and visual interpretation methods were combined to construct fine-scale mangrove and *S. alterniflora* maps using high-resolution satellite imagery from 2005 to 2019. We applied spatial analysis, centroid migration, and landscape indexes to analyze the spatio–temporal distribution changes of mangroves and *S. alterniflora* in these two ecotones over time. We used the landscape expansion index to investigate the *S. alterniflora* invasion process and expansion patterns. The annual change rates of mangrove and *S. alterniflora* areas in the Zhangjiang Estuary showed a continuous growth trend. However, the mangrove areas in the Dandou Sea showed a fluctuating trend of increasing, decreasing, and then increasing again, while *S. alterniflora* areas kept rising from 2005 to 2019. *Spartina alterniflora* showed larger annual change rates compared with mangroves, indicating rapid *S. alterniflora* invasion in the intertidal zones. The opposite centroid migration directions of mangroves and *S. alterniflora* and the decreasing distances between the mangrove and *S. alterniflora* centroids indirectly revealed the fierce competition between mangroves and *S. alterniflora* for habitat resources. Both regions saw a decrease in mangrove patch integrality and connectivity. The integrality of mangrove patches in the Zhangjiang Estuary was always higher than those in the Dandou Sea. We observed the growth stage (2011–2014) and outbreak stage (2014–2019) of *S. alterniflora* expansion in the Zhangjiang Estuary and the outbreak stage (2005–2009) and plateau stage (2009–2019) of *S. alterniflora* expansion in the Dandou Sea. The expansion pattern of *S. alterniflora* varies in time and place. Since the expansion of *S. alterniflora* in the outbreak stage is rapid, with a large annual change rate, early warning of *S. alterniflora* invasion is quite important for the efficient and economical removal of the invasive plant. Continuous and accurate monitoring of *S. alterniflora* is highly necessary and beneficial for the scientific management and sustainable development of coastal wetlands.

Keywords: mangroves; *Spartina alterniflora*; invasion; remote sensing; landscape expansion index; ecotone; spatial analysis; centroid migration

1. Introduction

Mangroves are woody trees or shrubs that grow mainly along sheltered coastlines within the tropic and subtropic latitudes [1]. As one of the world's most productive ecosystems, mangroves play crucial roles in protecting and stabilizing coastlines [1–5], providing feeding habitat for animals [6,7], maintaining ecological stability and biodiversity [8–10], mitigating global warming [11], etc. Despite having high ecological and socio-economic values, mangroves are being deforested or degraded [12] due to rapid agriculture reclamation, urban development, aquaculture encroachment, invasive plants, and other stresses caused by climate changes and human disturbances [13–17]. The protection and restoration of mangroves have attracted much attention [18–20]. Since the 1980s, significant efforts have been made to protect and manage mangroves [21,22]. The Chinese government has allocated enormous amounts of funds for mangrove restoration [23] and established many nature reserves to resist the impact of human activities on the environment for mangrove conservation [24].

Spartina alterniflora Loisel, the only coastal salt marsh species among the sixteen most serious invasive alien species listed by the State Environmental Protection Administration of China in 2003 [25], poses tremendous threats to the native biodiversity of mangrove ecosystems in China [21]. Its strong adaptability and reproductive capacity have enabled *S. alterniflora* to spread widely across Chinese coastal wetlands [26,27] and encroach on many nature reserves [28,29]. Researchers found that *S. alterniflora* invasion threatened the survival and breeding of mangroves [17], and mangroves are vulnerable to continuous *S. alterniflora* invasion, which results in the loss of ecological functions and services [30,31]. Prior research primarily focused on the mechanisms of how *S. alterniflora* invasion impacted mangroves at microscopic scales, such as the diversity and composition of sediment diazotrophic communities in different habitats and ecosystems' physical and chemical responses [32,33]. The spatial pattern differences of *S. alterniflora* invasion and its nearby mangrove distributions in different latitudes have been less studied.

Mangroves [12,21,22,34–41] and *S. alterniflora* [26,27,42–46] have been monitored extensively with remote sensing due to the advantages of large geographic coverage, access to inaccessible areas, and the capability to track long-term holistic changes. For example, Giri et al. [47] created the first global mangrove map for the year 2000 with Landsat images and a hybrid classification method combining supervised and unsupervised algorithms. Chen et al. [48] mapped the mangrove distribution in China for the year 2015 with a phenology-based classification method based on Landsat 7/8 and Sentinel-1A imagery time series in Google Earth Engine. Bunting et al. [49,50] generated the first time-series database of global mangrove forests using a two-iteration mangrove mapping approach based on ALOS and Landsat images. Jia et al. [12] built a dataset of mangrove distributions in China based on Landsat imagery from 1973 to 2015, with object-based image analysis (OBIA) and classification methods. Wang et al. [51] generated annual maps of mangroves, salt marshes, and tidal flats in China with a spatial resolution of 30 m, based on Landsat images from 1984 to 2018. Using 2-m resolution Ziyuan-3 and Gaofen-1 satellite images together with field data, Zhang et al. [52] presented a mangrove map of China for 2018. Jia et al. [34] produced a global mangrove dataset at a resolution of 10 m for the year 2020, using OBIA and random forest classification based on Sentinel-2 imagery. Liu et al. [26] generated the distribution map of *S. alterniflora* on the basis of Landsat-8 images from 2014 to 2016 using OBIA and support vector machine methods along China's mainland coast. Mao et al. [27] monitored *S. alterniflora* invasion in China from 1990 to 2015 with Landsat images. However, existing mangrove or *S. alterniflora* datasets were largely derived from medium spatial resolution satellite imagery, which lacked spatial details [34]. There are few studies on fine-scale, long-term, and multi-regional monitoring of mangroves and *S. alterniflora* simultaneously based on the same satellite datasets.

The dynamic expansion pattern research of *S. alterniflora* is important for the invasion mechanism study, contributing to the scientific eradication and management of *S. alterniflora* [53]. Previous studies mostly used multi-temporal *S. alterniflora* distribution maps based on remote sensing time series data to study the invasion characteristics of *S. alterniflora* in various regions [26,54]. Researchers used the landscape expansion index (LEI) to recognize the spatial expansion patterns of *S. alterniflora* [53,55,56]. To understand *S. alterniflora's* invasion mechanism, Wang et al. [55] and Yan et al. [53] showed three *S. alterniflora* expansion patterns, which were marginal expansion, external isolated expansion, and tidal creek-leading expansion, based on patch fractal dimension and LEI in Yancheng, Jiangsu, China. Using LEI, Li [56] investigated the expansion mode of *S. alterniflora* in Jiuduansha, Shanghai, China. However, these studies mostly focused on one local coastal region north of Zhejiang Province, where no mangroves survive. Little research has been performed on the differences in *S. alterniflora* expansion modes in mangrove–salt marsh ecotones of different latitudes.

The current work aims to evaluate spatial pattern differences of *S. alterniflora* invasion and its nearby mangroves in different latitudes. We selected the Dandou Sea and the Zhangjiang Estuary, two representative mangrove–salt marsh ecotones in the south and north of the Tropic of Cancer, as the study area. In both regions, *S. alterniflora* proliferates on the tidal flats, forming monoculture or mixed communities with mangroves [57]. The rapid spread of *S. alterniflora* poses a significant threat to the native mangrove ecosystems [31]. Based on multi-temporal Google Earth imagery, supplemented by Landsat imagery from 2005 to 2019, we used OBIA and visual interpretation methods [28] to monitor the spatial–temporal changes of mangroves and *S. alterniflora* from 2005 to 2019. Then, we applied LEI to investigate the spatial expansion patterns of *S. alterniflora* in these two ecotones. The results of this study can illuminate the process and pattern of *S. alterniflora* expansion into mangroves, offering valuable references for controlling *S. alterniflora* expansion.

2. Materials and Methods

2.1. Study Area

The Zhangjiang Estuary is situated in Zhangjiangkou National Mangrove Nature Reserve (117°24′07″–117°30′00″ E, 23°53′45″–23°56′00″ N), Yunxiao County, Fujian Province (Figure 1). It owns a natural mangrove community with the most mangrove species and the best growth status in the north of the Tropic of Cancer in China. The topography shows a northwest-to-southeast descending terrain [28]. The climate is a subtropical maritime monsoon. The average annual temperature is 21.2 °C, and the average annual precipitation is 1714.5 mm [29]. Tides are irregular semi-diurnal tides, and the average tidal range is 2.32 m. Intertidal vegetation includes native mangroves (*Kandelia obovate* Sheue, H.Y.Liu and J.W.H.Yong, *Avicennia marina* (Forsk.) Vierh., *Aegiceras corniculatum* L. (Blanco)) and invasive *S. alterniflora* [58]. *Spartina alterniflora* has invaded and occupied the mud flats, mangrove fringes, and mangrove canopy gaps [31].

The Dandou Sea (109°38′20″ E–109°40′48″ E, 21°30′40″ N–21°36′40″ N) is home to Shankou Mangrove National Nature Reserve in Hepu County, Beihai City, Guangxi Zhuang Autonomous Region (Figure 1). Located south of the Tropic of Cancer, it has the largest patch of pure *Rhizophora stylosa* Griff. forest and *Bruguiera gymnorrhiza* (L.) Savigny forest at its climax of succession; it is an important stopover site for migratory birds and a gene pool of mangroves in China. The climate is a subtropical maritime monsoon. The average annual temperature is 22.9 °C, and the average annual precipitation is 1573.4 mm. Tides are irregular diurnal tides, and the average tidal range is 2.45 m. The dominant mangrove species are *R. stylosa*, *B. gymnorhiza*, and *A. corniculatum*.

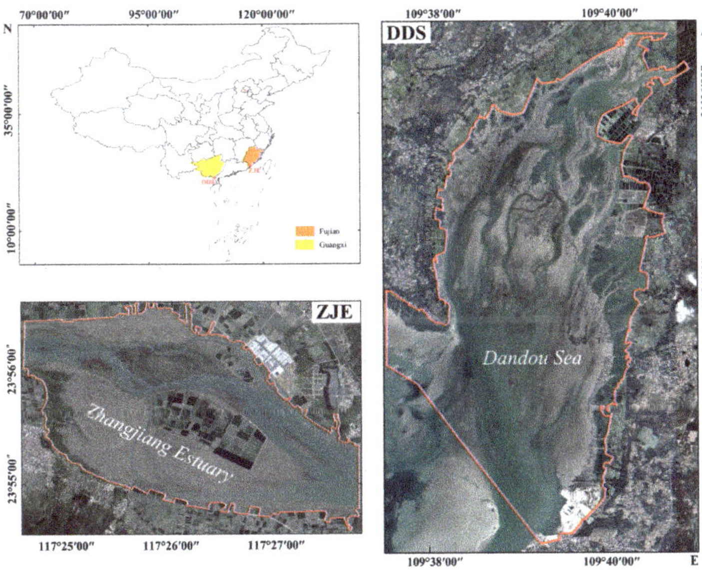

Figure 1. Map of the study area: the Zhangjiang Estuary is abbreviated as ZJE; the Dandou Sea is abbreviated as DDS.

2.2. Data Acquisition and Preprocessing

We collected high-resolution imagery from Google Earth 7.3.3, which offers public and free remote sensing images globally [28,59,60], from 2005 to 2019 in the two study areas (Table 1). We selected cloud-free imagery acquired during low tides in the autumn and winter seasons to ensure the accurate detection of mangroves and *S. alterniflora* [40]. As no cloud-free high-resolution Google Earth imagery at low tide was available in the Dandou Sea in 2005, we collected another cloud-free Landsat 5 Thematic Mapper (TM) image at low tide as a supplement. All the Google Earth images were geo-rectified, and the registration error was less than half a pixel.

Field surveys were conducted in November 2016 in the Zhangjiang Estuary and December 2019 in the Dandou Sea. Besides using a handheld GPS device (Guangzhou SOUTH Survey & Mapping Technology Co., Ltd., Guangzhou, China) to record the locations of mangroves and *S. alterniflora*, we mainly utilized unmanned aerial vehicles (DJI, Shenzhen, China) to collect very high spatial resolution photos to collect samples of mangroves and *S. alterniflora* [61]. The sample data were used to validate the accuracy of the detection results in the Zhangjiang Estuary in 2016 and the Dandou Sea in 2019.

Table 1. Characteristics of the selected remote sensing images in the two study areas.

Stuy Area	Source	Acquisition Date	Resolution (m)	Band	Tidal Level	Remark
ZJE	Google Earth	9 August 2005	0.55	3	Low	Classification
	Google Earth	24 December 2011	0.55	3	Low	Classification
	Google Earth	31 January 2014	0.55	3	Low	Classification
	Google Earth	28 February 2016	0.55	3	Low	Classification
	Google Earth	20 October 2019	0.55	3	Low	Classification
DDS	Landsat 5	21 January 2005	30	7	Low	Classification
	Google Earth	31 December 2005	0.56	3	High	Classification
	Google Earth	16 October 2009	0.56	3	Low	Classification
	Google Earth	22 October 2013	0.56	3	Low	Classification
	Google Earth	8 October 2015	0.56	3	Low	Classification
	Google Earth	8 December 2019	0.56	3	Low	Classification

2.3. Detection Method and Accuracy Assessment

We combined the OBIA and visual interpretation methods to detect mangroves and *S. alterniflora* from 2005 to 2019. Please refer to Liu et al. [28] for more details on the detection method. First, we performed the object-oriented method using eCognition 10 software. To improve the boundary detection results, we performed several pre-segmentation experiments for different segmentation scales and set the scale parameter to 100 for the Zhangjiang Estuary and the Dandou Sea. Then, we classified mangroves and *S. alterniflora* by visual interpretation. For the case of the Dandou Sea in 2005, we first obtained the classification results based on Google Earth imagery in 2005 and then modified the results using the Landsat-5 imagery acquired during low tide in 2005.

The classification result accuracies of the Zhangjiang Estuary in 2016, and the Dandou Sea in 2019, were assessed by ground survey samples. Owing to the lack of field survey data in other years, 800 independent points for every image were created by a random sampling method. We classified random points into *S. alterniflora*, mangroves, and other land cover types by consulting with experienced interpreters and local experts. We calculated confusion matrices for each year in the study areas and used the Kappa coefficient and overall accuracy to evaluate the accuracy of the mangrove and *S. alterniflora* detection results.

2.4. Landscape Indexes and Dynamic Metrics

Landscape indexes were used to evaluate the temporal and spatial variations of the mangrove and *S. alterniflora* landscape patterns [62,63]. Here, we selected patch number (NP), patch density (PD), largest patch index (LPI), patch shape index (LSI), area-weighted mean shape index (AWMSI), landscape division index (DIVISION), and aggregation index (AI) to assess the landscape characteristics, such as patch fragmentation, shape complexity, and patch connectivity for mangroves and *S. alterniflora* (Table 2). The landscape indexes were calculated with the FRAGSTATS v4.2.1 software. In addition, we used the annual change rate and annual expansion rate to analyze the dynamics of mangroves and *S. alterniflora*. The equations are listed as follows:

$$K = \frac{U_a - U_b}{U_b} \times \frac{1}{N} \times 100\% \quad (1)$$

$$U_a = U_b \times (1 + p)^N \quad (2)$$

where K represents the annual change rate; p denotes the annual expansion rate; U_a and U_b represent the area at the ending year and beginning year in each stage, respectively; and N denotes the time in years [28].

Table 2. The formula for the landscape indexes and the ecological significances.

Indicator	Formula	Unit	Code	Ecological Significance
Class Area	$\sum_{i=1}^{n} a_i$	ha	CA	CA is the total area of mangrove or *S. alterniflora* patches, which can influence the species richness, structure, and stability of mangrove or *S. alterniflora* ecosystems.
Number of Patches	n	#	NP	NP is the total number of mangrove or *S. alterniflora* patches, which can reflect the landscape fragmentation degree.
Patch Density	NP/A	#/ha	PD	PD represents the number of mangrove or *S. alterniflora* patches per unit area, reflecting the fragmentation and heterogeneity of the mangrove or *S. alterniflora* landscape.

Table 2. *Cont.*

Indicator	Formula	Unit	Code	Ecological Significance
Largest Patch Index	$\frac{max\left(a_{ij}\right)}{A} \times 100$	%	LPI	LPI reflects the concentration degree of patches, and indirectly describes the interference of anthropogenic activities.
Landscape Shape Index	$\frac{0.25P_{ij}}{\sqrt{a_{ij}}}$	-	LSI	LSI indicates the shape complexity of mangrove or *S. alterniflora* patches, and indirectly reflects the influence of human disturbance activities. The higher the value, the more scattered the patch distribution.
Area-weighted Mean Shape Index	$\sum_{i=1}^{m}[\frac{0.25P_{ij}}{\sqrt{a_{ij}}}\frac{a_i}{A}]$	-	AWMSI	AWMSI indicates the heterogeneity of mangrove or *S. alterniflora* patches. The higher the value, the more complex the shape of the patches.
Landscape Division Index	$[1-\sum_{i=1}^{m}\sum_{j=1}^{n}(\frac{a_{ij}}{A})(\frac{a_{ij}}{A})]$	-	DIVISION	DIVISION represents the separation degree of mangrove or *S. alterniflora* patches. The value is close to 1 (0) when the patch composition is complex (simple).
Aggregation Index	$[\sum_{i=1}^{m}\frac{g_{ii}}{maxg_{ii}}Q_i] \times 100$	%	AI	AI describes the connectivity among mangrove or *S. alterniflora* patches. The lower the value is, the more scattered the patches.
Splitting Index	$\frac{A^2}{\sum_1^m a_{i^2}}$	-	SPLIT	SPLIT represents the dispersion and aggregation degree of *S. alterniflora* patches. The higher the value, the more dispersed the patches.

2.5. Centroid Migration

Centroid migration of mangroves and *S. alterniflora* can reflect their spatial and temporal evolution features and relationships [40]. We calculated and analyzed the centroid migrations of mangroves and *S. alterniflora* here. The equation of centroid is listed as follows:

$$\begin{cases} X_C = \frac{\sum_{i=1}^{n}(C_iX_i)}{\sum_{i=1}^{n}C_i} \\ Y_C = \frac{\sum_{i=1}^{n}(C_iY_i)}{\sum_{i=1}^{n}C_i} \end{cases} \quad (3)$$

where X_c and Y_c represent the centroid coordinates of one land cover type weighted by its area, X_i and Y_i are the latitude and longitude coordinates for the centroid of the ith patch of one land cover type, C_i is the area of the ith patch, and n is the total number of patches of a certain land cover type.

2.6. *Spartina alterniflora* Expansion Pattern Analysis

We used LEI [64] to investigate the *S. alterniflora* expansion patterns at the patch level. The LEI equation is listed as follows:

$$\text{LEI} = \frac{A_P - A_O}{A_P + A_O} \quad (4)$$

where A_p represents the expansion area of the *S. alterniflora* patch and A_0 denotes the original area of the *S. alterniflora* patch. When $A_0 \neq 0$, and the range of LEI values is $(-1, 1)$, the expansion mode is marginal expansion, which indicates that the *S. alterniflora* patches expand outward from their original locations while the newly generated patches are adjacent to older patches [53]. When the expansion area is smaller than the original area of the *S. alterniflora* patch, $LEI < 0$, which represents the marginal expansion 1 mode. When the expansion area is equal to or larger than the original area of the *S. alterniflora* patch, $LEI \geq 0$, which denotes the marginal expansion 2 mode [55]. When the original area of the *S. alterniflora* patch is 0, $A_0 = 0$, LEI = 1, and the expansion mode is external expansion.

3. Results

3.1. Accuracy Assessment

Table 3 presents the accuracy assessments in the Zhangjiang Estuary and the Dandou Sea. In the Zhangjiang Estuary, the detection accuracy in 2016 was lower than that of other years, which was due to the seasonal-dependent environmental noises (algae) [61] on the mud flats (Figure 2). In the Dandou Sea, the detection accuracy in 2019 was lower than that of other years, which might be due to the tidal activities leading to turbid coastal waters [28]. The overall accuracies of all the detection results are greater than 90%, and all Kappa coefficients exceed 0.90, confirming consistency between our detection results and those from the validation data.

Table 3. Detection accuracies in Zhangjiang Estuary (ZJE) and Dandou Sea (DDS).

ZJE	**Year**	**2005**	**2011**	**2014**	**2016**	**2019**
	Overall Accuracy	95%	95%	96%	97%	92%
	Kappa	0.93	0.94	0.95	0.96	0.90
DDS	**Year**	**2005**	**2009**	**2013**	**2015**	**2019**
	Overall Accuracy	95%	96%	92%	94%	97%
	Kappa	0.94	0.95	0.90	0.93	0.96

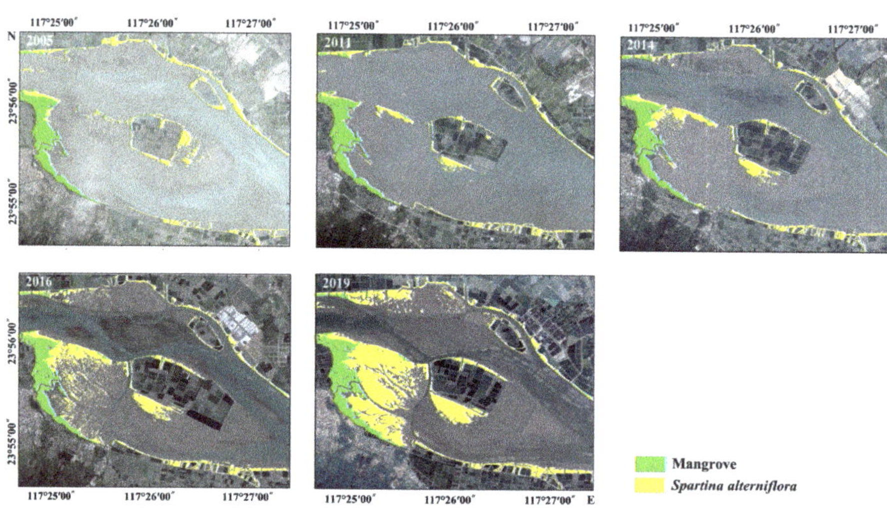

Figure 2. Distribution maps of mangroves and *S. alterniflora* in Zhangjiang Estuary from 2005 to 2019. Each mangrove and *S. alterniflora* detection result is superimposed on Google Earth imagery.

3.2. Spatial–Temporal Distribution Changes of Mangroves and *S. alterniflora*

Figure 2 shows the distribution maps of mangroves and *S. alterniflora* in the Zhangjiang Estuary from 2005 to 2019. Most mangroves in the Zhangjiang Estuary were concentrated in the west of the study area, while S. alterniflora grew on the seaward side of the mangroves or near aquaculture ponds. In the north of the study area, S. alterniflora were distributed in small patches and continued to expand on the mudflats towards the sea (Figure 2). From 2005 to 2019, the mangrove area in the Zhangjiang Estuary expanded, with a total area increase of 9.03 hm^2. The annual change rates of mangroves were 0.23% from 2005 to 2011, 0.54% from 2011 to 2014, 1.27% from 2014 to 2016, and 3.28% from 2016 to 2019, respectively (Table 4). From 2005 to 2019, the S. alterniflora expanded with a total area increase of 165.09 hm^2. The annual change rates of S. alterniflora were 0.42% from 2005 to 2011, 9.14% from 2011 to 2014, 18.68% from 2014 to 2016, and 30.92% from 2016 to 2019, respectively (Table 4). From 2014 to 2019, many small S. alterniflora patches on the mudflats gradually converged, forming larger S. alterniflora stands (Figure 2).

Table 4. Mangrove and *S. alterniflora* areas from 2005 to 2019, and the area changes during different stages in the Zhangjiang Estuary.

Year	Mangroves			*S. alterniflora*		
	Area/hm^2	Annual Change Rate/%	Annual Expansion Rate/%	Area/hm^2	Annual Change Rate/%	Annual Expansion Rate/%
2005	56.22	/	/	67.14	/	/
2011	57.00	0.23	0.23	68.83	0.42	0.42
2014	57.93	0.54	0.54	87.71	9.14	8.42
2016	59.40	1.27	1.26	120.48	18.68	17.2
2019	65.25	3.28	3.18	232.23	30.92	24.45

The centroid of mangroves in the Zhangjiang Estuary varied as the mangrove distributions changed. The mangrove centroid moved northeast by 0.04 km from 2005 to 2011, northwest by 0.008 km from 2011 to 2014, northeast by 0.064 km from 2014 to 2016, and southwest by 0.01 km from 2016 to 2019 (Figure 3). The centroid of S. alterniflora also shifted as S. alterniflora invaded the mudflats (Figure 2). The S. alterniflora centroid migrated southeast by 0.207 km from 2005 to 2011, westward by 0.321 km from 2011 to 2014, westward by 0.357 km from 2014 to 2016, and northwest by 0.382 km from 2016 to 2019 (Figure 3).

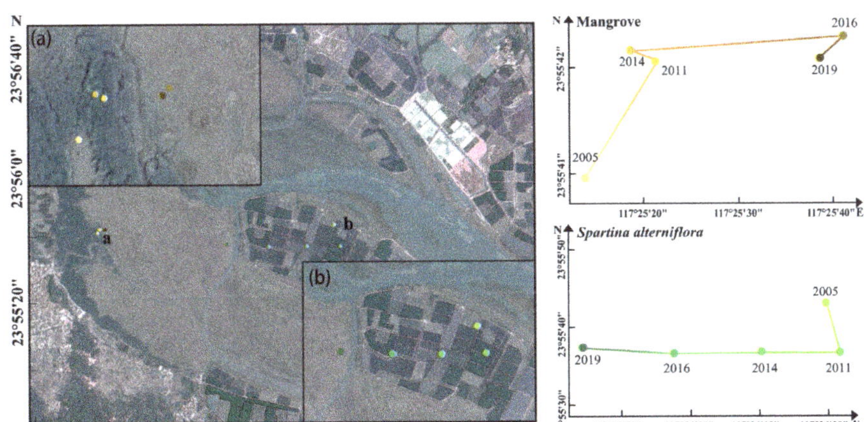

Figure 3. Centroid migration maps of mangroves and *S. alterniflora* in Zhangjiang Estuary from 2005 to 2019.

Figure 4 presents the distribution maps of mangroves and S. alterniflora in the Dandou Sea from 2005 to 2019. In this study area, mangroves were mainly found on the north and east sides or near the aquaculture ponds. S. alterniflora occupied the mud flats near mangroves in the north of the Dandou Sea and continuously expanded towards the sea. The mangrove area in the Dandou Sea increased from 475.78 hm^2 in 2005 to 493.24 hm^2 in 2009, decreased to 475.12 hm^2 in 2015, and then increased to 475.41 hm^2 in 2019 (Table 5). The mangrove area decreases were mainly due to anthropological activities, like land reclamations along the southern coast of the Dandou Sea (Figures 4 and 5), and the encroachment of *S. alterniflora* into mangroves also prohibited the mangrove expansion. From 2005 to 2019, the *S. alterniflora* spread with an area increase of 228.28 hm^2. The annual change rates of *S. alterniflora* were 23.25% from 2005 to 2009, 7.77% from 2009 to 2013, 0.40% from 2013 to 2015, and 0.32% from 2015 to 2019, respectively (Table 5). The *S. alterniflora* rapid spread was obvious along the eastern coast of the Dandou Sea from 2009 to 2013 (Figure 5).

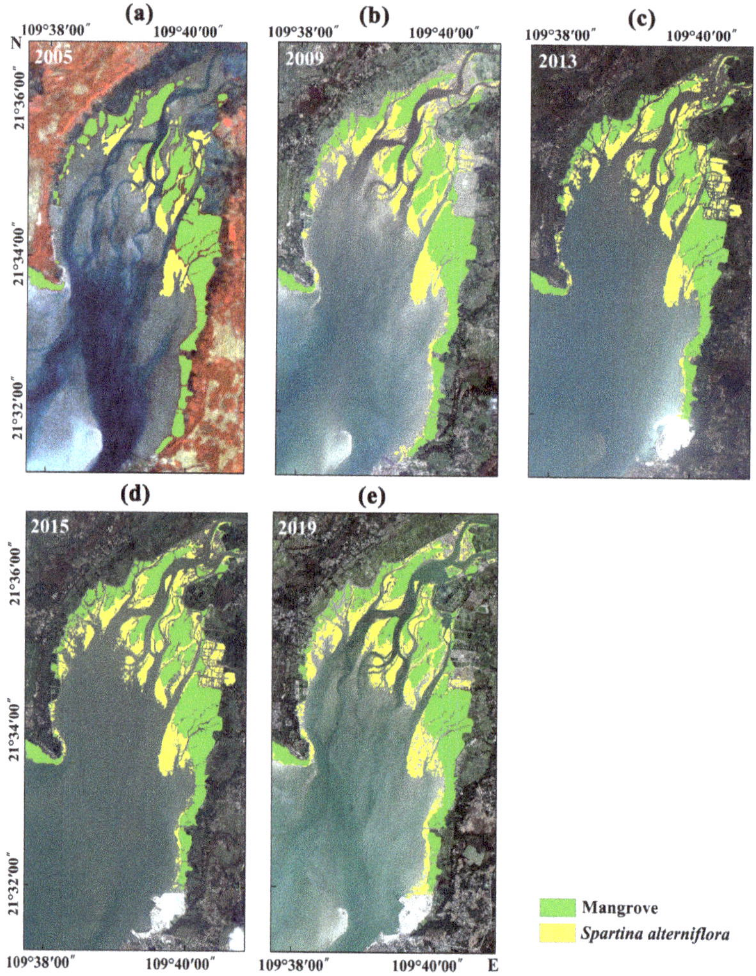

Figure 4. Distribution maps of mangroves and *S. alterniflora* in the Dandou Sea from 2005 to 2019. Each mangrove and *S. alterniflora* detection result is superimposed on Landsat TM (**a**) or Google Earth (**b**–**e**) data. The TM image is a false-color composite of band 4 in the red channel, band 3 in the green channel, and band 2 in the blue channel. The Google Earth image is a true-color composite.

Table 5. Areas of mangroves and *S. alterniflora* from 2005 to 2019 and the area changes during different stages in the Dandou Sea.

Year	Mangroves			*S. alterniflora*		
	Area/hm^2	Annual Change Rate/%	Annual Expansion Rate/%	Area/hm^2	Annual Change Rate/%	Annual Expansion Rate/%
2005	475.78	/	/	144.21	/	/
2009	493.24	0.92	0.91	278.34	23.25	17.87
2013	486.20	−0.36	−0.36	364.87	7.77	7.00
2015	475.12	−1.14	−1.15	367.76	0.40	0.40
2019	475.41	0.02	0.02	372.49	0.32	0.32

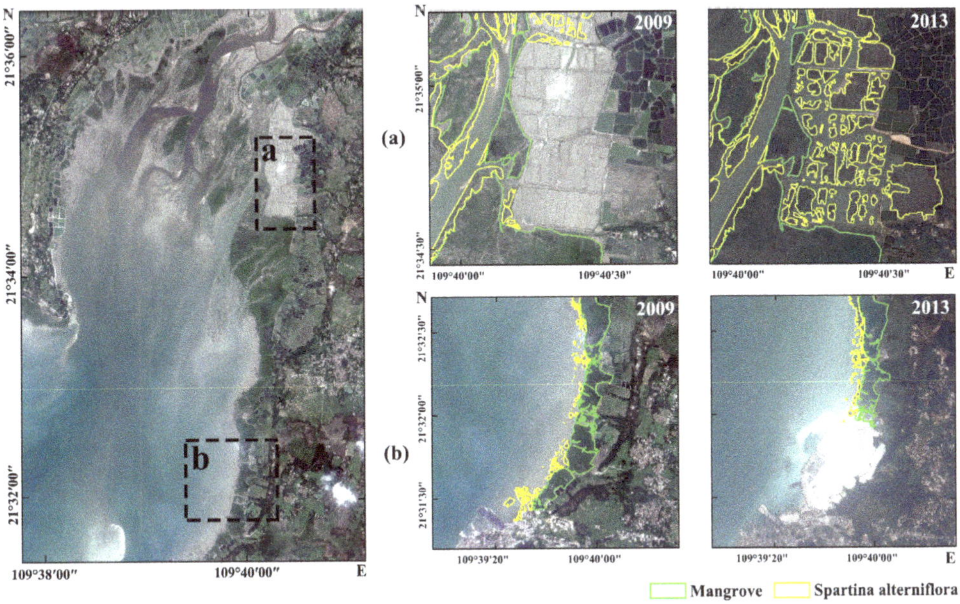

Figure 5. Human-induced mangrove distribution changes in the Dandou Sea.

The mangrove centroid in the Dandou Sea shifted towards the northeast by 0.128 km from 2005 to 2009, northwest by 0.242 km from 2009 to 2013, eastward by 0.014 km from 2013 to 2015, and eastward by 0.017 km from 2015 to 2019, respectively (Figure 6). The *S. alterniflora* centroid moved northwest by 0.223 km from 2005 to 2009, northeast by 0.278 km from 2009 to 2013, southwest by 0.062 km from 2013 to 2015, and southwest by 0.391 km from 2015 to 2019 (Figure 6).

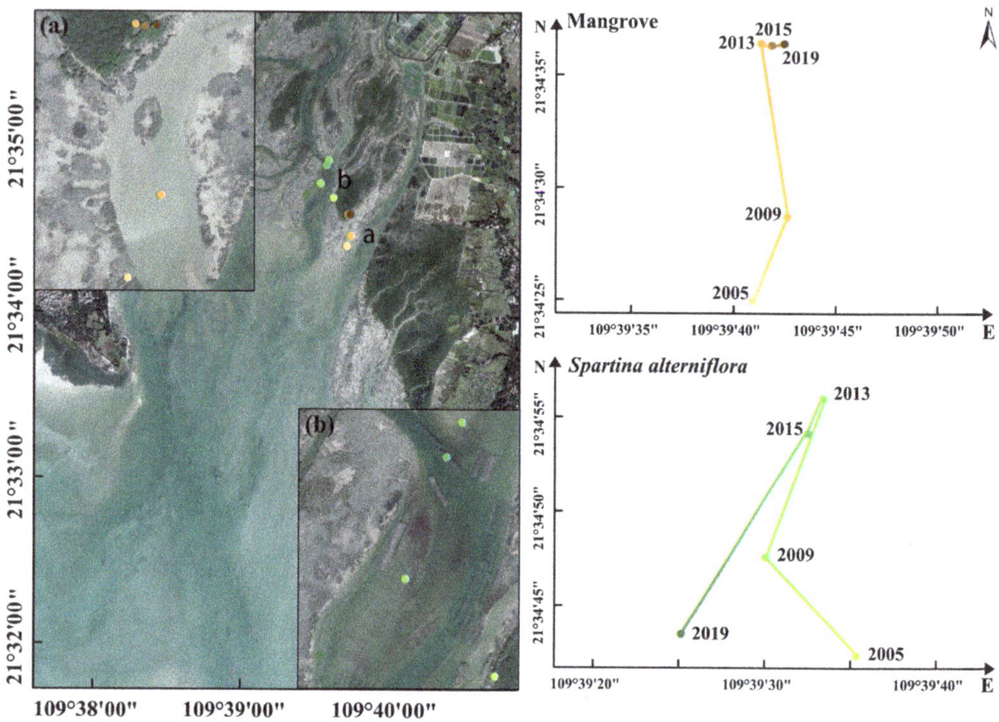

Figure 6. Centroid migration maps of mangroves and *S. alterniflora* in Dandou Sea from 2005 to 2019.

3.3. Changes of Mangrove and S. alterniflora Landscape Indexes

3.3.1. Mangroves

The landscape index results for mangroves and *S. alterniflora* in the Zhangjiang Estuary and the Dandou Sea are displayed in Figure 7. In the Zhangjiang Estuary, the total number of mangrove NP presented a continuous growth trend since 2014, peaking at 117 in 2019 and increasing by 46 from 2005 to 2019. This suggested that mangrove patches were increasingly fragmented in the Zhangjiang Estuary, and the patch distributions became more discrete. The mangrove PD fluctuated from 72.32 to 39.33, showing the mangrove heterogeneity. The mangrove LPI showed a trend of increasing and then decreasing, with the lowest point of 5.76 in 2019, reflecting the low concentration degree of the mangrove patches. The mangrove LSI fluctuated from 10.74 to 13.75, suggesting the scattered mangrove distributions. The mangrove AWMSI first increased and then decreased, reaching a peak at 3.08 in 2011. The mangrove DIVISION increased from 0.97 to 0.99, suggesting the complex mangrove patch composition. The mangrove AI decreased from 99.35 in 2005 to 99.26 in 2019, suggesting the decreasing mangrove connectivity in the Zhangjiang Estuary.

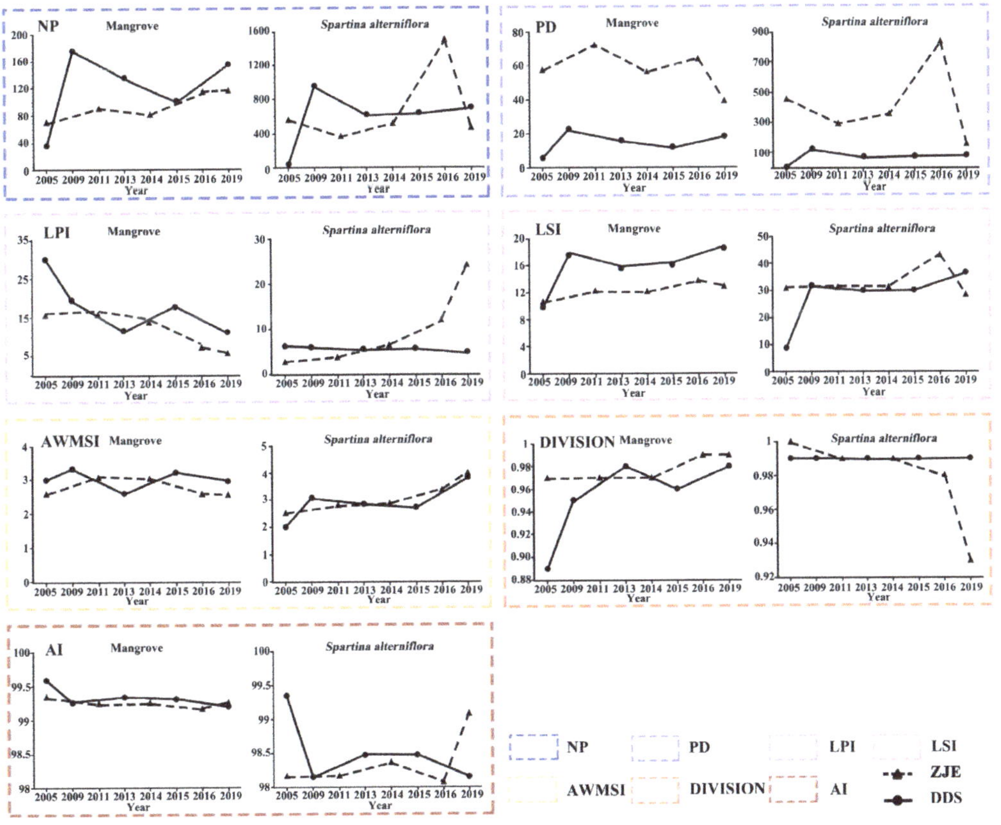

Figure 7. Temporal changes of mangrove and *S. alterniflora* landscape indexes in Zhangjiang Estuary (ZJE) and Dandou Sea (DDS). Here, NP denotes Number of Patches, PD denotes Patch Density, LPI denotes Largest Patch Index, LSI denotes Landscape Shape Index, AWMSI denotes Area-weighted Mean Shape Index, DIVISION denotes Landscape Division Index, and AI denotes Aggregation Index.

In the Dandou Sea, the total number of mangrove NP showed a fluctuating trend of increasing, decreasing. and then increasing again, peaking at 176 in 2009 (Figure 7). The trend of mangrove PD was similar to that of mangrove NP. The highest point of mangrove PD in the Dandou Sea was 22.81 in 2009, smaller than the mangrove PD values in the Zhangjiang Estuary. The mangrove LPI displayed a trend of decreasing, increasing, and then decreasing, which suggested the low concentration degree. The mangrove LSI displayed a trend of increasing, decreasing, and then increasing, peaking at 18.50 in 2019, which indicated that the mangrove distribution was scattered. The mangrove AWMSI fluctuated from 2.60 to 3.33, showing the mangrove heterogeneity. The mangrove DIVISION increased from 0.89 in 2005 to 0.98 in 2019, suggesting a more complex mangrove patch composition. The mangrove AI decreased from 99.59 in 2005 to 99.20 in 2019, suggesting decreasing mangrove connectivity in the Dandou Sea.

Taken together, the mangrove landscapes in both the Zhangjiang Estuary and the Dandou Sea were complex, with more fragmented patches, greater heterogeneity, and decreased connectivity. The mangrove patches in the Zhangjiang Estuary were more fragmented than those in the Dandou Sea.

3.3.2. *Spartina alterniflora*

In the Zhangjiang Estuary, the total number of *S. alterniflora* NP displayed a trend of decreasing, increasing, and then decreasing, peaking at 1508 in 2016 (Figure 7). The decline of *S. alterniflora* NP values from 2005 to 2011 was mainly because (1) the aquaculture pond constructions occupied the habitat areas of *S. alterniflora*, and (2) many small-sized *S. alterniflora* patches merged into larger ones. From 2014 to 2016, *S. alterniflora* spread rapidly, especially on the tidal mudflats west of the Zhangjiang Estuary, resulting in numerous small *S. alterniflora* patches scattered on the mudflats (Figure 2). From 2016 to 2019, those small patches continued to grow and then merged into larger patches, which resulted in decreased *S. alterniflora* NP values. This indicated that the period from 2011 to 2014 was the growth stage, and the years from 2014 to 2019 were the outbreak stage for the *S. alterniflora* expansion process in the Zhangjiang Estuary. The *S. alterniflora* PD fluctuated from 157.32 to 838.31, peaking at 838.31 in 2016, and the changing trend was consistent with that of NP. The *S. alterniflora* LPI continuously increased, peaking at 24.22 in 2019, indicating the increased concentration degree of the *S. alterniflora* patches. The *S. alterniflora* LSI first increased and then decreased, peaking at 43.23 in 2016, which also suggested the invasion outbreak of *S. alterniflora* expansion. The *S. alterniflora* AWMSI showed a continuous growth trend, indicating the increasing heterogeneity degree of *S. alterniflora* distributions in the Zhangjiang Estuary. The *S. alterniflora* DIVISION decreased dramatically from 1 to 0.93, suggesting the *S. alterniflora* patch composition was becoming simpler (Figure 2). The *S. alterniflora* AI dramatically increased from 98.08 in 2016 to 99.09 in 2019, suggesting improved connectivity among the *S. alterniflora* patches in the Zhangjiang Estuary.

In the Dandou Sea, the total number of *S. alterniflora* NP fluctuated from increasing to decreasing and then to increasing again, peaking at 960 in 2009 (Figure 7). The trend of *S. alterniflora* PD was similar to that of *S. alterniflora* NP. The highest point of *S. alterniflora* PD in the Dandou Sea was 124.42 in 2009, smaller than those in the Zhangjiang Estuary. This indicated that *S. alterniflora* patches in the Dandou Sea were less fragmented than those in the Zhangjiang Estuary. Also, the period from 2005 to 2009 was the outbreak stage, and the years from 2009 to 2019 were the plateau stage for the *S. alterniflora* expansion process in the Dandou Sea. The *S. alterniflora* LPI decreased from 6.22 in 2005 to 4.90 in 2019, indicating the *S. alterniflora* concentration degree decline in these years. The *S. alterniflora* LSI fluctuated from increasing to decreasing and then to increasing, peaking at 36.71 in 2019, which indicated the scattered distribution of *S. alterniflora*. The *S. alterniflora* AWMSI fluctuated from 2.01 to 3.82, showing the increased heterogeneity of *S. alterniflora* patches. The *S. alterniflora* DIVISION remained at 0.99, which indicated the severe fragmentation and complex patch composition of *S. alterniflora*. The *S. alterniflora* AI decreased from 99.35 in 2005 to 98.15 in 2019, suggesting the decreasing *S. alterniflora* connectivity in the Dandou Sea.

Taken together, the *S. alterniflora* landscapes in both the Zhangjiang Estuary and the Dandou Sea showed increasing heterogeneity. From 2005 to 2019, *S. alterniflora* in the Zhangjiang Estuary experienced the growth and outbreak stages and showed a higher concentration degree and improved connectivity, while those in the Dandou Sea experienced the outbreak and plateau stages and showed decreased connectivity. Comparing the landscape indexes of mangroves and *S. alterniflora*, we can see the differences in the landscape index trends in the two regions. In the Zhangjiang Estuary, from 2005 to 2019, the mangrove LPI showed a decreasing trend, while the *S. alterniflora* LPI showed an increasing trend. This indicated the possibility for *S. alterniflora* to be the dominant intertidal plant in the Zhangjiang Estuary in the future. In contrast, the *S. alterniflora* LPI in the Dandou Sea decreased over time. From 2009 onwards, the mangrove AI values of the two study areas were larger than the *S. alterniflora* AI values, indicating better connectivity of the mangrove patches than the *S. alterniflora* patches.

3.4. Expansion Mode of S. alterniflora from 2005 to 2019

As the spatial distribution changes of *S. alterniflora* varied in different areas from 2005 to 2019 (Figures 2 and 4), we further studied the expansion dynamics and analyzed the expansion pattern here. The *S. alterniflora* expansion modes included external expansion, marginal expansion 1, and marginal expansion 2 (Figure 8). In the Zhangjiang Estuary, from 2005 to 2011, *S. alterniflora* expanded by 371 patches (Table 6). The number of external expansion patches was 190, accounting for 51.21% of the total expansion patch number, suggesting that *S. alterniflora* expanded mainly through external expansion. A total of 140 patches, with a proportion of 37.74% and an area of 12.76 hm^2, were dominated by the marginal expansion 1 mode. From 2011 to 2014, *S. alterniflora* expanded by 502 patches. The number of external expansion patches was 356, accounting for 70.92% of the total expansion patches. However, the area of the marginal expansion patches was significantly higher than that of the external expansion patches. From 2014 to 2016, *S. alterniflora* expanded by 1476 patches. The number of *S. alterniflora* patches in external expansion mode accounted for 84.15% of the total. The area of marginal expansion patches was 12.14 hm^2 larger than those of the external expansion patches. From 2016 to 2019, the number of external expansion patches was slightly smaller than the total number of patches in marginal expansion 1 mode and marginal expansion 2 mode. The area of patches in marginal expansion 2 mode dramatically increased, while the area of external expansion patches declined. The AWMSI values of the patches (Table 6) suggested that patches in marginal expansion 2 mode were more complex than those in marginal expansion 1 mode, and patches in marginal expansion 1 mode also showed higher heterogeneity than those in external expansion mode. SPLIT values of *S. alterniflora* patches in external expansion mode were larger than those in marginal expansion 1 or marginal expansion 2 modes, indicating the external expansion mode could cause severe fragmentation of the patches. As displayed in Figure 8, we can clearly see that the period from 2011 to 2016 was the time when external expansion was the main *S. alterniflora* expansion mode on the mudflats southwest of the study region.

Table 6. Distribution statistics of LEI on different intervals in Zhangjiang Estuary.

Year	Expansion Pattern	LEI Interval Distribution	Number of Patches	Proportion of Total Number	Area/hm^2	AWMSI	SPLIT
2005–2011	Marginal expansion 1	LEI < 0	140	37.74%	12.76	2.60	689.4
	Marginal expansion 2	LEI ≥ 0	41	11.05%	18.95	3.74	38.7
	External expansion	LEI = 1	190	51.21%	5.38	2.37	1010.17
2011–2014	Marginal expansion 1	LEI < 0	114	22.71%	8.27	3.04	1167.52
	Marginal expansion 2	LEI ≥ 0	32	6.37%	19.48	5.55	32.88
	External expansion	LEI = 1	356	70.92%	5.76	1.58	1864.15
2014–2016	Marginal expansion 1	LEI < 0	118	7.99%	5.16	2.63	4909.22
	Marginal expansion 2	LEI ≥ 0	116	7.86%	21.63	4.37	122.51
	External expansion	LEI = 1	1242	84.15%	14.65	1.86	1117.57
2016–2019	Marginal expansion 1	LEI < 0	109	23.59%	10.1	3.18	8556.01
	Marginal expansion 2	LEI ≥ 0	125	27.06%	107.23	13.33	7.73
	External expansion	LEI = 1	228	49.35%	4.74	1.54	37,135.2

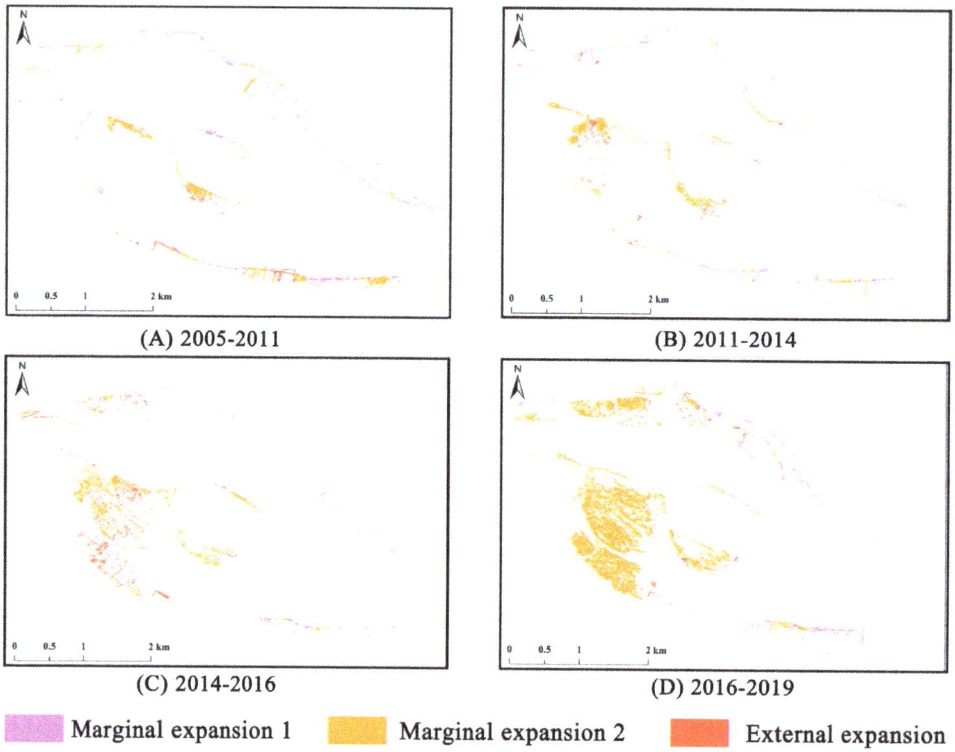

Figure 8. Spatial expansion patterns of *S. alterniflora* in Zhangjiang Estuary from 2005 to 2019.

In the Dandou Sea, from 2005 to 2009, *S. alterniflora* expanded by 1155 patches (Table 7). There were 930 external expansion patches, accounting for 80.52% of the total expansion patch number, with an area of 98.93 hm^2. It suggested *S. alterniflora* expanded mainly through external expansion in this period. There were also 207 patches in marginal expansion 1 mode, accounting for 17.92% of the total, with an area of 36.87 hm^2. From 2009 to 2013, there were 780 patches in marginal expansion 1 mode and 342 patches in marginal expansion 2 mode, accounting for 80.6% of the total, with an area of 82.17 hm^2 in total. There were only 270 patches dominated by external expansion mode, accounting for 19.4% of the total, indicating that *S. alterniflora* expanded mainly through marginal expansion in this period. From 2005 to 2013, both the AWMSI and SPLIT indexes of the marginal expansion 1 and marginal expansion 2 modes were higher than those of external expansion, indicating that the shape of marginal expansion patches showed higher heterogeneity and dispersion degree. From 2013 to 2015, the patch number in marginal expansion 1 mode was 1572, accounting for 80.91% of the total, with an area of 82.17 hm^2. The SPLIT index of external expansion was 4087.52, much larger than those in marginal expansion modes, indicating a severe fragmentation degree. From 2015 to 2019, the patch number in marginal expansion 1 mode was 970, accounting for 54.80% of the total, with an area of 36.26 hm^2. Both the number and area of the patches in external expansion mode were smaller than those in marginal expansion mode. The SPLIT index of patches in external expansion mode was larger than those of marginal expansion, indicating that the patches of external expansion were more dispersed. As demonstrated in Figure 9, the period from 2005 to 2009 was the time when external expansion was the dominant *S. alterniflora* expansion mode, while marginal expansion was the main *S. alterniflora* expansion mode from 2009 to 2019 in the Dandou Sea.

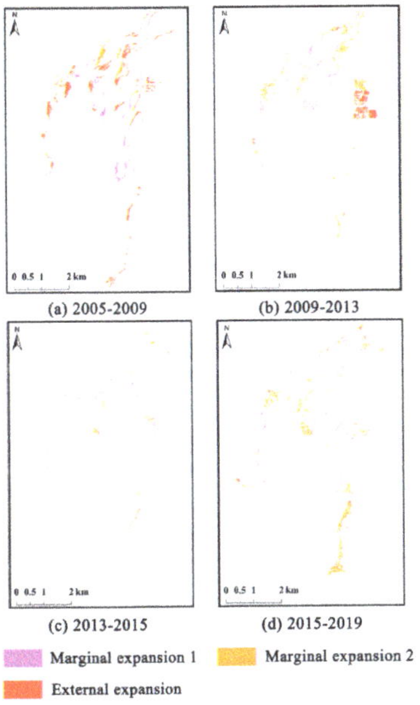

Figure 9. Spatial expansion patterns of *S. alterniflora* in Dandou Sea from 2005 to 2019.

Table 7. Distribution statistics of LEI on different intervals in Dandou Sea.

Year	Expansion Pattern	LEI Interval Distribution	Number of Patches	Proportion of the Total Number	Area/hm^2	AWMSI	SPLIT
2005–2009	Marginal expansion 1	LEI < 0	207	17.92%	36.87	2.82	334.78
	Marginal expansion 2	LEI ≥ 0	18	1.56%	18.94	2.57	273.91
	External expansion	LEI = 1	930	80.52%	98.93	2.48	128.22
2009–2013	Marginal expansion 1	LEI < 0	780	56.03%	34.56	3.3	1204.78
	Marginal expansion 2	LEI ≥ 0	342	24.57%	47.61	4.39	191.16
	External expansion	LEI = 1	270	19.40%	38.26	1.68	169.54
2013–2015	Marginal expansion 1	LEI < 0	1572	80.91%	29.85	2.52	507.38
	Marginal expansion 2	LEI ≥ 0	214	11.01%	11.15	2.64	400.76
	External expansion	LEI = 1	157	8.08%	3.84	1.61	4087.52
2015–2019	Marginal expansion 1	LEI < 0	970	54.80%	36.26	3.31	1000.46
	Marginal expansion 2	LEI ≥ 0	394	22.26%	57.91	5.15	90.3
	External expansion	LEI = 1	406	22.94%	12.16	1.88	3246.62

4. Discussion

4.1. Mangroves and *S. alterniflora* Mapping

In this study, the combined OBIA and visual interpretation methods were used to map mangroves and *S. alterniflora* in the Zhangjiang Estuary and the Dandou Sea from 2005 to 2019. We found that the high-resolution satellite imagery could detect the mangrove and *S. alterniflora* patches at a finer scale and accurately monitor the *S. alterniflora* invasion processes [28]. The overall accuracies in both the Zhangjiang Estuary and the Dandou Sea were larger than 90%, suggesting reliable classification results. The *S. alterniflora* areas in the Zhangjiang Estuary in 2005 and 2014, as measured by Liu et al. [28], were 66.76 hm^2 and 87.31 hm^2, respectively, which were close to the 67.14 hm^2 in 2005 and 87.71 hm^2 in 2014, respectively, observed in this study. In addition, the mangrove detection results from Zhong [65] (50.88 hm^2 in 2005 and 55.1 hm^2 in 2011) were consistent with our results

(56.22 hm^2 in 2005 and 57.00 hm^2 in 2011). As for the Dandou Sea, we compared our mangrove detection results in 2009, 2015, and 2019 with the mangrove datasets in 2010, 2015, and 2020 from Jia et al. [12] and found good spatial consistency (Figure A1). The main difference between our mangrove map in 2009 and the result from Jia et al. [12] in 2010 is shown in Figure A1, and we can see that some small mangrove patches were ignored by Jia et al. [12], which might be due to the coarse spatial resolution satellite imagery [12]. Furthermore, our mangrove map in 2019 and the mangrove detection results in 2018 from Zhang et al. [52] also demonstrated good spatial consistency (Figure A1), with the percentage of overlap greater than 95%. Pan et al. [66] monitored *S. alterniflora* in the Dandou Sea in 2013 and reported the area to be 372.11 hm^2 and the patch number to be 623. Their results were consistent with our study (*S. alterniflora* area 364.87 hm^2, patch number 642 in 2013). Thus, the long-term, fine-scale monitoring results of the mangrove–salt marsh ecotones in this study lay a good data basis for further landscape analysis.

4.2. Dynamic Changes of Mangroves and S. alterniflora

From 2005 to 2019, the annual change rates of mangrove and *S. alterniflora* in the Zhangjiang Estuary showed a continuous growth trend. Our results regarding the changes in the mangrove and *S. alterniflora* distribution area in the Zhangjiang Estuary are highly consistent with those of previous researchers [67,68]. In comparison, mangrove areas in the Dandou Sea first increased, then decreased, and then increased again, while *S. alterniflora* areas kept rising from 2005 to 2019. *S. alterniflora* in the two study regions showed larger annual change rates and annual expansion rates than mangroves, indicating the rapid *S. alterniflora* invasion in the intertidal zones.

From 2005 to 2019, in the Zhangjiang Estuary, the mangrove centroid experienced a trend of shifting seaward before 2016 and a process of migrating landward after 2016. The mangrove centroid migrated 77 m northeastward during the period. The centroid of *S. alterniflora* shifted southeastward before 2011 and then shifted westward after 2011. Additionally, the *S. alterniflora* centroid migrated 1019 m southeastward in total. As to the Dandou Sea, the mangrove centroid experienced a trend of shifting landward. The mangrove centroid migrated 358 m northeastward from 2005 to 2019. The *S. alterniflora* centroid shifted northward before 2013 and then shifted southward after 2013. The *S. alterniflora* centroid migrated 300 m northwestward from 2005 to 2019. It is noteworthy that the centroid migration directions of mangroves and *S. alterniflora* in the Zhangjiang Estuary were opposite, and the distance between the mangrove centroid and the *S. alterniflora* centroid was decreasing, changing from 2224 m in 2005 to 1159 m in 2019. In the Dandou Sea, the centroid migration directions of both mangroves and *S. alterniflora* shifted landward, while the mangrove centroid moved northeast and the *S. alterniflora* centroid shifted northwest, which was partly consistent with the opposite migration direction results in the Zhangjiang Estuary. In the Dandou Sea, the distance between the mangrove centroid and the *S. alterniflora* centroid was also decreasing, changing from 566 m in 2005 to 543 m in 2019. These indirectly reflect the fierce competition between mangroves and *S. alterniflora* for habitat resources in coastal areas. Also, the centroid distance changes between mangroves and *S. alterniflora* were larger in the Zhangjiang Estuary than those in the Dandou Sea, partly suggesting the more serious threat of *S. alterniflora* colonizing mangrove areas in the Zhangjiang Estuary.

The LSI, AWMSI, and DIVISION indexes indicate that mangrove segmentation is obvious and that the mangrove patch shapes become more complex and irregular. From 2005 to 2019, mangrove patches in the Dandou Sea exhibited higher complexity and fragmentation degrees compared with those in the Zhangjiang Estuary. The dramatic increasing trend of the *S. alterniflora* LPI indexes in the Zhangjiang Estuary suggested that *S. alterniflora* might continue to exert pressure on mangrove ecosystems in the near future.

4.3. *Spartina alterniflora* Invasion Process and Pattern

The *S. alterniflora* invasion process can be deduced from the area changes over time. A significant increase in the *S. alterniflora* area from 2016 to 2019 was notable in the Zhangjiang Estuary, with an annual change rate of 30.92%. Additionally, the *S. alterniflora* area changes revealed that *S. alterniflora* experienced a growth stage first and then an outbreak stage from 2014 to 2019. By comparison, the annual change rates of *S. alterniflora* decreased over time in the Dandou Sea, consistent with the results from Li et al. [69]. This indicated that *S. alterniflora* in the Dandou Sea experienced the outbreak stage from 2005 to 2009, followed by a decrease in expansion rate, and entered the plateau stage after 2009 mainly due to environmental limitations. *Spartina alterniflora* was reported to be first successfully planted in Fujian Province in 1981 and in Guangxi Zhuang Autonomous Region in 1979 [28]. The differences in planting time might be an important reason for the delayed *S. alterniflora* expansion outbreak stage in the Zhangjiang Estuary, compared with those in the Dandou Sea.

Researchers noted that the invasion of *S. alterniflora* in China exhibited characteristics of "point-source dispersal, multi-point outbreak" [70]. The *S. alterniflora* expansion processes in the Zhangjiang Estuary and the Dandou Sea also conformed to this rule. As displayed in Figure A2, *S. alterniflora* in the Zhangjiang Estuary first formed numerous small and irregular patches, subsequently, these scattered small patches on the mudflats rapidly expanded, coalescing into large contiguous clumps. Thus, although *S. alterniflora* can reproduce both asexually and sexually [71], the *S. alterniflora* clonal growth mainly contributed to area increase in the two study areas from 2005 to 2019. Additionally, the *S. alterniflora* outbreak stages were also obvious in Figure A2.

The expansion pattern of *S. alterniflora* varies in different regions and different periods. In terms of the patch number, the *S. alterniflora* expansion was dominated by external expansion from 2005 to 2016 in the Zhangjiang Estuary and from 2005 to 2009 in the Dandou Sea. In terms of the patch area, the *S. alterniflora* expansion was dominated by marginal expansion from 2005 to 2019 in the Zhangjiang Estuary and from 2009 to 2019 in the Dandou Sea. The shape of patches in marginal expansion mode was more complex compared to those in external expansion mode. A significant fragmentation trend of *S. alterniflora* patches in external expansion mode can be observed in both study areas.

Previous researchers mentioned that the *S. alterniflora* expansion could be divided into three stages, which were the growth stage, outbreak stage, and plateau stage [72]. However, owing to the lack of high spatial resolution satellite imagery before 2005, we did not observe the growth stage of *S. alterniflora* expansion in the Dandou Sea. Since the "Special Action Plan for Mangrove Protection and Restoration (2020–2025)" was implemented in 2020, we selected satellite data acquired before 2020 to prevent the impact of this large-scale action plan. So, we only observe the growth stage and outbreak stage of the *S. alterniflora* expansion in the Zhangjiang Estuary. However, by combining the results in the two areas, we can conclude the full *S. alterniflora* expansion process.

4.4. Implications for *S. alterniflora* Control and Management

China has implemented a series of control measures for *S. alterniflora* management to guarantee ecological security. The Biosecurity Law of the People's Republic of China came into force in 2021 and proposed to "strengthen the investigation, monitoring, early warning, control, assessment, removal and ecological restoration of invasive alien species". The "Wetland Protection Law of the People's Republic of China", which was officially implemented in 2022, stipulates that "the introduction and release of alien species into wetlands is prohibited". The "Special Action Plan for Mangrove Protection and Restoration (2020–2025)" points out that "the focus is to strengthen the prevention and control of harmful biological disasters such as *S. alterniflora*". By the end of 2022, the "Special Action Plan for the Prevention and Control of *Spartina alterniflora* (2022–2025)" was issued, marking *S. alterniflora* control as an important component of China's coastal wetland management and conservation efforts [73]. Following the deployment arrangements of the Chinese government, several provinces have launched a new round of *S. alterniflora* control campaigns based on previous control efforts. By the end of October 2023, China has completed the clearance of approximately 300 km^2, accounting for 86% of the annual target. Provinces such as Fujian, Shandong, Liaoning, and Hainan have completed the clearance of *S. alterniflora* in the entire region and have entered the stage of comprehensive management, protection, and ecological restoration [74].

According to our study, the expansion of *S. alterniflora* in the outbreak stage is rapid with a large annual change rate; the early warning of *S. alterniflora* invasion is highly significant for the efficient and economical removal of the invasive plant. With the national-scale implementation of *S. alterniflora* eradication campaigns, the *S. alterniflora* area will decrease or even disappear by 2025. However, the remaining roots in the soil and the seeds floating on the water make it challenging to exterminate *S. alterniflora* over the short term [61,75]. Continuous monitoring of *S. alterniflora* is essential to control secondary invasion and protect coastal wetlands [61].

5. Conclusions

As there are few studies on the spatial pattern differences of *S. alterniflora* invasion and the nearby mangrove distributions in different latitudes, we selected the Zhangjiang Estuary and the Dandou Sea, two representative mangrove–salt marsh ecotones in the north and south of the Tropic of Cancer, as the study areas. We combined OBIA and visual interpretation methods to map mangroves and *S. alterniflora* with high-resolution satellite imagery from 2005 to 2019. Based on the constructed long-term, fine-scale mangrove and *S. alterniflora* maps, we applied spatial analysis, centroid migration, and landscape indexes to analyze and compare the spatio–temporal distribution changes of mangroves and *S. alterniflora* in these two ecotones over time. The annual change rates of mangrove and *S. alterniflora* areas in the Zhangjiang Estuary showed a continuous growth trend. The mangrove areas in the Dandou Sea showed a fluctuating trend of increasing, decreasing, and then increasing again, while *S. alterniflora* areas kept rising from 2005 to 2019. *Spartina alterniflora* showed larger annual change rates and annual expansion rates compared with the mangroves, indicating a rapid *S. alterniflora* invasion in the intertidal zones. The opposite centroid migration directions of mangroves and *S. alterniflora* and the decreasing distances between the mangrove centroid and the *S. alterniflora* centroid revealed the fierce competition between mangroves and *S. alterniflora* for habitat resources. The LSI, AWMSI, and DIVISION landscape indexes suggest mangrove segmentation and complex patch shapes. From 2005 to 2019, mangrove patches in the Dandou Sea displayed higher complexity and fragmentation degrees compared with those in the Zhangjiang Estuary. The AI index suggested decreasing mangrove connectivity in both study areas. The impact of *S. alterniflora* invasion on mangroves was progressively serious as more mangrove canopy gaps were invaded and *S. alterniflora* patches near mangroves became larger. *S. alterniflora* invasion contributed to the severe fragmentation and degraded connectivity patterns of mangroves in both study regions. Thus, the control and management of *S. alterniflora* is urgent and imperative for scientific mangrove restoration and conservation.

The *S. alterniflora* invasion process and expansion mode were illuminated for the two study areas. From 2005 to 2019, we observed the growth stage and outbreak stage of *S. alterniflora* expansion in the Zhangjiang Estuary and the outbreak stage and plateau stage of *S. alterniflora* expansion in the Dandou Sea. The different planting times and latitudes might lead to the delayed *S. alterniflora* expansion outbreak stage in the Zhangjiang Estuary, in contrast with those in the Dandou Sea. The expansion characteristics of "point-source dispersal, multi-point outbreak" were obviously shown in these two regions. In terms of the patch number, the *S. alterniflora* expansion was dominated by external expansion from 2005 to 2016 in the Zhangjiang Estuary and from 2005 to 2009 in the Dandou Sea. In terms of the patch area, the *S. alterniflora* expansion was dominated by marginal expansion from 2005 to 2019 in the Zhangjiang Estuary and from 2009 to 2019 in the Dandou Sea. As *S. alterniflora* expansion in the outbreak stage is rapid, with a large annual change rate, early warning of *S. alterniflora* invasion is of significance for the efficient and economical removal of the invasive plant. Moreover, with the national-scale implementation of *S. alterniflora* eradication campaigns, continuous monitoring of *S. alterniflora* is essential to control secondary invasion and protect coastal wetlands. This study contributes to the understanding of the *S. alterniflora* invasion process and expansion mode in mangrove–salt marsh ecotones with different latitudes. These conclusions are highly beneficial for the management and sustainable development of coastal wetlands.

Author Contributions: Conceptualization, D.D. and H.H.; methodology, D.D. and Q.G.; software, Q.G.; validation, D.D., H.H. and Q.G.; formal analysis, D.D., H.H. and Q.G.; investigation, D.D. and Q.G.; resources, H.H.; data curation, Q.G.; writing—original draft preparation, D.D. and Q.G.; writing—review and editing, D.D. and H.H.; visualization, Q.G.; supervision, H.H.; project administration, H.H.; funding acquisition, H.H. All authors have read and agreed to the published version of the manuscript.

Funding: This work was supported by the Marine Economy Special Project of the Guangdong Province (GDNRC[2024]36); the Director's Foundation of the South China Sea Bureau of Ministry of Natural Resources (230206); the Science and Technology Project of Guangdong Forestry Administration (2024): Monitoring and Ecological Value Assessment of Coastal Wetland Resources in the Guangdong Province; the Science and Technology Project of Guangdong Forestry Administration (2023): Research on Carbon Storage Verification, Potential Assessment and Carbon Sink Trading Mechanism of Typical Coastal Wetlands.

Data Availability Statement: The authors confirm that the data supporting the findings of this study are available within the article.

Acknowledgments: The authors would like to thank the USGS for providing Landsat satellite images. Thanks to the reviewers and editors for their valuable suggestions on this paper.

Conflicts of Interest: The authors declare no conflicts of interest.

Appendix A

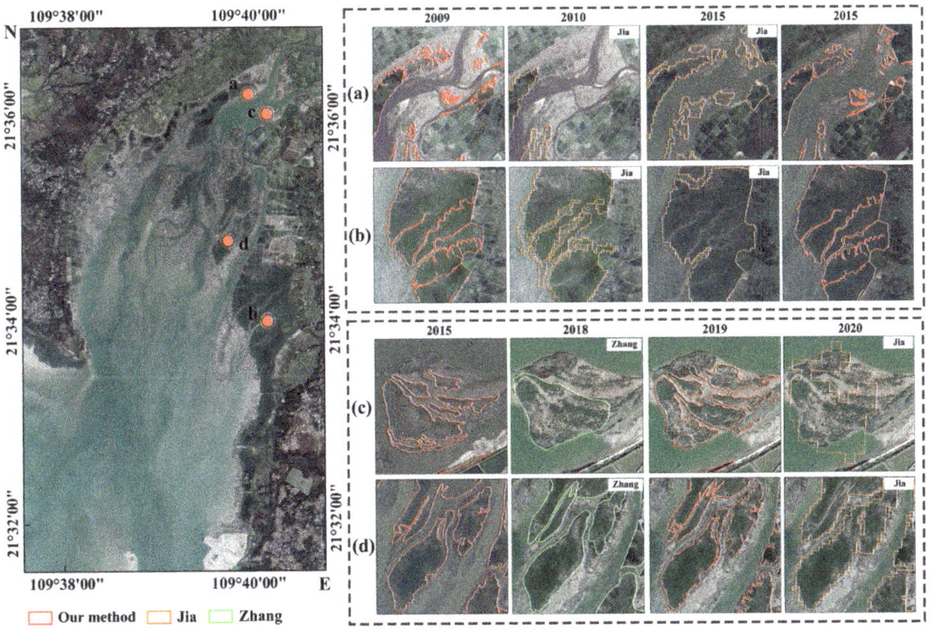

Figure A1. Comparison of mangrove patches from our study, Jia et al. [12] and Zhang et al. [52], in Dandou Sea.

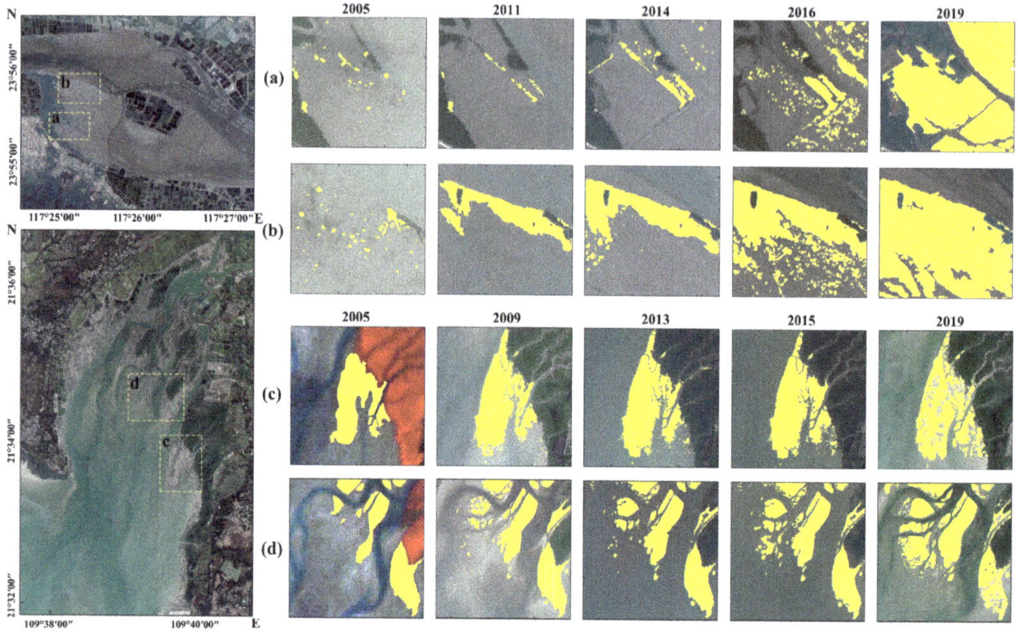

Figure A2. Examples of *S. alterniflora* expansions in Zhangjiang Estuary and Dandou Sea.

References

1. Kathiresan, K.; Bingham, B.L. Biology of mangroves and mangrove ecosystems. *Adv. Mar. Biol.* **2001**, *40*, 81–251.
2. Carlton, J.M. Land-building and stabilization by mangroves. *Environ. Conserv.* **1974**, *1*, 285–294. [CrossRef]
3. Marois, D.E.; Mitsch, W.J. Coastal protection from tsunamis and cyclones provided by mangrove wetlands—A review. *Int. J. Biodivers. Sci. Ecosyst. Serv. Manag.* **2015**, *11*, 71–83. [CrossRef]
4. Spalding, M.; Mcivor, A.; Tonneijck, F.; Tol, S.; Eijk, P.V. Mangroves for coastal defence. In *Guidelines for Coastal Managers & Policy Makers*; Wetlands International and The Nature Conservancy: Wageningen, The Netherlands, 2014.
5. Jia, M.; Wang, Z.; Luo, L.; Zhang, R.; Zhang, H. Global status of mangrove forests in resisting cyclone and tsunami. *Innov. Geo.* **2023**, *1*, 100024. [CrossRef]
6. Arceo-Carranza, D.; Chiappa-Carrara, X.; Chávez López, R.; Yáñez Arenas, C. Mangroves as feeding and breeding grounds. In *Mangroves: Ecology, Biodiversity and Management*; Rastogi, R.P., Phulwaria, M., Gupta, D.K., Eds.; Springer Singapore: Singapore, 2021; pp. 63–95, ISBN 9789811624933.
7. Lugendo, B.R.; Nagelkerken, I.; Kruitwagen, G.; Van Der Velde, G.; Mgaya, Y.D. Relative importance of mangroves as feeding habitats for fishes: A comparison between mangrove habitats with different settings. *Bull. Mar. Sci.* **2007**, *80*, 497–512.
8. Corte, G.N.; Checon, H.H.; Shah Esmaeili, Y.; Lefcheck, J.S.; Amaral, A.C.Z. Mangrove fragments as key coastal reservoirs of taxonomic and functional biodiversity. *Biodivers. Conserv.* **2021**, *30*, 1573–1593. [CrossRef]
9. Rahman, M.M.; Zimmer, M.; Ahmed, I.; Donato, D.; Kanzaki, M.; Xu, M. Co-benefits of protecting mangroves for biodiversity conservation and carbon storage. *Nat. Commun.* **2021**, *12*, 3875. [CrossRef]
10. Carugati, L.; Gatto, B.; Rastelli, E.; Lo Martire, M.; Coral, C.; Greco, S.; Danovaro, R. Impact of mangrove forests degradation on biodiversity and ecosystem functioning. *Sci. Rep.* **2018**, *8*, 13298. [CrossRef]
11. Alongi, D.M. Global significance of mangrove blue carbon in climate change mitigation. *Science* **2020**, *2*, 67. [CrossRef]
12. Jia, M.; Wang, Z.; Zhang, Y.; Mao, D.; Wang, C. Monitoring loss and recovery of mangrove forests during 42 years: The achievements of mangrove conservation in China. *Int. J. Appl. Earth Obs.* **2018**, *73*, 535–545. [CrossRef]
13. Kirwan, M.L.; Megonigal, J.P. Tidal wetland stability in the face of human impacts and sea-level rise. *Nature* **2013**, *504*, 53–60. [CrossRef] [PubMed]
14. Huber, L.C.; Sainge, M.N.; Feka, Z.N.; Kamara, R.A.; Kamara, A.; Sullivan, M.; Cuni-Sanchez, A. Human-driven degradation impacts on mangroves in southern Sierra Leone. *Trees For. People.* **2023**, *14*, 100445. [CrossRef]
15. Gilman, E.L.; Ellison, J.; Duke, N.C.; Field, C. Threats to mangroves from climate change and adaptation options: A review. *Aquat. Bot.* **2008**, *89*, 237–250. [CrossRef]
16. Biswas, S.R.; Choudhury, J.K.; Nishat, A.; Rahman, M.M. Do invasive plants threaten the Sundarbans mangrove forest of Bangladesh? *For. Ecol. Manag.* **2007**, *245*, 1–9. [CrossRef]

17. Lu, C.; Li, L.; Wang, Z.; Su, Y.; Su, Y.; Huang, Y.; Jia, M.; Mao, D. The national nature reserves in China: Are they effective in conserving mangroves? *Ecol. Indic.* **2022**, *142*, 109265. [CrossRef]
18. Romañach, S.S.; DeAngelis, D.L.; Koh, H.L.; Li, Y.; Teh, S.Y.; Barizan, R.S.R.; Zhai, L. Conservation and restoration of mangroves: Global status, perspectives, and prognosis. *Ocean Coast. Manag.* **2018**, *154*, 72–82. [CrossRef]
19. Domínguez-Domínguez, M.; Zavala-Cruz, J.; Rincón-Ramírez, J.A.; Martínez-Zurimendi, P. Management strategies for the conservation, restoration and utilization of mangroves in Southeastern Mexico. *Wetlands* **2019**, *39*, 907–919. [CrossRef]
20. Ngongolo, K.; Mtoka, S.; Mahulu, A. Challenges and opportunities for restoring the threatened mangroves. *J. Sci. Res. Rep.* **2015**, *5*, 352–360. [CrossRef]
21. Zhang, R.; Jia, M.; Wang, Z.; Zhou, Y.; Mao, D.; Ren, C.; Zhao, C.; Liu, X. Tracking annual dynamics of mangrove forests in mangrove National Nature Reserves of China based on time series Sentinel-2 imagery during 2016–2020. *Int. J. Appl. Earth Obs.* **2022**, *112*, 102918. [CrossRef]
22. Jia, M.; Liu, M.; Wang, Z.; Mao, D.; Ren, C.; Cui, H. Evaluating the effectiveness of conservation on mangroves: A remote sensing-based comparison for two adjacent protected areas in Shenzhen and Hong Kong, China. *Remote Sens.* **2016**, *8*, 627. [CrossRef]
23. Xu, S.; Xie, J.; Chen, M. Integrated management improves emerging coastal industries and ecological restoration with the participation of social capital. *Front. Mar. Sci.* **2023**, *9*, 1015262.
24. de Almeida, L.T.; Olímpio, J.L.S.; Pantalena, A.F.; de Almeida, B.S.; de Oliveira Soares, M. Evaluating ten years of management effectiveness in a mangrove protected area. *Ocean Coast. Manag.* **2016**, *125*, 29–37. [CrossRef]
25. Dong, D.; Wang, C.; Yan, J.; He, Q.; Zeng, J.; Wei, Z. Combing Sentinel-1 and Sentinel-2 image time series for invasive Spartina alterniflora mapping on Google Earth Engine: A case study in Zhangjiang Estuary. *J. Appl. Remote Sens.* **2020**, *14*, 044504.
26. Liu, M.; Mao, D.; Wang, Z.; Li, L.; Man, W.; Jia, M.; Ren, C.; Zhang, Y. Rapid invasion of Spartina alterniflora in the coastal zone of mainland China: New observations from Landsat OLI images. *Remote Sens.* **2018**, *10*, 1933. [CrossRef]
27. Mao, D.; Liu, M.; Wang, Z.; Li, L.; Man, W.; Jia, M.; Zhang, Y. Rapid Invasion of Spartina Alterniflora in the Coastal Zone of Mainland China: Spatiotemporal Patterns and Human Prevention. *Sensors* **2019**, *19*, 2308. [CrossRef]
28. Liu, M.; Li, H.; Li, L.; Man, W.; Jia, M.; Wang, Z.; Lu, C. Monitoring the invasion of Spartina alterniflora using multi-source high-resolution imagery in the Zhangjiang Estuary, China. *Remote Sens.* **2017**, *9*, 539. [CrossRef]
29. Gao, X.; Wang, M.; Wu, H.; Wang, W.; Tu, Z. Effects of Spartina alterniflora Invasion on the Diet of Mangrove Crabs (Parasesarma plicata) in the Zhangjiang Estuary, China. *J. Coast. Res.* **2018**, *34*, 106–113. [CrossRef]
30. Jackson, M.V.; Fuller, R.A.; Gan, X.; Li, J.; Mao, D.; Melville, D.S.; Murray, N.J.; Wang, Z.; Choi, C.-Y. Dual threat of tidal flat loss and invasive Spartina alterniflora endanger important shorebird habitat in coastal mainland China. *J. Environ. Manag.* **2021**, *278*, 111549. [CrossRef]
31. Zhang, Y.; Huang, G.; Wang, W.; Chen, L.; Lin, G. Interactions between mangroves and exotic Spartina in an anthropogenically disturbed estuary in southern China. *Ecology* **2012**, *93*, 588–597. [CrossRef]
32. Huang, X.; Feng, J.; Dong, J.; Zhang, J.; Yang, Q.; Yu, C.; Wu, M.; Zhang, W.; Ling, J. Spartina alterniflora invasion and mangrove restoration alter diversity and composition of sediment diazotrophic community. *Appl. Soil Ecol.* **2022**, *177*, 104519. [CrossRef]
33. Xia, S.; Wang, W.; Song, Z.; Kuzyakov, Y.; Guo, L.; Van Zwieten, L.; Li, Q.; Hartley, I.P.; Yang, Y.; Wang, Y. Spartina alterniflora invasion controls organic carbon stocks in coastal marsh and mangrove soils across tropics and subtropics. *Glob. Chang. Biol.* **2021**, *27*, 1627–1644. [CrossRef] [PubMed]
34. Jia, M.; Wang, Z.; Mao, D.; Ren, C.; Song, K.; Zhao, C.; Wang, C.; Xiao, X.; Wang, Y. Mapping global distribution of mangrove forests at 10-m resolution. *Sci. Bull.* **2023**, *68*, 1306–1316. [CrossRef] [PubMed]
35. Jiang, Y.; Zhang, L.; Yan, M.; Qi, J.; Fu, T.; Fan, S.; Chen, B. High-resolution mangrove forests classification with machine learning using worldview and uav hyperspectral data. *Remote Sens.* **2021**, *13*, 1529. [CrossRef]
36. Wang, L.; Jia, M.; Yin, D.; Tian, J. A review of remote sensing for mangrove forests: 1956–2018. *Remote Sens. Environ.* **2019**, *231*, 111223. [CrossRef]
37. Kuenzer, C.; Bluemel, A.; Gebhardt, S.; Quoc, T.V.; Dech, S. Remote sensing of mangrove ecosystems: A review. *Remote Sens.* **2011**, *3*, 878–928. [CrossRef]
38. Lu, Y.; Wang, L. The current status, potential and challenges of remote sensing for large-scale mangrove studies. *Int. J. Remote Sens.* **2022**, *43*, 6824–6855. [CrossRef]
39. Yin, D.; Wang, L.; Lu, Y.; Shi, C. Mangrove tree height growth monitoring from multi-temporal UAV-LiDAR. *Remote Sens. Environ.* **2024**, *303*, 114002. [CrossRef]
40. Lu, C.; Liu, J.; Jia, M.; Liu, M.; Man, W.; Fu, W.; Zhong, L.; Lin, X.; Su, Y.; Gao, Y. Dynamic analysis of mangrove forests based on an optimal segmentation scale model and multi-seasonal images in Quanzhou Bay, China. *Remote Sens.* **2018**, *10*, 2020. [CrossRef]
41. Giri, C. Observation and Monitoring of Mangrove Forests Using Remote Sensing: Opportunities and Challenges. *Remote Sens.* **2016**, *8*, 783. [CrossRef]
42. Tian, J.; Wang, L.; Yin, D.; Li, X.; Diao, C.; Gong, H.; Shi, C.; Menenti, M.; Ge, Y.; Nie, S.; et al. Development of spectral-phenological features for deep learning to understand Spartina alterniflora invasion. *Remote Sens. Environ.* **2020**, *242*, 111745. [CrossRef]
43. Okoye, O.K.; Li, H.; Gong, Z. Retraction of invasive Spartina alterniflora and its effect on the habitat loss of endangered migratory bird species and their decline in YNNR using remote sensing technology. *Ecol. Evol.* **2020**, *10*, 13810–13824. [CrossRef] [PubMed]

44. Li, N.; Li, L.; Zhang, Y.; Wu, M. Monitoring of the invasion of Spartina alterniflora from 1985 to 2015 in Zhejiang Province, China. *BMC Ecol.* **2020**, *20*, 7. [CrossRef] [PubMed]
45. Zhang, X.; Xiao, X.; Wang, X.; Xu, X.; Chen, B.; Wang, J.; Ma, J.; Zhao, B.; Li, B. Quantifying expansion and removal of Spartina alterniflora on Chongming island, China, using time series Landsat images during 1995–2018. *Remote Sens. Environ.* **2020**, *247*, 111916. [CrossRef] [PubMed]
46. Zhang, X.; Xiao, X.; Wang, X.; Xu, X.; Qiu, S.; Pan, L.; Ma, J.; Ju, R.; Wu, J.; Li, B. Continual expansion of Spartina alterniflora in the temperate and subtropical coastal zones of China during 1985–2020. *Int. J. Appl. Earth Obs.* **2023**, *117*, 103192. [CrossRef]
47. Giri, C.; Ochieng, E.; Tieszen, L.L.; Zhu, Z.; Singh, A.; Loveland, T.; Masek, J.; Duke, N. Status and distribution of mangrove forests of the world using earth observation satellite data. *Glob. Ecol. Biogeogr.* **2011**, *20*, 154–159. [CrossRef]
48. Chen, B.; Xiao, X.; Li, X.; Pan, L.; Doughty, R.; Ma, J.; Dong, J.; Qin, Y.; Zhao, B.; Wu, Z. A mangrove forest map of China in 2015: Analysis of time series Landsat 7/8 and Sentinel-1A imagery in Google Earth Engine cloud computing platform. *ISPRS J. Photogramm. Remote Sens.* **2017**, *131*, 104–120. [CrossRef]
49. Bunting, P.; Rosenqvist, A.; Lucas, R.M.; Rebelo, L.-M.; Hilarides, L.; Thomas, N.; Hardy, A.; Itoh, T.; Shimada, M.; Finlayson, C.M. The global mangrove watch—A new 2010 global baseline of mangrove extent. *Remote Sens.* **2018**, *10*, 1669. [CrossRef]
50. Bunting, P.; Rosenqvist, A.; Hilarides, L.; Lucas, R.M.; Thomas, N.; Tadono, T.; Worthington, T.A.; Spalding, M.; Murray, N.J.; Rebelo, L.-M. Global mangrove extent change 1996–2020: Global mangrove watch version 3.0. *Remote Sens.* **2022**, *14*, 3657. [CrossRef]
51. Wang, X.; Xiao, X.; Xu, X.; Zou, Z.; Chen, B.; Qin, Y.; Zhang, X.; Dong, J.; Liu, D.; Pan, L. Rebound in China's coastal wetlands following conservation and restoration. *Nat. Sustain.* **2021**, *4*, 1076–1083. [CrossRef]
52. Zhang, T.; Hu, S.; He, Y.; You, S.; Yang, X.; Gan, Y.; Liu, A. A fine-scale mangrove map of China derived from 2-meter resolution satellite observations and field data. *ISPRS Int. J. Geo Inf.* **2021**, *10*, 92. [CrossRef]
53. Yan, D.; Luan, Z.; Li, J.; Xie, S.; Wang, Y. Monitoring Spartina alterniflora Expansion Mode and Dieback Using Multisource High-Resolution Imagery in Yancheng Coastal Wetland, China. *Remote Sens.* **2023**, *15*, 3853. [CrossRef]
54. Wang, J.; Lin, Z.; Ma, Y.; Ren, G.; Xu, Z.; Song, X.; Ma, Y.; Wang, A.; Zhao, Y. Distribution and invasion of Spartina alterniflora within the Jiaozhou Bay monitored by remote sensing image. *Acta Oceanol. Sin.* **2022**, *41*, 31–40. [CrossRef]
55. Wang, J.; Liu, H.; Li, Y.; Liu, L.; Xie, F. Recognition of spatial expansion patterns of invasive Spartina alterniflora and simulation of the resulting landscape-changes. *Acta Ecol. Sin.* **2018**, *38*, 5413–5422. (In Chinese)
56. Li, J. Study on Population Dynamics and Diffusion Model of Invasive Plant Spartina alterniflora. Master's Thesis, North China University of Science and Technology, Tangshan, China, 2021. (In Chinese)
57. Shen, H.-K.; Zhao, B.; Chen, M.-Y.; Huang, R.-Y.; Yu, K.-F.; Liang, W. Changes of the area of Spartina alterniflora and mangroves in Guangxi Shankou Mangrove National Nature Reserve from 1995 to 2019. *J. Appl. Ecol.* **2022**, *33*, 397–404. (In Chinese)
58. Wang, F.; Sun, X.; Zhao, Y.; Wang, H.; Song, X.; Wei, S.; Chen, H. Does Spartina invasion affect the carbohydrate assimilation of polychaetes in mangroves? A case study in the Zhangjiang Estuary Mangrove National Nature Reserve. *J. Sea Res.* **2023**, *195*, 102435. [CrossRef]
59. Li, W.; Dong, R.; Fu, H.; Wang, J.; Yu, L.; Gong, P. Integrating Google Earth imagery with Landsat data to improve 30-m resolution land cover mapping. *Remote Sens. Environ.* **2020**, *237*, 111563. [CrossRef]
60. Liang, J.; Gong, J.; Li, W. Applications and impacts of Google Earth: A decadal review (2006–2016). *ISPRS J. Photogramm. Remote Sens.* **2018**, *146*, 91–107. [CrossRef]
61. Dong, D.; Huang, H.; Gao, Q. Tracking the Dynamics of *Spartina alterniflora* with WorldView-2/3 and Sentinel-1/2 Imagery in Zhangjiang Estuary, China. *Water* **2024**, *16*, 1780. [CrossRef]
62. Ahmadzadeh, R.; Dehdar Dargahi, M.; Khorasani, N.; Farsad, F.; Rahimibashar, M.R. Assessment of wetland landscape changes based on landscape metrics and trophic state index (case study: Anzali International Wetland). *Environ. Monit. Assess.* **2023**, *195*, 1206. [CrossRef]
63. Millán-Aguilar, O.; Nettel-Hernanz, A.; Hurtado-Oliva, M.Á.; Dodd, R.S.; Flores-Cárdenas, F.; Manzano-Sarabia, M. Landscape Metrics and Conservation Status of Five Mangrove Wetlands in the Eastern Gulf of California Margin. *J. Coast. Res.* **2019**, *36*, 102–194. [CrossRef]
64. Pengfei, W.; De-min, Z.; Huili, G. A new landscape expansion index: Definition and quantification. *Acta Ecol. Sin.* **2012**, *32*, 4270–4277. [CrossRef]
65. Zhong, L.; Lu, C.; Wang, Z.; Liu, M.; Chen, Y.; Lin, X.; Su, Y.; Jia, M. Health assessment of mangrove swamp in Zhangjiang Estuary based on GIS and RS. *Chinese J. Ecol.* **2019**, *38*, 2553–2563. (In Chinese)
66. Pan, L.; Shi, X.; Tao, Y.; Fan, H.; Mo, Z. Distribution and Expansion of Spartina alterniflora in Coastal Tidal Zone, Guangxi. *Wetland Sci.* **2016**, *14*, 464–470. (In Chinese)
67. Pan, W.; Chen, J.; Wang, Y. Analysis of spatio-temporal dynamical change and landscape characteristics of mangroves and *Spartina alterniflora* in Fujian based on satellite imageries from 1999 to 2018. *J. Ecol. Rural Environ.* **2020**, *36*, 1428–1436. (In Chinese)
68. Chen, G. Landscape pattern evolution and driving factors analysis of wetland in Zhangjiang Estuary Mangrove National Nature Reserve. Master's Thesis, Fujian Agriculture and Forestry University, Fuzhou, China, 2022. (In Chinese)
69. Li, L.; Liu, W.; Tao, Y.; Xu, X.; Fu, W.; Dong, J. Diffusion dynamics and driving forces of Spartina alterniflora in the Guangxi Shankou Mangrove Reserve. *Acta Ecol. Sin.* **2021**, *41*, 6814–6824. (In Chinese)

70. Deng, Z.; An, S.; Zhi, Y.; Zhou, C.; Chen, L.; Zhao, C.; Fang, S.; Li, H. Preliminary studies on invasive model and outbreak mechanism of exotic species, *Spartina alternilfora* Loisel. *Acta Ecol. Sin.* **2006**, *26*, 2678–2686. (In Chinese)
71. Liu, H.; Lin, Z.; Qi, X.; Zhang, M.; Yang, H. The relative importance of sexual and asexual reproduction in the spread of Spartina alterniflora using a spatially explicit individual-based model. *Ecol. Res.* **2014**, *29*, 905–915. [CrossRef]
72. Zhu, W.; Ren, G.; Wang, J.; Wang, J.; Hu, Y.; Lin, Z.; Li, W.; Zhao, Y.; Li, S.; Wang, N. Monitoring the invasive plant *Spartina alterniflora* in Jiangsu coastal wetland using MRCNN and Long-Time Series Landsat Data. *Remote Sens.* **2022**, *14*, 2630. [CrossRef]
73. Li, H.; Mao, D.; Wang, Z.; Huang, X.; Li, L.; Jia, M. Invasion of Spartina alterniflora in the coastal zone of mainland China: Control achievements from 2015 to 2020 towards the Sustainable Development Goals. *J. Environ. Manag.* **2022**, *323*, 116242. [CrossRef]
74. Xie, B.; Han, G. Control of invasive plant Spartina alterniflora: Concept, technology and practive. *Bull. Chin. Acad. Sci.* **2023**, *38*, 1924–1938. (In Chinese)
75. Min, Y.; Cui, L.; Li, J.; Han, Y.; Zhuo, Z.; Yin, X.; Zhou, D.; Ke, Y. Detection of large-scale Spartina alterniflora removal in coastal wetlands based on Sentinel-2 and Landsat 8 imagery on Google Earth Engine. *Int. J. Appl. Earth Obs.* **2023**, *125*, 103567. [CrossRef]

forests MDPI

Article

Naturalization of the Ornamental Plant *Crocus tommasinianus* Herb. (Iridaceae) in Forest Ecosystems: A Case Study from Poland

Renata Nowińska * and Aneta Czarna

Department of Botany, Faculty of Agronomy, Horticulture and Bioengineering, Poznań University of Life Sciences, Wojska Polskiego 28, 60-637 Poznań, Poland
* Correspondence: renata.nowinska@up.poznan.pl

Abstract: Highlights: Though not highly invasive, bulb and corm ornamental plants can escape cultivation and naturalize in new areas. Studying their naturalization is key to understanding their ecological impact and managing biodiversity. **Objectives:** This study aimed to document the first naturalization case of *Crocus tommasinianus* Herb. in Poland and assess the morphological variability of the naturalized population under different environmental conditions. Another objective was to identify diagnostic features in seed testa ornamentation to distinguish *C. tommasinianus* from related species (*C. vernus* (L.) Hill. and *C. scepusiensis* (Rehmann et Wol.) Borbás ex Kulcz.). **Methods:** The morphometric studies were performed within four subpopulations of *C. tommasinianus* differing in environmental conditions, determined with Ellenberg indices. Multivariate tests, ANOVA, and post-hoc tests were used to determine the morphometric diversity of specimens and to relate them to environmental factors. Seed micro-ornamentation was examined using a scanning electron microscope. **Results:** Light and temperature were negatively correlated, while moisture, soil pH, and nitrogen were positively correlated with many morphological traits. Plants spreading into forest ecosystems exhibited better-developed features (larger leaves and flowers) than those in former cultivation sites, indicating higher survival potential. The seed coat is papillate, with distinct differences in the shape, size, and secondary sculpture of the papillae compared to *C. vernus* and *C. scepusiensis*. Given that floristic studies often occur during the fruiting period of crocuses, testa ornamentation is crucial for identifying the studied species. **Conclusions:** The observed naturalization of *C. tommasinianus* demonstrates the high morphological plasticity of plants, which makes them capable of colonizing new areas, including forest habitats.

Keywords: alien species; early crocus; woodland crocus; phenotypic plasticity; diagnostic characters

1. Introduction

1.1. Problem of Alien and Invasive Plant Species in Europe and in Poland

Any live specimens (or plant parts that can reproduce) of a species, subspecies, or lower taxon introduced outside its natural range as a consequence of human intervention are classified as alien [1]. Species migration and changes in their distribution ranges, resulting from intentional and unintentional human activity, are a global problem that concerns all parts of the world to a greater or lesser extent [2–8]. It is estimated that the introduction of alien species to Central Europe peaked in the 19th century, and later, it was slightly slower [9]. Currently, Europe ranks second (after North America) in terms of the number of naturalized alien species, estimated at 4140. Most of them are species deriving from other parts of Europe and the temperate zone of Asia but also from Africa, North America, tropical Asia, and South America [4].

Considering exclusively the area of Poland, it can be concluded that the degree of anthropogenic transformation of flora is high, as indicated by the high proportion of alien species to native ones, namely 30% to 70%, respectively [10,11]. Among the alien species introduced since the end of the 15th century, those deriving from Europe prevail (35%), followed by those from North America (30%) and Asia (24%) [10]. About 4% of alien species naturalized in Poland are invasive, i.e., spreading dynamically in the natural environment and threatening biodiversity, as they transform habitats and outcompete native species [12]. It is very challenging to slow down the rate of the spread of invasive species, such as *Reynoutria japonica* Houtt, *R. sachalinensis* (F. Schmidt) Nakai, *R.* ×*bohemica* Chrtek et Chrtkova, *Impatiens parviflora* DC, *I. glandulifera* Royle, *Heracleum sosnowskyi* Manden., *H. mantegazzianum* Sommier et Levie, *Solidago canadensis* L., and *S. gigantea* Aiton. It requires individualized and complex actions linked with high costs [13–16]. In the case of non-invasive alien species, which spread slower and less strongly affect the environment, the major threat associated with them is so-called homogenization, i.e., a reduction in the floristic uniqueness of individual regions and the loss of local, specific features within ecosystems [17–21]. That is why one of the key tasks in biodiversity protection, related to the spread of alien species, is currently the early detection of the spread of each alien species and its later monitoring.

1.2. Horticulture and Bulbous Plants in the Context of the Spread of Alien Species

Horticulture—linked with the import of ornamental plants, e-commerce, and plant cultivation—is currently the most important way of introducing new plant species to areas where they did not occur before [22,23]. Most of the introduced and cultivated species remain only in a place of cultivation, but some start to spread, reproduce, and establish in the wild [22]. This phenomenon, called the *naturalization* of alien species, allows the creation of stable populations in new environments, with or without direct human activity [24]. The number of alien plants escaping from cultivation into native ecosystems is steadily increasing; currently, at least 75% and 93% of the global naturalized alien flora are grown in private and botanical gardens, respectively [25].

An important role in the global ornamental plant trade is played by bulb and corm plants. They belong to over 800 genera, but only a few of them dominate in the trade: *Tulipa* L., *Lilium* L., *Narcissus* L., *Gladiolus* L., *Hyacinthus* L., *Crocus* L., and *Iris* L. [26], as well as *Scilla* L., *Ornithogalum* L., *Galanthus* L., and *Allium* L. Bulb and corm geophytes, are not highly invasive plants [27], but many species tend to escape from cultivation and become naturalized in natural and semi-natural ecosystems. All the above-mentioned genera are currently naturalized outside their natural ranges of distribution, and the size of their secondary ranges is usually not larger than their primary ranges [28]. In Poland, only two of them are classified as naturalized—*Tulipa sylvestris* L. and *Crocus vernus* (L.) Hill [10]—although the list of naturalized bulb and corm plants will probably be soon extended to include other taxa.

The naturalization success of many ornamental plants results from the preference of a majority of gardeners for plants that are easy to cultivate [29], i.e., have a wide ecological spectrum. In the case of bulb, corm, and tuber plants, their advantage is also the ability to reproduce both vegetatively, by means of new bulbs or tubers, and sexually (with few exceptions) through seed dispersal. The geophytes that escape from cultivation by means of seeds in cemeteries and in former manor parks in Poland include *Ornithogallum umbellatum* L., *O. nutans* L., *O. boucheanum* Asch., *Scilla sibirica* Haw., *Chionodoxa sardensis* Barr. et Sudg., *Ch. Forbesii* Baker, *Eranthis hyemalis* (L.) Salisb., and *Tulipa sylvestris* [30–34]. For example, *Ornithogallum umbellatum* was found on roadsides, meadows, and forest edges, near allotments (small gardens), and even in a national park [35]. Investigations in Lower Saxony [35] showed that due to vegetative reproduction, some populations of *Tulipa sylvestris* have survived as a cultural relic for at least 250 years in places of their former cultivation (e.g., historical parks and churchyards), forming populations composed of over 10,000 individuals. The spontaneous dispersal of crop plants can be additionally supported by human activity. Ornamental geophytes can escape to the natural environment with discarded plant matter from gardens and composted waste [22]. This applies, for example, to species of the genera *Crocus*, *Galanthus*, and *Scilla*. In contrast, the increasing range and number of localities of the invasive *Allium paradoxum* are partly due to its intentional dispersal by wild food lovers [36].

1.3. *Crocus tommasinianus*—A Cultivated, Alien, Naturalized Species

The genus *Crocus* includes 261 species [37], which are grouped into two subgenera, two sections, and 15 series [38–41]. Species from this genus are native to the Mediterranean basin (North Africa, Iberian Peninsula), South and Central Europe, and West Asia, reaching as far east as western China, with the largest number of species in the Balkan Peninsula and Asia Minor [42]. As a result of naturalization, the genus *Crocus* now also occurs in North America [28].

In the past, many *Crocus* species were introduced to botanic gardens and then distributed [43]. Today, numerous species of this genus are cultivated as ornamental plants [44], but the most popular in cultivation are *Crocus chrysanthus* (Herb.) Herb., *Crocus flavus* Weston, *Crocus sieberi* J. Gay, *Crocus tommasinianus* Herb. and *Crocus vernus*, together with hundreds of varieties and hybrids derived from them [45]. Due to their bright colors and multitude of varieties, they are valued for their decorative appeal. They are also relatively easy to grow [43], which makes them popular among both experienced gardeners and amateurs. Crocuses adapt to almost all soils, respond well to organic fertilizers, and tolerate sunny, semi-shady, and shady locations [46].

In Central Europe, most cultivated *Crocus* species bloom in early spring, after a period of dormancy in winter [44]. One of the smallest [46] and earliest blooming species [47] is Tommasini's crocus, *Crocus tommasinianus*. Two synonymous English names are used to refer to this species, which reflect its characteristics well: *early crocus* [48] and *woodland crocus* [49]. Sometimes, the less formal name *snow crocus* is also used [50]. *C. tommasinianus* can be observed in winter—at the end of February or in early March—and sometimes in spring until early April. It typically begins flowering 2–3 weeks earlier than the very commonly cultivated *C. vernus* [47], although the cross-pollination of both species and spontaneous introgression are occasionally observed [47,51].

C. tommasinianus forms corms with a distinctly reticulate tunic from which three to five linear leaves grow 4–8 mm wide. The flowers are large and slender, 2–4 cm long and 4–5 cm in diameter, with a perianth tube diameter of about 2–3 mm. Their color ranges from deep violet or purple to snow white, but the perianth tube is always clear white. The style is shortly divided into three branches, each of them fimbriate at the apex; the stigma does not exceed the stamens. The plant height is 7–10 cm [38,42,52,53].

This species naturally occurs only on the Balkan Peninsula: from southern Hungary and northwestern Bulgaria to the West Balkans, namely Croatia, Bosnia-Herzegovina, Montenegro, and Serbia [28,38,54], where it is found in forests and other shaded areas [55,56]. It is a particularly useful winter or early spring flowering species in designed landscapes [57], often grown in gardens and parks. Under favorable conditions, it multiplies quickly by tubers and by self-seeding, so it can escape from cultivation [58,59]. Its naturalization has been recorded in Europe (Belgium, Germany, Great Britain, The Netherlands) and in the eastern part of North America (Delaware, New York, Virginia) [28]. The process of escaping from cultivation was interestingly described by [47], based on observations in London, where colonies of fair size spread especially along shady, lightly wooded parkland verges, where the soil is not heavily compacted.

Since the naturalization of this species was not recorded earlier in Poland [28,60], the main goal of our study was to document the first naturalized case of *C. tommasinianus* and to analyze the causes and possible consequences of this process.

Another goal was to determine the morphological variability of the local population depending on environmental conditions, in the context of species expansion from the place of introduction to a forest community.

The basis of every species monitoring effort is species identification. In the case of crocuses (especially from the *Verni* series, to which *C. tommasinianus* belongs), identification is difficult due to the close relationship of other species, related to polyploidization [38,40,61–63]. This results in high interspecific morphological similarity and the high intraspecific morphological variability of macroscopic features. In such cases, searching for microscopic features is very justified. In the case of the *Verni* series, the features analyzed so far concerned the anatomical structure of the leaves [64–66]. There are some studies on seeds of *C. tommasinianus* and *C. vernus*, but so far, they have only considered the primary sculpture of the testa without characterizing the secondary sculpture [67,68]. Seed micro-ornamentation patterns are constant and not very susceptible to environmental influences, making them more reliable for systematic purposes than other macro- and micromorphological features [69]. For these reasons, the third aim of the study was to search for new diagnostic features of *C. tommasinianus* based on the primary and secondary testa sculpture and comparison with seeds of two closely related species from the *Verni* series occurring in Poland: *C. vernus* and *C. scepusiensis* (Rehmann and Woł.) Borbás (not yet studied for testa micro-ornamentation).

2. Materials and Methods

2.1. Description of the New Location

Field research was conducted in 2023 in the Wielkopolska National Park (WNP, western Poland; 52°16′09.5″ N 16°47′82.0″ E; Figure 1). WNP climate is temperate, transitional between oceanic and continental, with mean annual temperature of 8.4 °C and mean annual precipitation of 521 mm, for the years 1951—[70].

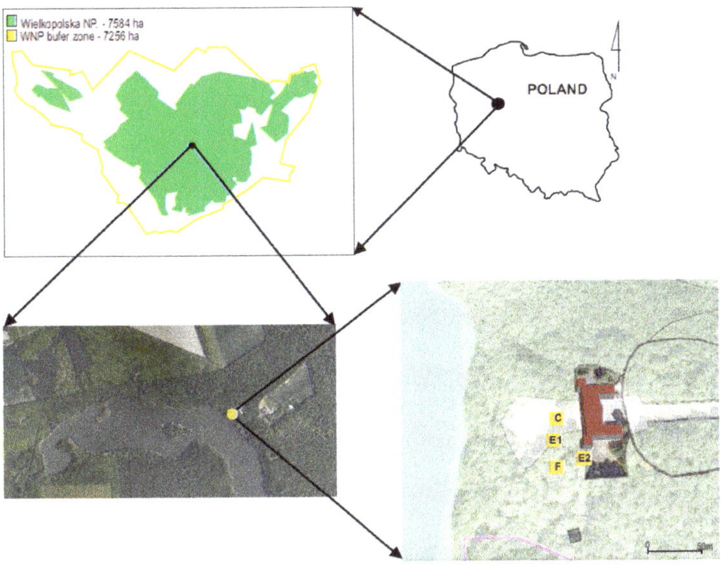

Figure 1. Location map of the study area. Abbreviations: WNP—Wielkopolska National Park, C—clearing, F—oak forest (*Potentillo albae-Quercetum*), E1—ecotone1, E2—ecotone2.

Crocus tommasinianus was found here to be spreading from a former cultivation site (in a clearing near a palace building) to a nearby forest. The local population was first observed at the described site in 1994, but it was not identified as a species at that time (A. Czarna—unpublished data). Recently, the species has been determined at the flowering stage, based on the key and descriptions included in the works of Mathew et al. [38] and Dostál [53]—Figure 2.

(a) (b)

Figure 2. *Crocus tommasinianus* in the Wielkopolska National Park: (**a**) plant habit; (**b**) group of specimens (photo: A. Czarna, 9 March 2024).

In the study area, the species covers about 500 m^2. Currently, four groups (i.e., subpopulations) can be distinguished: (1) in a clearing (place of former cultivation), (2) in a nearby oak forest, (3) at the edge of the oak forest and the clearing (hereinafter referred to as ecotone1) and far away from buildings, and (4) at the edge of the oak forest and the clearing, but in a site partly shaded by the palace building (hereinafter referred to as ecotone2)—Figure 1.

In a representative part of each subpopulation (i.e., representing its typical density and distribution of specimens), a plan of the distribution of specimens was made in a 1 m × 1 m plot. In these places, in areas of 10 m^2, phytosociological relevés characterizing the local vegetation were also taken. The species lists were prepared separately for three layers of vegetation: trees (A), shrubs (B), and herbaceous (C). The names of species were given after Mirek [60]. The Braun-Blanquet cover-abundance scale was used to show the shares of the recorded species [71].

Phytosociological data are often used to infer environmental characteristics [72]. Therefore, based on the prepared relevés and Ellenberg's indices [73], average values of temperature, light, soil moisture, pH, and soil nutrients were calculated for the areas occupied by individual subpopulations.

2.2. Morphometric Measurements of Individuals

To determine whether habitat conditions affect the diversity of *C. tommasinianus*, the morphological features of 30 individuals from each subpopulation were measured. The number of leaves emerging from the corm was counted; the length of the longest leaf, the length of the flower (i.e., the sum of the total length of the perianth tube and the longest tepal), the length of the outer tepal, the width of the outer tepal, the length of the anther and filament, as well as the length of the stigma and style were measured. The measurements were taken using a 10 cm VIN caliper with an accuracy of 0.05 mm.

2.3. Carpological Examination

To determine whether *C. tommasinianus* is carpologically distinguishable from two closely related species from the *Verni* series (*C. vernus* and *C. scepusiensis*), their seeds were analyzed. Carpological material for *C. tommasinianus* and *C. scepusiensis* came from the Botanical Garden of Adam Mickiewicz University in Poznan (Poland), and that for *C. vernus* came from the Index Seminum of the University of Ljubljana Botanical Garden (Slovenia). Three features (seed length, width, and elaiosome length) of 30 fully developed seeds of each species were measured under a stereoscopic microscope PZO MST 132 with an eyepiece micrometer scale (Polskie Zakłady Optyczne PZO, Warsaw, Poland). Seed-coat sculpture was studied using scanning electron microscope (SEM) micrographs prepared with a Zeis EVO 10 microscope (Carl Zeiss Microscopy GmbH, Jena, Germany).

2.4. Statistical Analyses

Principal component analysis (PCA) on the correlation matrix was used to examine the combined effects of environmental factors (i.e., average values of Ellenberg indicators) on morphological characters of *Crocus tommasinianus*. The analysis included morphometric measurements of all the studied specimens without prior division into subpopulations. The scree test was used to select the main components that significantly explained the variability of the set [74].

Discriminant function analysis (DFA) was used to determine which morphological characters most effectively differentiate the studied subpopulations. Before analysis, the quantitative characters were standardized to compare characters that were previously recorded in different units. The data were not transformed because they were normally or near-normally distributed (checked with a normality plot combined with the Kolmogorov–Smirnov test). The models were built using a backward stepwise procedure. The contribution of particular characters to the discrimination between groups was interpreted based on standardized coefficients.

Differences in morphometric and carpological characters between subpopulations were tested using ANOVA and Tukey's HSD test.

3. Results

3.1. Description of the New Location

The terrain where *Crocus tommasinianus* occurs is undulating, with a slope of 10%–30%. Rusty brown soils were formed here on clayey sands [75]. The plants grew in a clearing in the adjacent forest ecosystem. In the clearing with a meadow ecosystem, the cover of the herbaceous layer is about 60%. The dominant species are *Festuca pratensis* Huds. and *F. heterophyllal* Lam., with an admixture of dicotyledonous plants, among which the most abundant are *Hieracium pilosella* L. and *Trifiolium repens* L. (Appendix A). The forest ecosystem represents the habitat type of moderately moist deciduous forest, phytosociologically closest to an oak forest (Ass. *Potentillo albae-Quercetum* Libb. 1933, All. *Potentillo albae-Quercion petraeae;* O. *Quercetalia pubescenti-petraeae*). The forest stand is formed by approximately 100-year-old sessile oaks (*Quercus petraea* (Matt.) Liebl.) with an admixture of *Carpinus betulus* L., *Acer platanoides* L., *Fraxinus excelsior* L., *Ulmus laevis* Pall., and *Pinus sylvestris* L. The average density of the tree layer is about 50%. The shrub layer is poorly developed, covering up to 5%. The herbaceous layer coverage is about 60%, with no dominant species, although *Poa nemoralis* L. and *Quercus petraea* are abundant in some places. In the forest edge zones (referred to here as ecotones), a separate plant community is not formed. ecotone1 and ecotone2 differ in tree density (40% and 15%, respectively), the number of species in the herbaceous layer (36 and 11, respectively), and their coverage (50% and 65%, respectively). In the ecotones, certain ornamental bulbous species occur with a low degree of cover: *Chionodoxa sardensis*, *Galanthus nivalis* L., and *Scilla sibirica*.

The mean Ellenberg indices, calculated separately for each site of *C. tommasinianus*, were as follows: light (L) 5.00–7.61; temperature (T) 5.00–5.79; moisture (F) 4.07–5.40; soil pH (R) 5.22–7.00; nitrogen content (N) 3.80–5.83. The habitat conditions were quite similar in the forest and ecotones (especially ecotone1) and markedly different in the clearing, where the L value was significantly higher in the clearing, while the nitrogen content (N) was significantly lower (Figure 3).

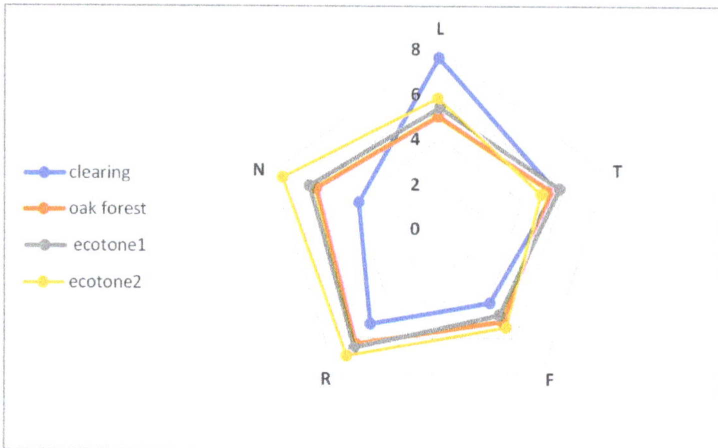

Figure 3. Mean Ellenberg indicator values for light (L), temperature (T), moisture (F), soil pH (R), and nitrogen (N) calculated for sites of *Crocus tommasinianus* occurrence.

In 1994, the population of *Crocus tommasinianus* was larger, but it was limited to the clearing where it was cultivated, covering about 200 m^2. The species did not show any tendency to spread at that time. Since then, the population has spread to the adjacent forested areas, now occupying about 500 m^2, with only about 30 m^2 remaining at the original cultivation site. In 2023, the density of this species (Figure 4) was the highest at the edges of the forest (about 100–140 individuals/m^2) and the lowest in the clearing (fewer than 20 individuals/m^2). The share of flowering individuals was very high—at least 95%—regardless of the place of occurrence.

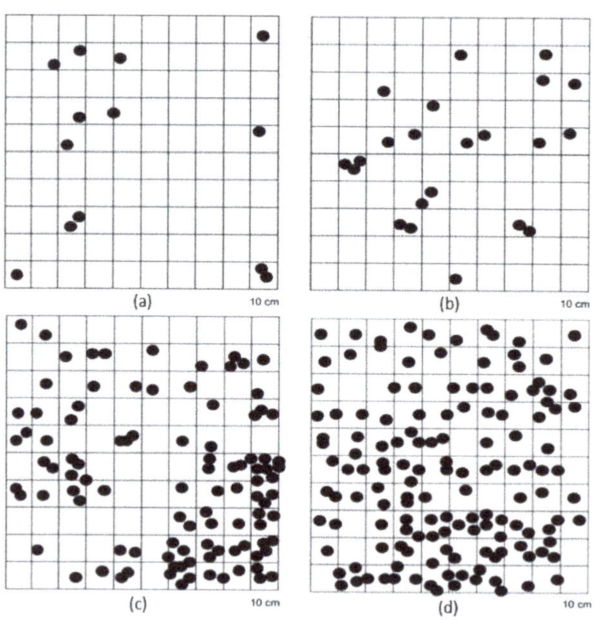

Figure 4. Distribution of *Crocus tommasinianus* specimens on 1 m × 1 m plots: (**a**) clearing, (**b**) oak forest, (**c**) ecotone1, (**d**) ecotone2.

3.2. Morphometric Measurements of Individuals

The examined specimens markedly varied morphologically. Individual specimens had three to five leaves, with a mean of 3.90. At the time of flowering, the mean length of the longest leaf was 5.94 cm, with a range from 2 cm to 14.5 cm. Flower length varied from 4.5 cm to 16.5 cm (mean 8.49 cm), while the outer tepal length varied from 2.2 cm to 4.5 cm (mean 3.10 cm). Among the sexual parts of the flowers, the filament length was the most variable, ranging from 4 mm to 14 mm, with a mean of 8.35 mm (Table 1).

Table 1. Descriptive statistics of the analyzed features of *Crocus tommasinianus*.

Character	Mean ± SD	Min.	Max.
Number of leaves	3.90 ± 0.59	3.0	5.0
Length of longest leaf (cm)	5.94 ± 2.82	2.0	14.5
Flower length (cm)	8.49 ± 2.48	4.5	16.5
Outer tepal length (cm)	3.10 ± 0.44	2.2	4.5
Outer tepal width (cm)	1.03 ± 0.19	0.6	1.5
Stigma length (mm)	3.90 ± 0.80	2.0	6.0
Style length (cm)	1.97 ± 0.42	0.8	3.0
Anther length (cm)	1.27 ± 0.17	0.8	1.7
Filament length (mm)	8.35 ± 1.74	4.0	14.0

Principal component analysis (PCA) was used to examine the relationships between Ellenberg indicators and morphological characters. The first two principal components explained almost 60% of the total variance (Figure 5). Light (L) was negatively correlated with all the analyzed characters, most strongly with the length of the longest leaf (−0.54) and flower length (−0.47). Temperature (T) was negatively correlated with all characters except style length; the strongest correlations were with the length of the longest leaf (−0.61) and number of leaves (−0.56). Moisture (F), soil pH (R), and nitrogen content (N) were very strongly ($r > 0.90$) positively correlated with each other. Among the last three environmental traits, moisture had the strongest influence on the morphological characters, particularly on the length of the longest leaf (0.65), flower length (0.51), and outer tepal length (0.49).

Based on the stepwise discriminant function analysis (DFA), seven out of nine morphological features significantly differentiated the studied subpopulations (Wilks' lambda = 0.05038; $F(27,316) = 20.859$, $p < 0.001$). The first two discriminant functions explained 96% of the total variability in the tested material and had a significant impact on the discrimination (chi-square tests; $p < 0.001$). Almost 60% of the total discriminatory power was contributed by the first function, which primarily separated the forest subpopulation from ecotone1 and the clearing subpopulations (Figure 6).

The first function was most heavily affected by the length of the longest leaf as well as the flower length (Table 2). The second function explained nearly 40% of the total variability and primarily separated the subpopulation in ecotone2 from the remaining subpopulations, with this distinction being driven primarily by various flower characters (flower length, outer tepal length, style length, and filament length), as well as the number of leaves.

The results of the analysis of variance (ANOVA, Table 2) and the post-hoc tests (Figure 7) confirmed that the characters related to the size and robustness of the plants (number of leaves, longest leaf length, flower length) were crucial for distinguishing the described subpopulations.

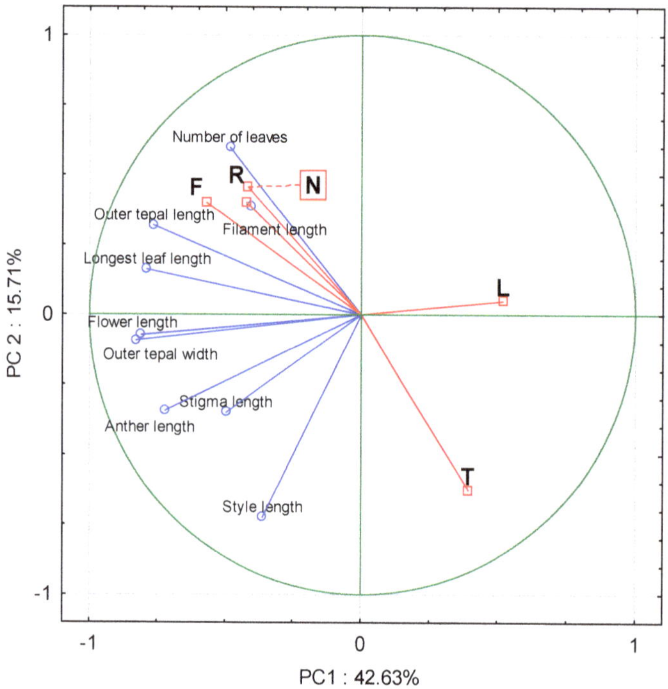

Figure 5. Principal component analysis (PCA) biplot of morphological characters of *Crocus tommasinianus* and mean Ellenberg indicator values for light (L), temperature (T), moisture (F), soil pH (R), and nitrogen (N).

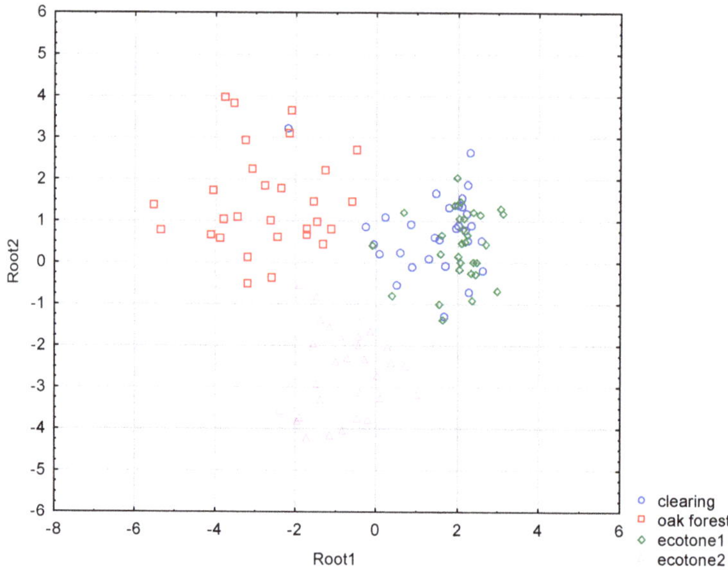

Figure 6. Results of discriminant function analysis (DFA) based on 9 morphological features of specimens from the studied subpopulations of *Crocus tommasinianus*.

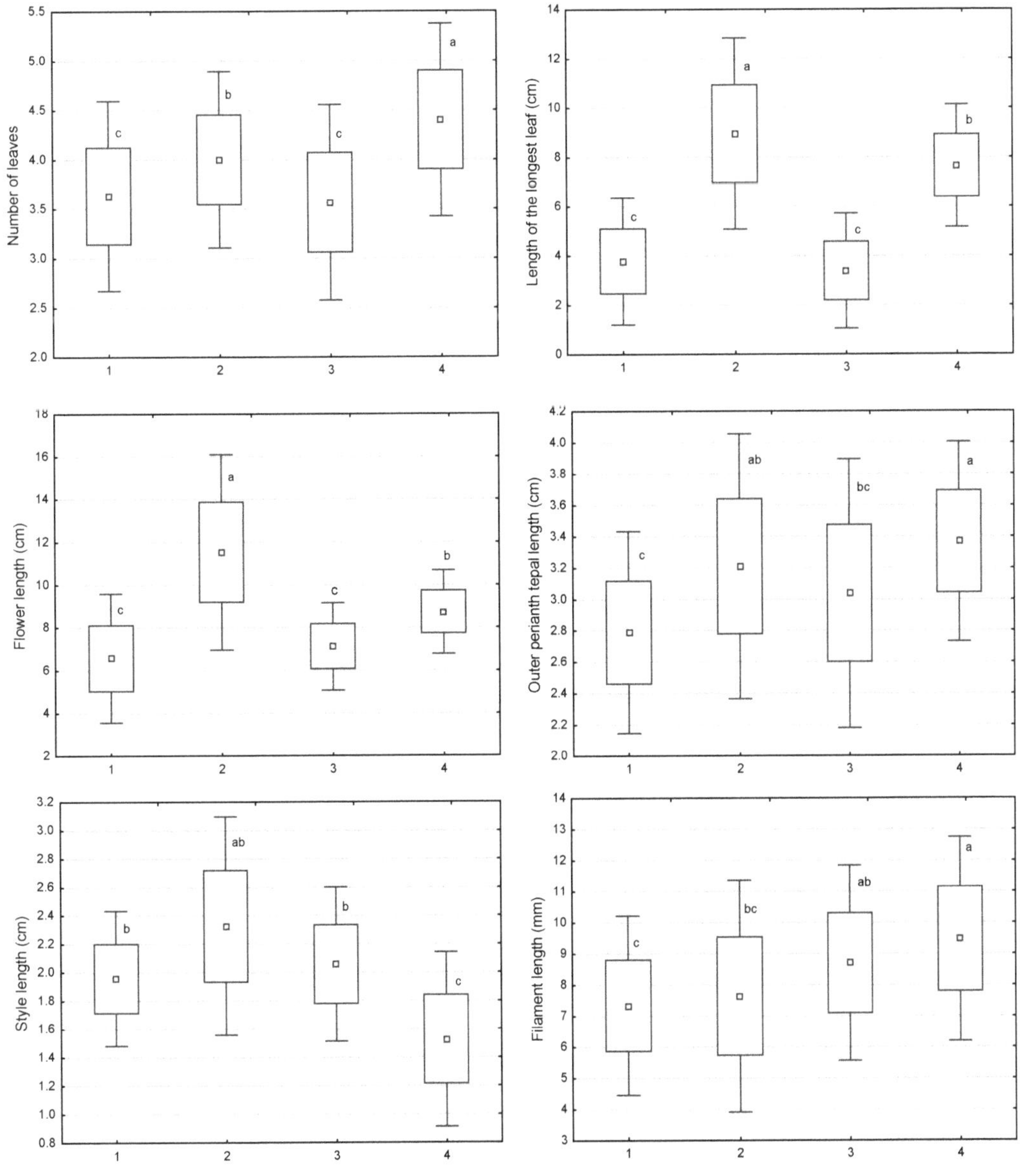

Figure 7. The mean (point), mean ± SD (standard deviation, box), and mean ± 1.96*SD (whisker) for the morphological characters of *Crocus tommasinianus* identified by discriminant function analysis. The horizontal axis shows the groups (subpopulations) from different habitats: 1—clearing; 2—oak forest; 3—ecotone1; 4—ecotone2. Different lowercase letters indicate subpopulations that differ in a given character (Tukey's HSD test, $p < 0.05$).

Table 2. Results of discriminant function analysis (DFA) based on morphological characters of *Crocus tommasinianus*. The highest values of standardized coefficients are bolded.

Character	Wilk's Lambda	Partial Lambda	p	Standardized Coefficients		ANOVA F	p
				CAN1	CAN2		
Number of leaves	0.05	0.92	0.03	−0.09	**−0.32**	17.48	<0.001
Length of longest leaf (cm)	0.09	0.59	<0.001	**−0.84**	−0.29	107.62	<0.001
Flower length (cm)	0.06	0.82	<0.001	**−0.46**	**0.40**	62.47	<0.001
Outer tepal length (cm)	0.06	0.91	0.02	0.33	**−0.30**	12.95	<0.001
Outer tepal width (cm)	0.05	0.97	0.32	−0.15	0.01	12.10	<0.001
Stigma length (mm)	0.06	0.84	<0.001	0.34	−0.35	3.10	0.03
Style length (cm)	0.09	0.54	<0.001	0.03	**0.94**	36.99	<0.001
Anther length (cm)	0.05	0.96	0.26	−0.04	0.16	10.00	<0.001
Filament length (mm)	0.07	0.77	<0.001	0.22	**−0.59**	7.63	<0.001
Eigenvalues				3.59	2.51		
Cumul. total variance expl. (%)				0.57	0.96		

3.3. Carpological Examinaton

Crocus tommasinianus has brown seeds that are rounded, 2.43–3.00 mm long, and 1.63–2.24 mm wide, with an elaiosome length of 0.34–0.61 mm. The seed coat micro-ornamentation is papillate, with long papillae that taper towards the top, are flattened on both sides, and are slightly bent at the top. All papillae are more or less the same size, about 40 μm long and 10 μm wide at the base with wrinkled secondary sculpture (Figure 8, Table 3).

Table 3. Seed size in three species of the genus *Crocus* occurring in Poland. Different lowercase letters indicate species that differ in a given characteristic (Tukey's HSD test, $p < 0.05$).

Crocus Species		Seed Length (mm)	Seed Width (mm)	Elaiosome Length (mm)
C. vernus	Mean ± SD	2.87 ± 0.24 [ab]	2.13 ± 0.22	0.42 ± 0.06
	Min–Max	2.13–3.15	1.71–2.43	0.30–0.53
C. scepusiensis	Mean ± SD	3.06 ± 0.26 [a]	2.10 ± 0.10	0.47 ± 0.11
	Min–Max	2.77–3.69	1.90–2.28	0.30–0.65
C. tommasinianus	Mean ± SD	2.78 ± 020 [b]	2.08 ± 0.16	0.48 ± 0.09
	Min–Max	2.43–3.00	1.63–2.24	0.34–0.61

Crocus vernus has brown seeds that are rounded, 2.13–3.15 mm long, and 1.71–2.43 mm wide, with an elaiosome length of 0.30–0.53 mm. The seed coat micro-ornamentation is papillate; the papillae are equally long (about 25 μm), slightly tapering towards the top. Their width at the base is about 20 μm. Their dorsal side is convex, and the ventral side is concave. The base of the papillae is often visibly sunken. As for the secondary sculpture, the cuticle surface is slightly longitudinally wrinkled at the edges of the papillae, as well as on their concave side.

Crocus scepusiensis has brown seeds that are rounded, 2.77–3.69 mm long, and 1.90–2.28 mm wide, with an elaiosome length of 0.30–0.65 mm. The micro-ornamentation seed coat is papillate; the papillae are equally long, with an extended base and a slightly tapering top. They are about 30 μm long and 20 μm wide at the base. Their dorsal side is convex, and the ventral side is concave. As for the secondary sculpture, the cuticle surface is irregularly, longitudinally, and deeply wrinkled at the edges of the papillae and on the concave side, with wax platelets near the base.

A statistical analysis of seed size showed significant differences only in seed length: *C. tommasinianus* seeds are significantly shorter than *C. scepusiensis* seeds but do not differ significantly in length from *C. vernus* seeds.

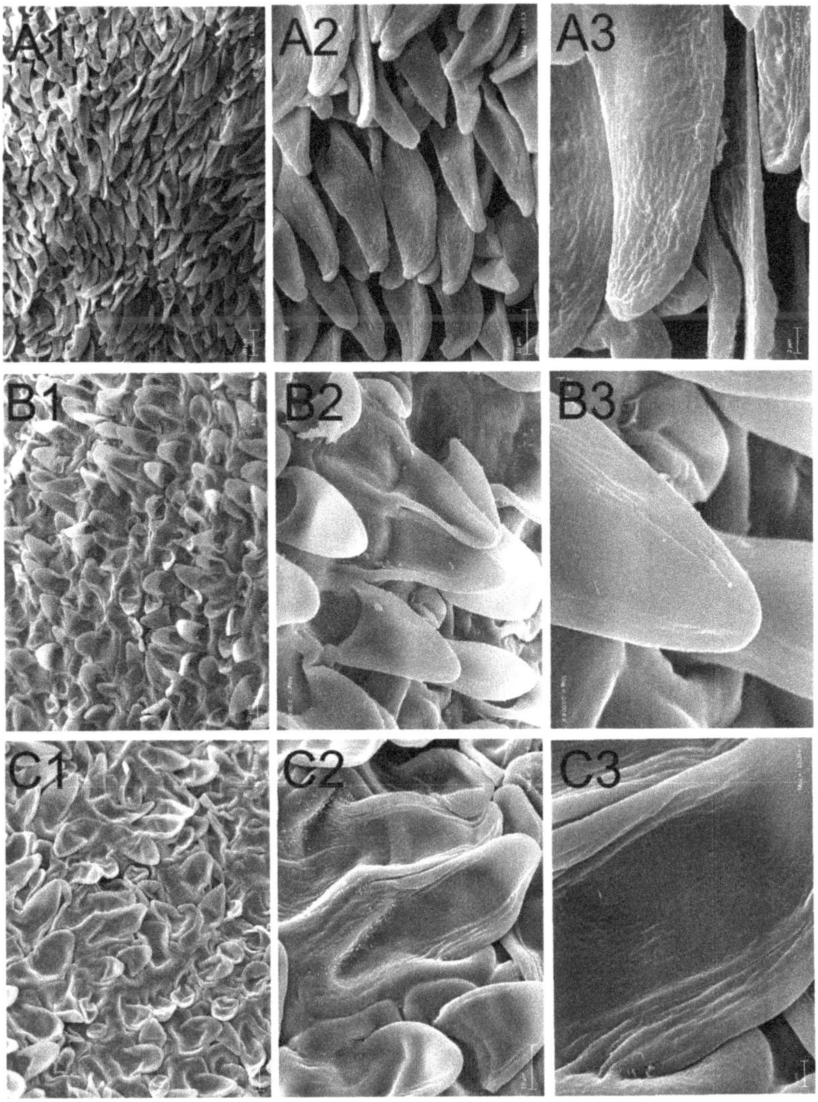

Figure 8. Seed sculpture of examined taxa (SEM): *Crocus tommasinianus* (**A**), *Crocus vernus* (**B**) and *Crocus scepusiensis* (**C**). Magnifications: ×1000 (**1**), ×3500 (**2**), and ×10,000 (**3**).

4. Discussion

So far, only two species from the genus *Crocus* have been recorded in the wild in Poland: *C. scepusiensis* Rehmann et Wol. and *C. vernus* Hill [60], both of the *Verni* series. The first species is native and occurs in the southern part of Poland, particularly in the mountain ranges of the Western Carpathians [76]. In contrast, *C. vernus* is an alien species that often escapes from cultivation and becomes naturalized [10]. Our research indicates that a third species, *C. tommasinianus*, can now be included in the flora of Poland. According to the geographical-historical classification of floras [77], this species should be considered as an alien (anthropophyte) introduced and naturalized after the 15th century (metaphyte, kenophyte).

Although there have been no previous reports of *C. tommasinianus* escaping from cultivation in Poland, the described site is not unique within the country. Searching for other locations of *C. tommasinianus* in Poland, we examined the iNaturalist database [78]. Of the 36 locations identified by observers as *C. tommasinianus*, 10 are classified as reliable, and 3 are regarded as probable. Some of the confirmed locations were marked by observers as wild-growing individuals, although this is not always corroborated by the visible surroundings in the photographs. The vast majority of probably wild specimens are found in urban or rural areas near cultivation sites. However, at least one site seems to be situated in a forest, as indicated by the abundant presence of *Ficaria verna* Huds.

4.1. Description of the New Location

Crocus tommasinianus has been cultivated at the study site for at least 30 years. During that period, the distribution of individuals has changed markedly, reflecting the transformation from a cultivated species into a naturalized one. At the original introduction site (i.e., the clearing), the number of individuals is currently very small. The population has dispersed into nearby forest areas. We assume that the small number of individuals in the clearing is at least partly due to the unsuitable location for cultivation of this species, as *C. tommasinianus* is moderately shade-tolerant [38,53]. In the study area, it is more abundant under the canopy of trees, and all of the studied morphological characters are more or less negatively correlated with light indicator values. Due to its similarity to the commonly cultivated and sold *C. vernus*, it may have been mistakenly planted as *C. vernus*, which is known to prefer sunlight.

Another factor that affects the subpopulation of *C. tommasinianus* in the clearing is mowing. For all crocus species cultivated on lawns, it is recommended to delay mowing until the end of May to avoid damaging the leaves during photosynthesis, allowing the corms to accumulate the necessary nutrients for flowering in the following year [79,80]. Since *C. tommasinianus* is one of the smallest cultivated crocus species [46], mowing might not significantly damage its leaves. An experiment [24] on the growth and development of *C. tommasinianus* 'Ruby Giant' on a lawn composed of highly competitive warm-season turfgrasses showed that strong competition, herbicides, and lawn mowing did not affect the cultivar's bulb survival and flowering, suggesting that typical lawn management practices are not detrimental to the bulbs. However, the cultivar 'Ruby Giant' used in the experiment is sterile [81]. The population observed in our study sets fruits (capsules) that ripen in summer. During ripening, the pedicel elongates significantly, raising the capsules above the ground surface. Mowing can damage or destroy unripe fruit, thus reducing the number of viable seeds, which likely affects dispersal within the clearing. This effect does not occur in the ecotone or the forest. As indicated by [82,83], generative reproduction significantly contributes to the increase in the population size of *C. tommasinianus*. This species disperses its seeds primarily through autochory, with further dissemination aided by ants (myrmecochory) attracted by the elaiosomes on the seeds. This two-phase dispersal (diplochory) is exceptionally effective, allowing relatively long-range seed dispersal and reducing seed mortality [84]. The adaptation of seed traits for ant-mediated dispersal is widespread among plants inhabiting temperate deciduous and mixed forests [85].

In the case of bulb and corm plants, we must take into account their vegetative propagation by the formation of daughter bulbs and corms. Although in the study area, we did not conduct detailed research on the proportions of sexual and vegetative reproduction in this population and subpopulations, such observations will be initiated in the coming years as an element of monitoring of this species. In 2024, traces of rooting by wild boars were found in ecotone2, i.e., in the place where *C. tommasinianus* was the most abundant. Wild boars could search for tubers of *Ficaria verna* (which grows there) but also for corms of *C. tommasinianus*. Crocuses can be part of the food base of wild boars, as reported earlier for *C. heufelianus* Herb. [55] and *C. ilvensis* Peruzzi and Carta [86]. Published studies [87,88] indicate that rooting by wild boars causes the fragmentation of underground organs of forest geophytes, such as *Ficaria verna* or *Dentaria bulbifera* L., thereby promoting the dispersion of these organs and intensifying their multiplication. As a result of rooting by wild boars, the density of geophytes increases. It seems that the activity of wild boars can also have a positive influence on asexual reproduction and population density in this crocus species. If wild boars root there repeatedly, as is often the case [89], then further expansion of the alien species can be expected.

When assessing the influence of this species on native flora and plant communities, we concluded that according to the classification proposed by Faliński [90,91], it currently enriches the species composition, i.e., the new species does not outcompete the coexisting species and does not replace other species found earlier in the plant community. It seems, however, that if its density increases in the future, then its competitiveness may also increase. The geophytes that develop in spring in forest ecosystems may slow down the development of other plants in the herb layer, mostly annuals, as reported, e.g., in the case of the influence of *Ficaria verna* on *Impatiens parviflora* DC. [92]. A specific form of competition may be also allelopathy, which is understood in this context as inhibition of the development of other plants by means of chemical substances (secondary metabolites) leached from plants by rain, exuded from roots, or volatilized or released during decomposition of dead plant matter [93]. Although there is no available publication on the possible allelopathic properties of *C. tommasinianus*, such research was conducted on the closely related *C. sativus* L. Due to the high phylogenetic conservation of the composition of secondary metabolites [94], such a comparison may be justified. Extracts from the leaves, corms, or flowers of *C. sativus* have allelopathic properties, hampering seed germination and leading to a decrease in the biomass and leaf area of various plant species (both field weeds and cultivated plants) [95–98]. Thus, similar effects can be expected in the case of *C. tommasinianus*. A meta-analysis of 16,810 allelopathic effects obtained from 384 studies showed that native plants suffer more from leachates of naturalized alien plants than from leachates of other native plants [99], so allelopathy may aid alien plant species in colonizing new plant communities.

4.2. Morphometric Measurements of Individuals

Crocuses undeniably exhibit substantial morphological variation, even within a single population [100]. Discriminant function analysis revealed that individual subpopulations differed in features contributing to the larger size of individuals, such as the number of leaves, the length of the longest leaf, as well as flower length and the length of the outer tepal; individuals growing in moderate shade were clearly larger. This is likely because plants can fully exploit their growth potential in better conditions, e.g., in partial shade for *C. tommasinianus*. Seed dimensions, however, tend to be more conservative and less variable in response to environmental conditions [101], as confirmed by the data in Table 3. Ecological studies conducted on various species show that the conditions under which a plant matures affect the number of seeds produced and their germination capacity rather than seed size [101].

4.3. Carpological Examination

Seed sculpture is also a very conservative feature [69]. Our study shows that *C. tommasinianus* clearly differs from *C. vernus* and *C. scepusiensis* in terms of both ornamentation and secondary testa sculpture. Since many floristic studies are conducted during the fruiting period of crocuses rather than during flowering, the diagnostic usefulness of testa ornamentation for species identification is crucial. This is particularly important in the case of the pair *C. tommasinianus* and *C. vernus*. Both species are widely cultivated throughout Poland, and both may escape from cultivation, so their ranges overlap. These species are morphologically very similar, so outside the flowering period, they are extremely difficult to distinguish.

5. Conclusions

The naturalization of alien plant species, their further expansion, and even invasion are most often the result of a combination of several factors: their inherent potential, supported by human activity, and certain random events. For these reasons, the future of alien species is often difficult to predict. The documentation of the first cases of alien species naturalization is very important, as it provides an opportunity to implement preventive and protective strategies to minimize the risks associated with their introduction.

For the first time, we observed the naturalization of the ornamental *Crocus tommasinianus* in Poland. This shade-tolerant species, initially cultivated in a location situated near forest ecosystems, has spread and successfully reproduced in a nearby oak forest. This may lead to an increase in population size and further expansion of the species in the future. In this study, we examined the morphological plasticity of this species in response to environmental conditions outside its cultivated area. We identified the probable causes of its spread in the forest ecosystem and the reduction in the number of individuals at its former cultivation site. We also analyzed the current and potential future impact of this species on the forest ecosystem and other coexisting organisms.

Due to the lack of data on the ecology of *Crocus tommasinianus* in naturalized sites in Central Europe, this area will be used for monitoring studies. Long-term monitoring will track the species' dispersal dynamics, the proportion of vegetative and sexual reproduction in different habitat conditions, morphological variability, and potential interactions with other species. This approach will capture the effects of seasonal and annual changes on the naturalization of *Crocus tommasinianus*, providing a deeper understanding of the species' long-term dynamics. A limitation of this study is that it covers only an area of about 500 m^2. Considering the climatic and soil variability in Poland, future sampling from a larger number of locations would provide a better understanding of *C. tommasinianus* adaptation to different ecological conditions.

Author Contributions: Conceptualization, A.C. and R.N.; methodology, R.N. and A.C.; formal analysis, R.N.; investigation, A.C. and R.N.; writing—original draft preparation, R.N. and A.C.; writing—review and editing, R.N. and A.C.; visualization R.N. and A.C. All authors have read and agreed to the published version of the manuscript.

Funding: This publication was financed by the Polish Minister of Science and Higher Education as part of the Strategy of the Poznan University of Life Sciences for 2024–2026 within the field of improving scientific research and development work in priority research areas.

Data Availability Statement: The data presented in this study are available on request from the corresponding author.

Acknowledgments: We would like to express our gratitude to Maria Morozowska for her assistance with the study of seed coat micro-ornamentation. We also sincerely thank the editor and the anonymous reviewers for their valuable comments and suggestions, which have significantly improved the quality of this paper.

Conflicts of Interest: The authors declare no conflicts of interest.

Appendix A

Table A1. Plant Communities with *Crocus tommasinianus* subpopulation studied. Abbreviations: 1—clearing; 2—oak forest (*Potentillo albae-Quercetum*); 3—ecotone1; 4—ecotone2. For individual species, the values of the Ellenberg indices are provided for light (L), temperature (T), moisture (F), soil pH (R), and nitrogen content (N); x—no data.

Position of Locality	1	2	3	4	Ellenberg Indicators				
					L	T	F	R	N
Date	26.03.2021	26.03.2021	26.03.2021	26.03.2021					
Coordinates	52°16′11.0″ N 16°47′81.0″ E	52°16′08.4″ N 16°47′81.2″ E	52°16′11.6″ N 16°47′82.2″ E	52°16′09.5″ N 16°47′84.5″ E					

Table A1. *Cont.*

Position of Locality	**1**	**2**	**3**	**4**	**Ellenberg Indicators**				
					L	**T**	**F**	**R**	**N**
Sample collection area [m^2]	10	10	10	8					
Cover of tree layer a [%]	0	35	40	15					
Cover of shrub layer b [%]	0	5	2	0					
Cover of herbaceous layer c [%]	65	60	50	65					
Number of species	22	21	38	11					
***Crocus tommasinianus* Herb.**	**1.2**	**1.1**	**2.2**	**3.4**					
ChO. *Arrhenatheretalia* Pawł. 1928									
Achillea millefolium L.	r	.	.	.	8	x	4	x	5
Dactylis glomerata L.	.	.	.	r	7	x	5	x	6
Taraxacum officinale Web.	+	.	r		8	x	5	x	7
ChCl. *Molinio-Arrhenatheretea* T.Tx. 1937									
Festuca pratensis Huds.	2.2	.	.	.	8	x	6	x	6
Plantago lanceolata L.	+	.	.	.	6	x	x	x	x
Rumex acetosa L.	r	.	.	.	8	x	x	x	5
ChCl. *Querco-Fagetea* Br.-Bl. et Vlieg. 1937									
Acer platanoides L. (a)	.	1.1	1.1	1.1	(5)	6	x	x	x
Acer platanoides L. (b)	r	+	+	.					
Aegopodium podagraria L.	.	1.1	.	+	5	x	6	7	8
Anemone nemorosa L.	.	.	+	.	x	x	x	5	x
Fraxinus excelsior L. (a)	.	.	1.1	.	(4)	5	x	7	7
Fraxinus excelsior L. (c)		r	.	.					
Lonicera xylosteum L.	.	+	.	.	5	5	5	7	x
Poa nemoralis L.	.	2.2	1.1	.	5	x	5	5	3
ChCl. *Nardo-Callunetea* Prsg 1949									
Luzula campestris (L.) DC.	.	r	.	.	7	x	4	3	2
Pilosella officinarum Vaill.	1.1	.	.	.	7	x	4	x	2
ChCl. *Koelerio-Corynephoretea* Klika in Klika et Novak, 1941									
Helichrysum arenarium (L.) Moench	+	.	.	.	7	6	3	5	1
Sedum sexangulare L.	+	.	.	.	7	5	7	8	1
ChAll. *Vicio lathyroides-Potentillion* Brzeg in Brzeg et M. Woj. 1996									
Hypochoeris radicata L.	r	.	.	.	8	5	5	4	3
Potentilla argentea L.	r	.	.	.	9	x	2	3	1
Vicia lathyroides L.	+	.	.	.	8	7	2	3	2
ChAll. *Carpinion* Issl. 1931 em. Oeberd. 1953									
Carpinus betulus L. (a)	.	1.1	.	1.1	(4)	6	x	x	x
ChAll. *Fagion* R.Tx. et Diem. 1936									
Fagus sylvatica L. (c)	.	.	r	.	(3)	5	5	x	x
ChO. *Fagetalia* Pawł. in Pawł., Sokoł. et Wall. 1928									
Corydalis intermedia (L.) Mérat	.	r	+	.	3	5	5	7	7
Ficaria verna Huds.		1.1	+	1.1	4	5	6	7	7
ChAss. *Lolio-Cynosuretum* R.Tx. 1947									
Trifiolium repens L.	1.1	.	.	.	8	x	x	x	7
ChAll. *Alliarion* Oberd. (1957) 1962									
Geranium robertianum L.	.	+	+	.	4	x	x	x	7
ChAss. *Alliario-Chaerophylletum temuli* (Kreh 1935) Lohm. 1949									

Table A1. *Cont.*

Position of Locality	**1**	**2**	**3**	**4**	**Ellenberg Indicators**				
					L	**T**	**F**	**R**	**N**
Chaerophyllum temulum L.	.	.	+	.	5	6	5	x	8
Hedera helix L.	.	+	+	.	(4)	5	5	x	x
ChO. *Glechometalia* R.Tx. in R.Tx. et Brun.-Hool 1975			.						
Geum urbanum L.	.	+	.	.	4	5	5	x	7
ChCl. *Festuco-Brometea* Br.Bl. et R.Tx. 1943			.						
Ajuga genevensis L.	.	r	+	.	8	x	4	7	2
Euphorbia cyparissias L.	.	.	r	.	8	x	3	x	3
ChAll. Alliarion Oberd. (1957) 1962									
Impatiens parviflora DC.		.	+	.	4	6	5	x	6
ChO. *Glechometalia* R.Tx. in R.Tx. et Braun-Hool 1975									
Alliaria petiolata (M. Bieb.) Cavara and Grande	.	.	+	.	5	6	5	7	9
Glechoma hedaracea L.		+	+	.	6	5	6	x	7
ChAss. *Hordeo-Brometeum* (Allorge 1922) Lohm. 1950			.						
Bromus sterilis L.		.	+	.	7	7	4	x	5
ChAll. *Ulmenion minoris* Oberd. 1953									
Ulmus laevis Pall. (a)		.	.	1.1	(4)	6	8	7	7
ChCl. *Stellarietea mediae* R.Tx., Lohm. et Prsg. 1950			.						
Stellaria media (L.) Vill.		.	r	+	6	x	4	7	8
Others									
Aesculus hippocastanum L. (a)	.	.	1.1	.	.	.	.	.	.
Allium vineale L.	r	.	+	.	5	7	4	x	7
Arenaria serpyllifolia L.	.	.	+	.	9	5	4	x	x
Carex spicata Huds.	.	.	r	.	7	5	5	x	6
Chionodoxa sardensis Barr. et Sudg.	r	.	r	.					
Conyza canadensis (L.) Cronquist	.	.	+	.	8	x	4	x	4
Euonymus alatus (Thunb.) Siebold (b)	.	+	.	.	.	.	.	.	.
Festuca trachyphylla Sibth.	2.2	.	.	.	.	.	.	.	.
Gagea pratensis (Pers.) Dumort.	.	.	+	.	.	.	.	.	.
Galanthus nivalis L.	r	.	+	.	5	7	x	7	7
Geranium molle L. 'Alba'	+	.	.	.	7	6	3	5	4
Medicago lupulina L.	r	.	.	.	7	5	4	8	x
Myosotis stricta Link ex Roem. and Schult.	.	.	+	.	8	x	3	4	2
Ornithogalum nutans L.	.	.	+	.	.	.	.	.	.
Poa annua L.	.	.	.	r	7	x	6	x	8
Prunus domestica L. subsp. *syriaca* (Borkh.) Janch. (b)	.	.	+	.	.	.	.	.	.
Pyrus pyraster (L.) Burgsd. (c)	.	r	.	.	.	.	.	.	.
Quercus petraea (Matt.) Liebl. (a)	r	2.2	2.2	.	(6)	6	5	x	x
Scilla sibirica Haw.	.	.	r	1.1	.	.	.	.	.
Scilla sibirica Haw. 'Alba'	.	.	+	+	.	.	.	.	.
Veronica chamaedrys L.	+	.	r	.	6	x	4	x	x
Veronica sublobata M. A. Fisch.	.	1.1	1.1		6	6	5	7	7
Vinca minor L.	.	1.1	+	.	4	6	5	x	6
Viola suavis M. Bieb.	.	.	r	.	.	.	.	.	.

References

1. Regulation (EU) No 1143/2014 of the European Parliament and of the Council of 22 October 2014 on the Prevention and Management of the Introduction and Spread of Invasive Alien Species. Available online: https://eur-lex.europa.eu/eli/reg/2014/1143/oj (accessed on 25 August 2024).
2. Pyšek, P.; Richardson, D.M. The biogeography of naturalization in alien plants. *J. Biogeogr.* **2006**, *33*, 2040–2050. [CrossRef]
3. Aguin-Pombo, D. Biological Invasions and Global Trade: An Urgent Issue. In *Natural Resources, Sustainability and Humanity: A Comprehensive View*; Aguin-Pombo, D., Ed.; Springer: Dordrecht, The Netherlands, 2012; pp. 83–99.
4. Van Kleunen, M.; Dawson, W.; Essl, F.; Pergl, J.; Winter, M.; Weber, E.; Pyšek, P. Global exchange and accumulation of non-native plants. *Nature* **2015**, *525*, 100–103. [CrossRef] [PubMed]
5. Early, R.; Bradley, B.A.; Dukes, J.S.; Lawler, J.J.; Olden, J.D.; Blumenthal, D.M.; Gonzalez, P.; Grosholz, E.D.; Ibáñez, I.; Miller, L.P.; et al. Global threats from invasive alien species in the twenty-first century and national response capacities. *Nat. Commun.* **2016**, *7*, 12485. [CrossRef] [PubMed]
6. Zenni, R.D.; Essl, F.; García-Berthou, E.; McDermott, S.M. The economic costs of biological invasions around the world. *NeoBiota* **2021**, *67*, 1–9. [CrossRef]
7. Seebens, H.; Essl, F.; Hulme, P.E.; van Kleunen, M. Development of pathways of global plant invasions in space and time. In *Global Plant Invasions*; Clements, D.R., Upadhyaya, M.K., Joshi, S., Shrestha, A., Eds.; Springer: Cham, Switzerland, 2022; pp. 35–50. [CrossRef]
8. Lenzner, B.; Latombe, G.; Schertler, A.; Seebens, H.; Yang, Q.; Winter, M.; Weigelt, P.; van Kleunen, M.; Pyšek, P.; Pergl, J.; et al. Naturalized alien floras still carry the legacy of European colonialism. *Nat. Ecol. Evol.* **2022**, *6*, 1723–1732. [CrossRef]
9. Sudnik-Wójcikowska, B. *Rośliny Synantropijne*; Multico, Oficyna Wydawnicza: Warsaw, Poland, 2011.
10. Tokarska-Guzik, B. *The Establishment and Spread of Alien Plant Species (Kenophytes) in the Flora of Poland*; Wydawnictwo Uniwersytetu Śląskiego: Katowice, Poland, 2005.
11. Jackowiak, B. Man-made changes in the flora and vegetation of Poland: Current review. *Diversity* **2023**, *15*, 618. [CrossRef]
12. Tokarska-Guzik, B.; Dajdok, Z.; Zając, M.; Zając, A.; Urbisz, A.; Danielewicz, W.; Hołdyński, C. *Rośliny Obcego Pochodzenia w Polsce*; Generalna Dyrekcja Ochrony Środowiska: Warsaw, Poland, 2012.
13. Obidziński, A.; Kołaczkowska, E.; Otręba, A. *Metody Zwalczania Obcych Gatunków Roślin Występujących na Terenie Puszczy Kampinoskiej*; Kampinoski Park Narodowy: Izabelin, Poland; Wydawnictwo BioDar: Kraków, Poland, 2016.
14. Słowiński, K.; Grygierzec, B.; Synowiec, A.; Tabor, S.; Araniti, F. Preliminary study of control and biochemical characteristics of giant hogweed (Heracleum sosnowskyi Manden.) treated with microwaves. *Agronomy* **2022**, *12*, 1335. [CrossRef]
15. Švec, P.; Perglová, I.; Fröhlich, V.; Laštovička, J.; Seidl, J.; Růžičková, K.; Horáková, I.; Lukavský, J.; Ferko, M.; Štych, P.; et al. Perseverance of management is needed–Efficient long-term strategy of Reynoutria management. *NeoBiota* **2024**, *94*, 261–288. [CrossRef]
16. Szymura, M.; Szymura, T.H.; Wolski, K. Invasive Solidago species: How large area do they occupy and what would be the cost of their removal? *Pol. J. Ecol.* **2016**, *64*, 25–34. [CrossRef]
17. Winter, M.; Schweiger, O.; Klotz, S.; Nentwig, W.; Andriopoulos, P.; Arianoutsou, M.; Basnou, C.; Delipetrou, P.; Didžiulis, V.; Hejda, M.; et al. Plant extinctions and introductions lead to phylogenetic and taxonomic homogenization of the European flora. *Proc. Natl. Acad. Sci. USA* **2009**, *106*, 21721–21725. [CrossRef]
18. Yang, Q.; Weigelt, P.; Fristoe, T.S.; Zhang, Z.; Kreft, H.; Stein, A.; Seebens, H.; Dawson, W.; Essl, F.; König, C.; et al. The global loss of floristic uniqueness. *Nat. Commun.* **2021**, *12*, 7290. [CrossRef] [PubMed]
19. Qian, H.; Guo, Q. Linking biotic homogenization to habitat type, invasiveness, and growth form of naturalized alien plants in North America. *Divers. Distrib.* **2010**, *16*, 119–125. [CrossRef]
20. Daru, B.H.; Davies, T.J.; Willis, C.G.; Meineke, E.K.; Ronk, A.; Zobel, M.; Pärtel, M.; Antonelli, A.; Davis, C.C. Widespread homogenization of plant communities in the Anthropocene. *Nat. Commun.* **2021**, *12*, 6983. [CrossRef] [PubMed]
21. Šibíková, M.; Jarolímek, I.; Hegedüšová, K.; Májeková, J.; Mikulová, K.; Slabejová, D.; Škodová, I.; Zaliberová, M.; Medvecká, J. Effect of planting alien *Robinia pseudoacacia* trees on homogenization of Central European forest vegetation. *Sci. Total Environ.* **2019**, *687*, 1164–1175. [CrossRef]
22. Heywood, V.H.; Brunel, S. *Code of Conduct on Horticulture and Invasive Alien Plants*; Council of Europe Publ: Strasbourg, France, 2008.
23. Fry, C. *The Plant Hunters: The Adventures of the World's Greatest Botanical Explorers*; Univ. Chicago Press: Chicago, IL, USA, 2013.
24. Richardson, M.D.; McCalla, J.; Buxton, T.; Lulli, F. Incorporating early spring bulbs into dormant warm-season turfgrasses. *HortTechnology* **2015**, *25*, 228–232. [CrossRef]
25. van Kleunen, M.; Essl, F.; Pergl, J.; Brundu, G.; Carboni, M.; Dullinger, S.; Early, R.; González-Moreno, P.; Groom, Q.J.; Hulme, P.E.; et al. The changing role of ornamental horticulture in alien plant invasions. *Biol. Rev.* **2018**, *93*, 1421–1437. [CrossRef]
26. Alam, A.; Iqbal, M.; Vats, S. Cultivation of Some Overlooked Bulbous Ornamentals—A Review on its Commercial Viability. *Rep. Opin.* **2013**, *5*, 9–34.
27. Invasive Species Specialist Group ISSG. The Global Invasive Species Database. Version 2015.1. Available online: https://www.iucngisd.org/gisd/ (accessed on 24 September 2024).
28. Plants of the World Online (POWO). Available online: https://powo.science.kew.org (accessed on 25 February 2024).

29. Hoste, I. Ornamentals and Invasive Plants: If You Choose One You Get the Other as Well. In *Mini Symposium of Aliens and Invasive Species*; National Botanic Garden of Belgium, Bouchout Castle: Meise, Belgium, 2011.
30. Czarna, A.; Klimko, M.; Janyszek, S. Vascular Flora and Vegetation of the Former Manor Park in Radojewo (Wielkopolska Region, Poland). *Rocz. AR Pozn.* **2009**, *388*, 37–47.
31. Czarna, A. Vascular Flora in the Park at Jeziory in the National Park of Wielkopolska. *Rocz. AR Pozn.* **2010**, *389*, 39–44.
32. Czarna, A. Vascular Plant Flora in the Cytadela Cemeteries in Poznań (Poland). *Acta Agrobot.* **2016**, *69*, 1–17. [CrossRef]
33. Czarna, A. Vascular Plants in the Cemetery of the Meritorious (Cmentarz Zasłużonych) in Poznań (Poland). *Ann. UMCS* **2016**, *71*, 59–73. [CrossRef]
34. Nowińska, R.; Czarna, A.; Czekalski, M.; Morozowska, M. Vascular Flora of Selected Palace Parks in the Wielkopolska Region. *Steciana* **2016**, *20*, 137–157. [CrossRef]
35. Żukowski, W.; Latowski, K.; Jackowiak, B.; Chmiel, J. Rośliny Naczyniowe Wielkopolskiego Parku Narodowego. In *Prace Zakładu Taksonomii Roślin Uniwersytetu im. Adama Mickiewicza w Poznaniu*; Bogucki Wydawnictwo Naukowe: Poznań, Poland, 1995.
36. Drobnik, J. *Allium paradoxum* from Asia to Europe: Ornamental, Invasive, Edible, and Medicinal. *Hum. Ecol.* **2023**, *51*, 559–567. [CrossRef]
37. Rukšāns, J. *The World of Crocuses, the First Supplement*; The Latvian Academy of Sciences: Riga, Latvia, 2023.
38. Mathew, B.F. *Crocus* L. In *Flora Europaea, Vol. 5: Alismataceae to Orchidaceae (Monocotyledones)*; Tutin, T.G., Heywood, V.H., Burges, N.A., Moore, D.M., Valentine, D.H., Walters, S.M., Webb, D.A., Eds.; Cambridge University Press: Cambridge, UK, 2010; pp. 92–97.
39. Petersen, G.; Seberg, O.; Thorsøe, S.; Jørgensen, T.; Mathew, B. A phylogeny of the genus Crocus (Iridaceae) based on sequence data from five plastid regions. *Taxon* **2008**, *57*, 487–499. [CrossRef]
40. Surányi, G.; Máthé, C.; Mosolygó, Á.; Borbély, G.; Vasas, G. Analysis of genetic diversity in crocuses with Carpathian Basin origin using AFLP-markers. *Acta Biol. Hung.* **2010**, *61*, 149–155. [CrossRef]
41. Alsayied, N.F.; Fernández, J.A.; Schwarzacher, T.; Heslop-Harrison, J.S. Diversity and relationships of *Crocus sativus* and its relatives analysed by inter-retroelement amplified polymorphism (IRAP). *Ann. Bot.* **2015**, *116*, 359–368. [CrossRef]
42. Rukšans, J. *The World of Crocuses*; Latvian Academy of Sciences: Riga, Latvia, 2017.
43. Rand, E.S. *Popular Flowers, and How to Cultivate Them*; Hurd and Houghton: New York, NY, USA, 1876.
44. Snowarski, M. Flora Polski, Atlas Roślin.pl. Available online: https://atlas-roslin.pl/index.html (accessed on 25 February 2024).
45. Harris, S. *Crocus species* (Iridaceae). *Oxford University Plants 400. Department of Plant Sciences*; Oxford University. Available online: https://herbaria.plants.ox.ac.uk/bol/plants400/Profiles/cd/crocus (accessed on 18 October 2024).
46. Szilagyi, C.; Băla, M.; Toța, C. The behaviour of some crocus species upon different fertilizers and substrates. *J. Name* **2018**, *22*, 98–101.
47. Wurzell, B. Spring flowering crocuses. *BSBI News* **1992**, *60*, 36–38.
48. The Complete List of Taxon Names from the BSBI's Database. Available online: https://bsbi.org/taxon-lists (accessed on 14 October 2024).
49. McAwoy, M. *Non-Native Plants of Delaware*; Species Conservation and Research Program, Delaware Department of Natural Resources and Environmental Control, Division of Fish and Wildlife: Dover, DE, USA, 2024.
50. Tuttle, J. Plant spring blooming bulbs now. In *Ask a Master Gardener*; University of Minnesota Extension: St Paul, MN, USA, 2023; pp. 1–6.
51. Brighton, C.A. Cytological problems in the genus Crocus (Iridaceae): I. Crocus vernus aggregate. *Kew Bull.* **1976**, *31*, 33–46. [CrossRef]
52. Fedorov, A.A. Flora Partis Europaeae URSS. 1979; Volume 4.
53. Dostál, J. *Nová Květena ČSSR*; Academia: Prague, Czech Republic, 1989; 1548p.
54. Prišžter, S.Z. Uj säfrányfaj (*Crocus tommasinianus* HERB.) Magyarországon. *Bot. Közlemények* **1964**, *51*, 183–186.
55. Mihaly, A.; Kricsfalusy, V. Population biology and ecology of *Crocus heuffelianus* Herb. (Iridaceae) in Ukraine. *Linz. Biol. Beiträge* **1997**, *29*, 641–681.
56. Ževrnja, N.; Vladović, D. The genus Crocus L. in the flora of Svilaja Mountain. *Nat. Croat. Period. Musei Hist. Nat. Croat.* **2005**, *14*, 363–368.
57. Hitchmough, J.; Fieldhouse, K. *Plant User Handbook: A Guide to Effective Specifying*; Blackwell Science: Oxford, UK, 2004. [CrossRef]
58. Ogle, C.C.; La Cock, G.D. Additional records and observations of monocotyledons naturalised or casual in Manawatu Ecological Region, New Zealand. *Perspect. Biosecurity* **2019**, *4*, 7–31.
59. Hill, M.O.; Preston, C.D.; Shanklin, J.D. Geographical patterns in the flora of Cambridgeshire (vc 29). *Br. Ir. Bot.* **2020**, *2*, 285–308. [CrossRef]
60. Mirek, Z.; Piękoś-Mirkowa, H.; Zając, A.; Zając, M. *Vascular Plants of Poland: An Annotated Checklist*; W. Szafer Institute of Botany, Polish Academy of Sciences: Kraków, Poland, 2020.
61. Harpke, D.; Carta, A.; Tomović, G.; Ranđelović, V.; Ranđelović, N.; Blattner, F.R.; Peruzzi, L. Phylogeny, karyotype evolution and taxonomy of Crocus series Verni (Iridaceae). *Plant Syst. Evol.* **2015**, *301*, 309–325. [CrossRef]
62. Mosolygó, Á.; Sramkó, G.; Barabás, S.; Czeglédi, L.; Jávor, A.; Molnár, A.V.; Surányi, G. Molecular genetic evidence for allotetraploid hybrid speciation in the genus *Crocus* L. (Iridaceae). *Phytotaxa* **2016**, *258*, 121–136. [CrossRef]

63. Raca, I.; Blattner, F.R.; Waminal, N.E.; Kerndorff, H.; Ranđelović, V.; Harpke, D. Disentangling Crocus series Verni and its polyploids. *Biology* **2023**, *12*, 303. [CrossRef]
64. Kandemir, N. Comparative leaf anatomy of some endemic Crocus L. taxa from Turkey. *Bangladesh J. Bot.* **2011**, *40*, 155–162. [CrossRef]
65. Raca, I.; Ljubisavljević, I.; Jušković, M.; Ranđelović, N.; Ranđelović, V. Comparative anatomical study of the taxa from series Verni Mathew (*Crocus* L.) in Serbia. *Biol. Nyssana* **2017**, *8*, 15–22. [CrossRef]
66. Raca, I.; Jovanovic, M.; Ljubisavljevic, I.; Juskovic, M.; Randelovic, V. Morphological and leaf anatomical variability of *Crocus* cf. *heuffelianus* Herb. (Iridaceae) populations from the different habitats of the Balkan Peninsula. *Turk. J. Bot.* **2019**, *43*, 645–658. [CrossRef]
67. Kerndorff, H.; Pasche, E.; Harpke, D. The Genus Crocus (Liliiflorae, Iridaceae): Life Cycle, Morphology, Phenotypic Characteristics, and Taxonomically Relevant Parameters. *Stapfia* **2015**, *103*, 27–65.
68. Karaismailoğlu, M.C.; Şik, L.; Gemicioğlu, A.; Erol, O. Seed Structure of Some Taxa of the Genus Crocus L. (Iridaceae) Series Crocus. *Turk. J. Bot.* **2018**, *42*, 722–731. [CrossRef]
69. Song, Y.X.; Peng, S.; Mutie, F.M.; Jiang, H.; Ren, J.; Cong, Y.Y.; Hu, G.W. Evolution and taxonomic significance of seed micromorphology in Impatiens (Balsaminaceae). *Front. Plant Sci.* **2022**, *13*, 835943. [CrossRef] [PubMed]
70. Dyderski, M.K.; Jagodziński, A.M. Impact of invasive tree species on natural regeneration species composition, diversity, and density. *Forests* **2020**, *11*, 456. [CrossRef]
71. Braun-Blanquet, J. *Pflanzensoziologie. Grundzüge der Vegetationskunde*, 3rd ed.; Springer: Vienna, Austria, 1964.
72. Di Biase, L.; Tsafack, N.; Pace, L.; Fattorini, S. Ellenberg Indicator Values Disclose Complex Environmental Filtering Processes in Plant Communities along an Elevational Gradient. *Biology* **2023**, *12*, 161. [CrossRef]
73. Ellenberg, H. *Zeigerwerte der Gefäßpflanzen Mitteleuropas*; Verlag Erich Goltze KG: Göttingen, Germany, 1974.
74. Cattell, R.B. The Scree Test for the Number of Factors. *Multivar. Behav. Res.* **1966**, *1*, 245–276. [CrossRef]
75. Chudzicki, M. *Operat Ekosystemów Leśnych Wielkopolskiego Parku Narodowego—Opis Taksacyjny cz. 2. Projekt Planu Ochrony Wielkopolskiego Parku Narodowego*; Biuro Urządzania Lasu i Geodezji Leśnej Oddział w Poznaniu: Poznań, Poland, 2013.
76. Szafer, W.; Kulczyński, S.; Pawłowski, B. *Rośliny Polskie*; Volume I-II; PWN: Warsaw, Poland, 1988.
77. Kornaś, J. Geograficzno-historyczna klasyfikacja roślin synantropijnych. *Mater. Zakładu Fitosocjologii Stosow. UW* **1968**, *25*, 33–41.
78. iNaturalist. *Crocus tommasinianus, C. vernus* . iNaturalist. Available online: https://www.inaturalist.org (accessed on 23 September 2024).
79. Krause, J. *Niskie, Ale Pięknie Kwitnące Rośliny Cebulowe. Krokus, Szafirek*; PWRiL: Warszawa, Poland, 1987.
80. Knebel, C.H. *Rośliny Cebulowe*; Klub dla Ciebie: Warsaw, Poland, 2006.
81. Wendebourg, T. *Zwiebelpflanzen für den Garten*; Verlag Eugen Ulmer: Stuttgart, Germany, 2004.
82. Adams, K. *Krokusy, Tulipany, Lilie*; Świat Książki: Warsaw, Poland, 2014.
83. Johansson, B.G. Crocus på Gotland. *Gotlands Bot. Förening* **1991**, *11*, 28–32.
84. Wall, S.V.; Longland, W.S. Diplochory and the evolution of seed dispersal. In *Seed Fate: Predation, Dispersal and Seedling Establishment*; Forget, P.M., Lambert, J.E., Hulme, P.E., Vander Wall, S.B., Eds.; CABI Publishing: Wallingford, UK, 2005; pp. 297–314. [CrossRef]
85. Gorb, E.; Gorb, S. *Seed Dispersal by Ants in a Deciduous Forest Ecosystem: Mechanisms, Strategies, Adaptations*; Kluwer Academic Publishers: Dordrecht, The Netherlands, 2003.
86. Peruzzi, L.; Carta, A. *Crocus ilvensis* sp. nov. (Sect. *Crocus, Iridaceae), Endemic to Elba Island (Tuscan Archipelago, Italy). Nord. J. Bot.* **2011**, *29*, 6–13. [CrossRef]
87. Biały, K. The Effect of Boar (*Sus scrofa*) Rooting on the Distribution of Organic Matter in Soil Profiles and the Development of Wood Anemone (*Anemone nemorosa* L.) in the Oak-Hornbeam Stand (*Tilio-Carpinetum*) in the Białowieza Primeval Forest. *Forestry* **1996**, *38*, 77–88. [CrossRef]
88. Faliński, J.B. Interaction: Wild Boar Rooting—Participation of Geophytes in the Herb Layer of Oak-Linden-Hornbeam Forest. *Phytocoennosis* **2002**, *5*, 70.
89. Podgórski, T.; Lusseau, D.; Scandura, M.; Sönnichsen, L.; Jędrzejewska, B. Long-Lasting, Kin-Directed Female Interactions in a Spatially Structured Wild Boar Social Network. *PLoS ONE* **2014**, *9*, e99875. [CrossRef] [PubMed]
90. Faliński, J.B. Stadia Neofityzmu i Stosunek Neofitów do Innych Komponentów Zbiorowiska. *Mater. Zakładu Fitosocjologii Stosow. Uniw. Warsz.* **1968**, *25*, 15–31.
91. Faliński, J.B. Inwazje w Świecie Roślin: Mechanizmy, Zagrożenia, Projekt Badań. *Phytocoen. N.S.* **2004**, *16*, 1–32.
92. Piskorz, R.; Klimko, M. Współwystępowanie Niecierpka Drobnokwiatowego *Impatiens parviflora* DC. i wybranych roślin lasu dębowo-grabowego w Wielkopolskim Parku Narodowym. *Sylwan* **2007**, *151*, 43–58.
93. Rice, E.L. *Allelopathy*; Acad. Press: New York, NY, USA, 1984.
94. Wink, M. Evolution of Secondary Metabolites from an Ecological and Molecular Phylogenetic Perspective. *Phytochemistry* **2003**, *64*, 3–19. [CrossRef]
95. Barkhordari, K.; Sorooshzadeh, A.; Mokhtassi, B.A. Allelopathic Effect of Extraction Solution of Leaves and Corms of Saffron (*Crocus sativus*) in Phenological Stages on Seed Germination of Jimson Weed (*Datura stramonium*). *Modares J. Biotechnol.* **2018**, *9*, 233–239.

96. Mikolajchuk, V.; Panfilova, A.; Korkhova, M.; Drobitko, A. Allelopathic Activity of Water-Soluble and Volatile Secretions of *Crocus sativus* L. (Iridaceae) Flowers in the Northern Black Sea Region. *J. Biotechnol.* **2022**, *9*, 233–239. [CrossRef]
97. Hosseini, M.; Rizvi, S.J.H. A Preliminary Investigation on Possible Role of Allelopathy in Saffron (*Crocus sativus* L.). In *II International Symposium on Saffron Biology and Technology*; Kafi, M., Koocheki, A., Eds.; ISHS: Leuven, Belgium, 2006; pp. 75–79.
98. Feizi, H.; Salari, A.; Gharar, F. Study of the Allelopathic Effect of Saffron (*Crocus sativus* L.) Organs' Aqueous Extract on the Seed Germination and Seedling Growth of Sugar Beet and Safflower at Different Concentrations. *J. Med. Spice Plants* **2018**, *22*, 156–161.
99. Zhang, Z.; Liu, Y.; Yuan, L.; Weber, E.; van Kleunen, M. Effect of Allelopathy on Plant Performance: A Meta-Analysis. *Ecol. Lett.* **2021**, *24*, 348–362. [CrossRef]
100. Dolatyari, A.; Abolhasani, M.T.; Ardalani, F.; Rukšāns, J. A taxonomic revision of the genus Crocus (Iridaceae) in Iran. *Nord. J. Bot.* **2024**, e04270. [CrossRef]
101. Fenner, M.; Thompson, K. *The Ecology of Seeds*; Cambridge University Press: Cambridge, UK, 2005; Available online: https://www.bsbi.org/ (accessed on 17 October 2014).

forests

MDPI

Article

Decreased P Cycling Rate and Increased P-Use Efficiency after *Phyllostachys edulis* (Carrière) J. Houz. Expansion into Adjacent Secondary Evergreen Broadleaved Forest

Shuwang Song [1,2], Lin Wang [1], Zacchaeus G. Compson [3], Tingting Xie [1], Chuyin Liao [1], Dongmei Huang [4], Jun Liu [1], Qingpei Yang [1] and Qingni Song [1,*]

[1] Jiangxi Provincial Key Laboratory of Improved Variety Breeding and Efficient Utilization of Native Tree Species, Jiangxi Agricultural University, Nanchang 330045, China; shuwangsong@nenu.edu.cn (S.S.); wangllin962464@163.com (L.W.); xtting1999@126.com (T.X.); 13542773342@163.com (C.L.); liujun@jxau.edu.cn (J.L.); qingpeiyang@jxau.edu.cn (Q.Y.)

[2] Key Laboratory of Geographical Processes and Ecological Security of Changbai Mountains, School of Geographical Sciences, Northeast Normal University, Ministry of Education, Changchun 130024, China

[3] Department of Biological Sciences, Advanced Environmental Research Institute, University of North Texas, Denton, TX 76205, USA; zacchaeus.greg.compson@gmail.com

[4] School of Humanities and Public Administration, Jiangxi Agricultural University, Nanchang 330045, China; hdmae99@163.com

* Correspondence: songqingni@jxau.edu.cn

Abstract: (1) Background: Moso bamboo (*Phyllostachys edulis* (Carrière) J. Houz.) expansion has seriously altered the species composition and structure of adjacent forest ecosystems in subtropical regions. However, the shift in phosphorus (P) biogeochemical cycling has yet to be assessed, which is a critical gap considering the great variation in ecophysiological properties between invasive bamboo and the displaced native tree species. (2) Methods: We investigated and compared expansion-induced changes in P pools (plant, litter, and soil) and P fluxes (plant uptake and litterfall return) using paired sampling of the bamboo-dominated forest (BDF) and secondary evergreen broadleaved forest (EBF) at Jiangxi province's Dagang Mountain National Forest Ecological Station. (3) Results: Both the P storage of the plants and litter were significantly greater by 31.8% and 68.2% in the BDF than in the EBF, respectively. The soil total P and available P storage were 28.9% and 40.4% lower, respectively, in the BDF than in the EBF. Plant P uptake was 15.6% higher in the BDF than in the EBF, and the annual litter P return was 26.1% lower in the BDF than in the EBF due to higher P resorption efficiency for moso bamboo compared with evergreen broadleaved tree species. The ecosystem P cycling rate was reduced by 36.1% in the BDF compared with the EBF. (4) Conclusions: Moso bamboo expansion slowed the broadleaved forest ecosystem's P cycle rate, likely because moso bamboo has higher P-use efficiency, reserving more P in its tissues rather than returning it to the soil. The results from this study elucidate an understudied element cycle in the context of forest succession, demonstrating the ecosystem consequences related to bamboo invasion.

Keywords: *Phyllostachys edulis* expansion; P biogeochemical cycling; resorption efficiency; evergreen broadleaved forest; subtropical China

1. Introduction

Phosphorus (P) is a vital structural and functional component in plants. It exists in living cells as phosphate, is used in synthesizing essential biomolecules such as ATP, nucleic acids, and proteins, and directly impacts plant growth and development [1]. The P cycle is one of the most complex cycles in forest communities, and it is responsible for a series of key biogeochemical processes [2]. The pools and fluxes of the P cycle are easily affected by the community composition through various processes, including plant uptake, biomass distribution, plant P-use efficiency, and leaf litter decomposition [3,4]. Under natural conditions, the P cycle is in a dynamic equilibrium in forest ecosystems. However, biological invasion or community succession can affect the ecological processes associated with the P cycle, which in turn can disrupt the P cycle's equilibrium in the original forest environment [5,6].

Approximately 70% of China's bamboo forest area is made up of the native gigantic bamboo known as moso bamboo (*Phyllostachys edulis* (Carrière) J. Houzeau), which is extensively dispersed in subtropical climates. [7,8]. The rapid growth and aggressive clonal reproduction of this bamboo species provide it with great potential for expansion into nearby secondary forests. [9]. In recent decades, such expansion has been regularly noted in both native and imported regions (such as China and Japan), which has contributed to the global rise in the cover of bamboo forests [10]. The phenomenon of bamboo expansion has raised increasing attention from both the society community and the public due to a series of cascading effects, such as reduced biodiversity and increased soil erosion [11,12]. Evergreen broadleaved forests, the native vegetation of subtropical China, play a pivotal role in the maintenance of forest productivity, biodiversity, and ecological balance [13,14]. Currently, native evergreen broadleaved forests are suffering from serious interference from the expansion of moso bamboo, since they are secondary forests that have regenerated from previous destruction by natural disasters or human activities [10,15].

The spread of moso bamboo displaced broadleaved tree species, leading to the formation of monocultures. [16,17]. This shift in the community composition of forests may lead to changes in the pools and fluxes of the P cycle for invaded ecosystems because of the differential organ P content for moso bamboo compared to trees [18,19] but also because broadleaved trees and moso bamboo have different biomass allocations [20]. Furthermore, it has been demonstrated that moso bamboo expansion would change the quality of leaf litter, the soil microbial community's habitat [21], and the litter's subsequent decomposition process [22,23]. Moso bamboo's fine roots also differ from those of broadleaved trees in that they use a flexible strategy for absorbing nutrients. [24]. Therefore, moso bamboo expansion may also lead to changes in the plant P uptake and litterfall P return to the soil in the recipient ecosystem. Wu et al. (2018) demonstrated that following the expansion of moso bamboo into nearby coniferous forests, the total soil P pool decreased and the available soil P pool increased. [25]. Yang et al. (2024) found that moso bamboo expansion into broadleaved forests decreased the soil HCl-P and $CaCl_2$-P content but increased the total soil P pool [26]. However, these previous studies only focused on one or partial sections of the forest P cycle rather than the overall cycle. Consequently, it is necessary for us to provide an overall evaluation of the P cycle in the context of bamboo expansion.

In this paper, we selected typical moso bamboo-dominated forests (hereafter "BDF") and secondary evergreen broadleaved forests (hereafter "EBF") to thoroughly examine how the expansion of moso bamboo affects the P cycle in the forest ecosystem. Our specific objectives were to (1) quantify the P storage in plant tissues, litter, and soil, (2) examine plant P uptake and litter P return to the soil, and (3) determine the P-use efficiency in the BDF and EBF. Given the large amount of the root biomass for moso bamboo relative to the trees that make up the BDF, our hypothesis was that the expansion of bamboo would bring about (a) higher P uptake and P storage in plant tissues, (b) a lower P return from the leaf litter and a lower soil P pool in the recipient BDF ecosystem, and (c) a lower forest P cycling rate and higher P-use efficiency. This study not only provides important contextual information for assessing how the expansion of moso bamboo affects broadleaved evergreen forests ecologically but also has implications for evaluating the impact of vegetation alterations on the biogeochemical P cycle in the presence of plant invasion.

2. Methods

2.1. Study Site

This study was carried out at Dagangshan National Forest Ecological Station, situated in a mountainous region in Jiangxi Province, South China (27°30′–27°50′ N, 114°30′–114°45′ E). This region has a middle subtropical climate with monsoon activity. It has humid, rainy summers and cold, dry winters. The mean annual temperature ranges from 15.8 to 17.7 °C, with January and July having the lowest and highest temperatures of −8.3 °C and 39.9 °C, respectively. The annual mean precipitation is 1591 mm, with 80% of that falling between March and August. Based on the Chinese soil classification system, the soils in this region are categorized as mountain red and yellow varieties [27]. The climate and soil environments provide suitable potential habitats for bamboo growth. Moso bamboo is also widely distributed at this site, and it has aggressively expanded into areas historically occupied by diverse native EBF [20,28]. Therefore, this site is an ideal place to explore the changes in the P pools and fluxes in the context of conversion from EBF into BDF forest. We randomly selected three paired stands straddling unexpanded EBF (greater than 80% coverage with *Castanopsis fargesii* Franch., *Castanopsis sclerophylla* (Lindl.) Schottky, *Symplocos theophrastifolia* Siebold & Zucc, and *Schima superba* Gardner & Champ., making up the majority of the canopy layer), and BDF (moso bamboo predominates with some indigenous broadleaved trees, including *S. theophrastifolia*, *C. fargesii*, and *C. sclerophylla*) in August 2011. The moso bamboo invaded the EBF around thirty years ago, becoming the BDF [20]. Occasionally, the BDF's dead and old (>10 a) bamboos were taken for housing or fuel. No fertilization was performed at each site. In every stand, we set up a 20 m × 20 m plot with similar terrain, soil parent material, and historical plant community structure. This design enabled us to infer that the changes in the current P biogeochemistry cycle resulted from moso bamboo expansion. The stand density, height, and basal areas for EBF observed by Song et al. [20] were 1031 stems ha^{-1}, 13.1 m, and 22.3 m^2 ha^{-1}, respectively. Those corresponding indicators in the BDF were 5379 stems ha^{-1}, 12.9 m, and 46.2 m^2 ha^{-1}, respectively [20].

2.2. P Pool and Flux Estimation

We sampled fresh and senescent leaves, branches, stems, and coarse roots of the main plant species (*C. fargesii*, *C. sclerophylla*, *S. theophrastifolia*, *S. superba*, *Quercus chenii* Nakai, and moso bamboo) to determine the P concentrations in these pools of each stand in August 2011. For every species, we gathered three mature, healthy trees from each plot. The sampled trees had a breast diameter of 5–10 cm and a height of 5–10 m. Nine samples for each tissue of each species were collected from the BDF and EBF types. As the nutrient concentration of moso bamboo organs is significantly influenced by the culm age [29], we sampled aboveground tissues from the current-year bamboo shoots and those older than 1 year separately. All the samples were oven-dried at 70 °C in the laboratory and then pulverized and ground using a ball mill (JXFSTPRP-48L, Shanghai Jingxin Industrial Development Co., Ltd., Shanghai, China). There were three root samples collected from 0–20 cm of soil in each plot to monitor the P concentration. The treatment of the fine root samples was the same as that of the plant samples. Fine root biomass was obtained in the study of Liu et al. [24], whose plots and plant sampling times were the same with this study. The molybdate-blue method was used to determine the total plant P content, which involved digesting 0.25 g of sample in 5 mL of concentrated H_2SO_4 with a catalyst combination ($CuSO_4$, K_2SO_4). The leaf P resorption efficiency (PRE) was calculated according to Killingbeck [30].

We collected three 0–20 cm soil samples each month from 2011 to 2012 in each plot to determine the soil P content. In a ventilation room, there were 432 soil samples (3 soil samples × 3 plots × 2 stand types × 12 months × 2 years = 432) that were air-dried to a constant weight. The roots and stones were removed. The samples were then ground in a mortar and passed through a 2 mm sieve. The total P content was determined using the molybdate-blue method. The soil available P fractions were measured sequentially using a modified Hedley procedure [31,32], including the following summarized steps. (1) Resin-P_i: 0.5 g air-dried soil was added to 30 mL of deionized water plus 0.4 $NaHCO_3$ anion exchange resin in a nylon bag; this solution was then added to a 50 mL capped plastic centrifuge tube and incubated on a shaker for 16 h at 25 °C. P was then recovered from the resin by shaking it for 1 h in 50 mL of 0.5 mol L^{-1} HCl. (2) $NaHCO_3$-P_i and

$NaHCO_3$-P_o: The residue from step (1) was incubated on a shaker in 30 mL of 0.5 mol L^{-1} $NaHCO_3$ solution (pH = 8.5) for 16 h at 25 °C, and each mixture was centrifuged at 12,000× *g* for 10 min [32]. The supernatant was then filtered through a 0.45 μm Millipore filter, and the inorganic P concentration in the filtrate was colorimetrically determined using the molybdate-blue method [33]. The extracts were digested in concentrated H_2SO_4 and $HClO_4$ to measure the total P content. The $NaHCO_3$-P_o content in the extracts was calculated by the total P concentration minus the $NaHCO_3$-P_i concentration. The soil P content was the mean of the concentration of each month in each stand.

The formulas listed below were used to calculate the P pools [34]:

$$\text{Plant P pool} = \text{Biomass} \times \text{Tissue P content}; \quad (1)$$

$$\text{Litter P pool} = \text{Litter storage} \times \text{Litter P content}; \quad (2)$$

$$\text{Soil total P (STP) pool} = \text{STP content} \times \text{Bulk density} \times \text{Soil depth}; \quad (3)$$

$$\text{Available P (AP) pool} = \text{AP content} \times \text{Bulk density} \times \text{Soil depth}. \quad (4)$$

In these formulae, the values of the aboveground and belowground biomass for the trees and bamboo were obtained from Song et al. (2017) [20]. The values of the standing aboveground litter were obtained from Song et al. (2016) (EBF: 2.26 t ha^{-1}; BDF: 2.87 t ha^{-1}) [35], and Liu et al. (2013) provided those for the belowground litter (rhizome, coarse and fine roots) dry weight (EBF: 3.2 t ha^{-1}; BDF: 13.15 t ha^{-1}) [24].

The following formulae were used to estimate the main P fluxes [36]:

$$\text{Plant P uptake} = \text{Net primary production (NPP)} \times \text{P content}; \quad (5)$$

$$\text{Litter P return} = (\text{L+F}) \times \text{P content}; \quad (6)$$

$$\text{P cycling rate} = \text{Litter P return/P uptake}. \quad (7)$$

The data for the NPP were obtained from the reference of Song et al. (EBF: 19.8 t ha^{-1} yr^{-1}; BDF: 30.0 t ha^{-1} yr^{-1}) [20]. L was the annual aboveground litterfall production obtained from Song et al. (2016) (EBF: 9.09 t ha^{-1} yr^{-1}; BDF: 5.82 t ha^{-1} yr^{-1}) [35], and F was the annual growth of the fine root obtained from Liu et al. (2013) (EBF: 1.31 t ha^{-1} yr^{-1}; BDF: 9.46 t ha^{-1} yr^{-1}) [24].

The ecosystem P-use efficiency fluxes were estimated using the following equation [37]:

$$\text{P-use efficiency} = \text{NPP/annual P uptake} \quad (8)$$

All of these standing storage estimations for P are given in terms of the oven-dry weight per hectare. The P cycling rate is calculated annually as the weight per hectare.

2.3. Statistical Analysis

We used the analysis of variance (ANOVA) to detect differences in the pools, fluxes, and P-cycles between the EBF and BDF. Before the ANOVA, the homogeneity of variance for the data was tested using Levene's test, and the data that failed to pass the test were converted into homogeneity by the way of logarithmic, square root, or reciprocal. The P content of the plant tissues (stems, branches, leaves, and roots) and the P resorption efficiency between the moso bamboo and other trees were compared using the least significant difference test (LSD). For all the tests, we applied a p of 0.05 to infer statistical significance. Origin 2017 was used to construct the graphics, and SPSS 24.0 was used for all the statistical analyses.

3. Results

3.1. P Content

3.1.1. P Content in Plants

The P content of bamboo shoots and trees was found in the following order: leaves > branches > stems and coarse roots. The P content of all the plant tissues (except the leaves) for moso bamboo was significantly higher between the current shoots and the old bamboo shoots (>1a; Table 1). The P content of the bamboo leaves was higher than that of the trees. The other tissues of the old bamboo shoots contained a similar P content compared to the corresponding tissues of the broadleaved trees.

Table 1. Mean P content of various plant tissue types (g kg^{-1}, dry matter) in bamboo and broadleaved trees.

Species	Leaves	Branches	Stems	Coarse Roots	Rhizomes
Bamboo shoot (≤1a)	1.50 ± 0.12 ^{a}	0.80 ± 0.25 ^{a}	1.35 ± 0.12 ^{a}	1.02 ± 0.32 ^{a}	1.64 ± 0.41 ^{a}
Bamboo shoot (>1a)	1.23 ± 0.30 ^{a}	0.35 ± 0.13 ^{b}	0.27 ± 0.19 ^{b}	0.34 ± 0.15 ^{b}	0.45 ± 0.11 ^{b}
Castanopsis fargesii	0.76 ± 0.09 ^{b}	0.52 ± 0.16 ^{b}	0.34 ± 0.06 ^{b}	0.20 ± 0.12 ^{b}	n.a.
Castanopsis sclerophylla	0.69 ± 0.07 ^{b}	0.55 ± 0.20 ^{b}	0.39 ± 0.15 ^{b}	0.36 ± 0.09 ^{b}	n.a.
Symplocos theophrastifolia	0.64 ± 0.16 ^{b}	0.47 ± 0.28 ^{b}	0.39 ± 0.09 ^{b}	0.19 ± 0.12 ^{b}	n.a.
Schima superba	0.73 ± 0.11 ^{b}	0.43 ± 0.19 ^{b}	0.22 ± 0.07 ^{b}	0.34 ± 0.11 ^{b}	n.a.
Quercus chenii	0.73 ± 0.21 ^{b}	0.41 ± 0.18 ^{b}	0.22 ± 0.07 ^{b}	0.34 ± 0.07 ^{b}	n.a.

The data are given as the mean ± SD, with SD representing the standard deviation (n = 9). Significant differences between different species are indicated by lowercase letters in the same column at $p = 0.05$. The abbreviation n.a. denotes the absence of data.

3.1.2. P Content in Litter and P Resorption Efficiency

The litter P concentration for the BDF was significantly lower than that of the EBF. Moso bamboo had the lowest P concentration (0.36 g kg^{-1}) among all the species in the BDF and EBF stands. The leaf litter P content of the trees, except for *C. sclerophylla* in the BDF, was lower in the EBF. With respect to the P resorption efficiency (PRE), there were also statistical differences among the species. The PRE of moso bamboo was 70.6%, which equated to 9.0% to 41.1% higher than the PRE of the trees, regardless of whether they were in the EBF or the BDF (Table 2).

Table 2. Leaf litter P content and P resorption efficiency (PRE) for main species in the bamboo-dominant forest (BDF) and secondary evergreen broadleaved forest (EBF).

Stands/Species	P (g kg^{-1})	PRE (%)
EBF	0.53 ± 0.19 ^{A}	
C. fargesii	0.52 ± 0.13 ^{ab}	32.0 ± 6.3 ^{b}
C. sclerophylla	0.46 ± 0.12 ^{ab}	33.3 ± 3.8 ^{b}
S. theophrastifolia	0.58 ± 0.05 ^{a}	9.0 ± 5.6 ^{d}
S. superba	0.50 ± 0.18 ^{ab}	28.6 ± 7.8 ^{c}
Q. chenii	0.61 ± 0.20 ^{a}	28.7 ± 4.4 ^{c}
BDF	0.38 ± 0.21 ^{B}	
moso bamboo	0.36 ± 0.10 ^{b}	70.6 ± 8.2 ^{a}
C. fargesii	0.48 ± 0.08 ^{ab}	21.4 ± 4.4 ^{c}
C. sclerophylla	0.61 ± 0.21 ^{a}	21.4 ± 5.6 ^{c}
S. theophrastifolia	0.46 ± 0.07 ^{ab}	11.3 ± 2.7 ^{d}
S. superba	0.43 ± 0.11 ^{ab}	41.1 ± 6.3 ^{b}
Q. chenii	0.44 ± 0.11 ^{ab}	26.0 ± 5.6 ^{c}

The data are the means ± SD. SD is the standard deviation. n = 9 is for litter quality and P resorption efficiency. Uppercase letters in the same column indicate significant differences between two stands. Lowercase letters in the same column indicate significant differences among species at $p = 0.05$.

3.1.3. Soil P Content

Along with the conversion of EBF to BDF, the total soil P content decreased by 21.9%, with the value decreasing from 0.32 g kg^{-1} in the EBF to 0.25 g kg^{1} in the BDF. The P content of each component (Resin-P, $NaHCO_3$-P_i, and $NaHCO_3$-P_o) was also significantly reduced. Furthermore, the total available soil P content (Resin-P + $NaHCO_3$-P_i + $NaHCO_3$-P_o) decreased by 32.8% (Table 3).

Table 3. Soil chemical properties for bamboo-dominant forest (BDF) and secondary evergreen broadleaved forest (EBF).

	EBF	BDF
Total P (g kg^{-1})	0.32 ± 0.03 [a]	0.25 ± 0.02 [b]
Resin-P (mg kg^{-1})	2.0 ± 0.6 [a]	1.0 ± 0.8 [b]
$NaHCO_3$-Pi (mg kg^{-1})	3.3 ± 2.2 [a]	1.3 ± 0.6 [b]
$NaHCO_3$-Po (mg kg^{-1})	20.0 ± 2.3 [a]	14.7 ± 3.6 [b]
Available P (mg kg^{-1})	25.3 ± 3.1 [a]	17.0 ± 4.6 [b]

The data are given as the mean ± SE, with SE representing the standard error (n = 3). Significant differences between BDF and EBF are indicated by lowercase letters at $p = 0.05$.

3.2. Ecosystem P Pool

3.2.1. Plant P Pool

Compared to the EBF (66 kg P ha^{-1}), the plant P pool in the BDF (87 kg P ha^{-1}) was 31.8% greater ($F = 26.05$, $p = 0.027$; Figure 1). The two stands showed different P spatial allocation patterns (Figure 1a). The plants of the EBF and the BDF had equally distributed P in the aboveground tissues ($F = 1.09$, $p = 0.355$); however, compared with the EBF, the underground plant P pool in the BDF was 193.5% higher than that of the EBF ($F = 72.22$, $p = 0.001$). Additionally, bamboo accounted for 72.0% P in the BDF (61 kg P ha^{-1}), while P storage in the trees decreased by 40 kg ha^{-1} from the EBF to the BDF (Figure 1b).

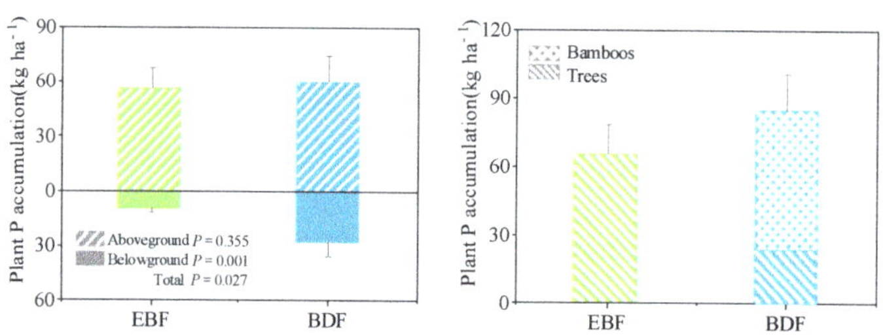

(**a**) (**b**)

Figure 1. The P allocation for the plant pool (kg P ha^{-1}) of secondary evergreen broadleaved forest (EBF) and bamboo-dominant forest (BDF). (**a**) spatial allocation patterns of P accumulation; (**b**) species allocation patterns of P accumulation. The data are given as the mean ± SD, with SD representing the standard deviation (n = 3).

3.2.2. Litter P Pool

The conversion of EBF to BDF resulted in a significant rise in the total litter P pool (2.2 kg P ha^{-1} compared to 3.7 kg P ha^{-1}, respectively; $F = 6.83$, $p = 0.040$). The largest contribution of belowground litter to the BDF is shown in Figure 2, where it was 127.2% greater than that of the EBF. With regard to the aboveground P component, there was no significant difference between the litter P storage pool in the BDF and EBF.

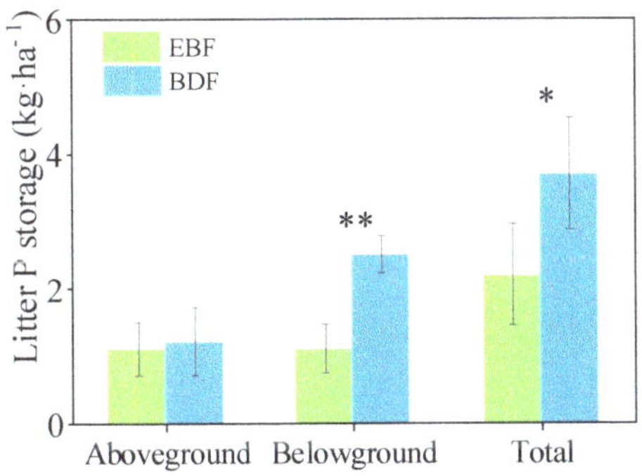

Figure 2. P allocation of litter pool (kg P ha^{-1}) in the secondary evergreen broadleaved forest (EBF) and bamboo-dominant forest (BDF). The data are given as the mean ± SD, with SD representing the standard deviation (n = 3; * $p < 0.05$; ** $p < 0.01$).

3.2.3. Soil P Pool

The total soil P pool and available P pool were both significantly lower in the BDF than in the EBF ($F = 18.21$, $p = 0.045$; $F = 29.53$, $p < 0.001$; respectively), with a decrease of 40.4% for the available soil P pool in the EBF (34 kg P ha^{-1}) compared to the BDF (57 kg P ha^{-1}) (Figure 3).

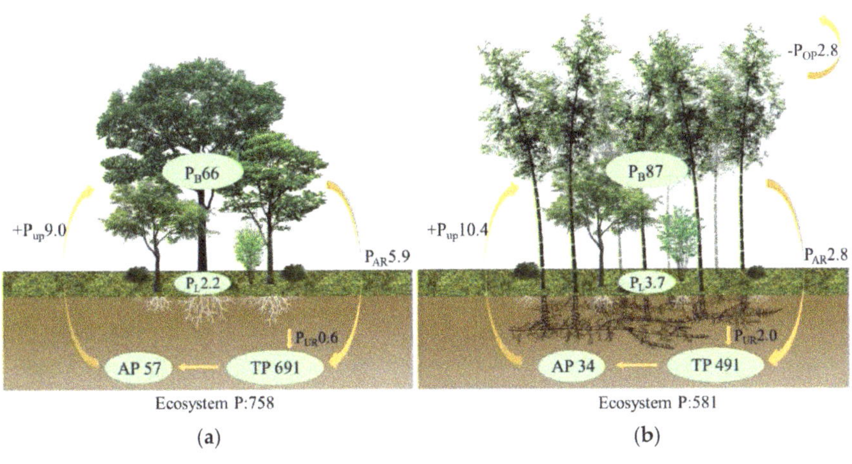

Figure 3. The P pools (kg ha^{-1}) and P fluxes (kg ha^{-1} a^{-1}) for secondary evergreen broadleaved forest (EBF) and bamboo-dominant forest (BDF). (**a**) evergreen broadleaved forest, EBF, and (**b**) bamboo-dominant forest, BDF. P pools included plant biomass (P_B), standing litter (P_L), and soil total P (TP) and available P (AP) within 0–20 cm. P fluxes included annual plant P uptake (+P_{UP}) and P return by litter (P_{AR} is the return by aboveground litterfall, P_{UR} is the P return from the underground fine root mortality), and P_{OP} is the logging output of BDF.

3.3. *Ecosystem P Fluxes*

3.3.1. Ecosystem P Cycle Rate

The annual P uptake of the belowground pool was 1.0 kg P ha^{-1} yr^{-1} in the EBF and 3.9 kg P ha^{-1} yr^{-1} in the BDF, which was 74.4% lower for the EBF ($F = 266.77$, $p = 0.004$). P uptake by the aboveground pool of the EBF was 23.2% higher than that of the BDF pool

($F = 9.47$, $p = 0.037$). In terms of the total amount, there was no significant difference in P uptake between the BDF and the EBF ($F = 5.96$, $p = 0.061$) (Figure 4).

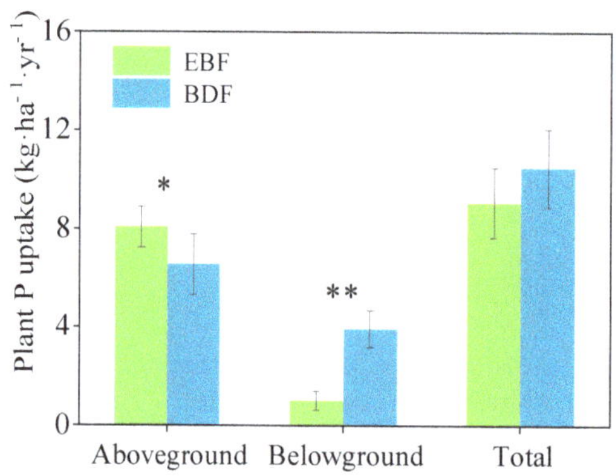

Figure 4. Annual P uptake for secondary evergreen broadleaved forest (EBF) and bamboo-dominant forest (BDF). The data are given as the mean ± SD, with SD representing the standard deviation (n = 3; * $p < 0.05$; ** $p < 0.01$).

The pool of leaf litter P return to soil was significantly lower in the BDF (2.8 kg P ha^{-1} yr^{-1}) than in the EBF (5.9 kg P ha^{-1} yr^{-1}). In contrast, the belowground root return was higher in the BDF (2.0 kg P ha^{-1} yr^{-1}) than in the EBF (0.6 kg P ha^{-1} yr^{-1}) ($F = 11.24$, $P = 0.041$), representing a decrease of 26.2% for the BDF (Figure 3). The ecosystem P cycling rate (calculated as the ratio of P return to P uptake) was 36.1% lower in the BDF than in the EBF, with values of 0.46 and 0.72 in the BDF and EBF, respectively (Figure 5). Ultimately, about 2.8 kg P ha^{-1} yr^{-1} of P was produced by the BDF as a result of harvesting old or dead bamboo.

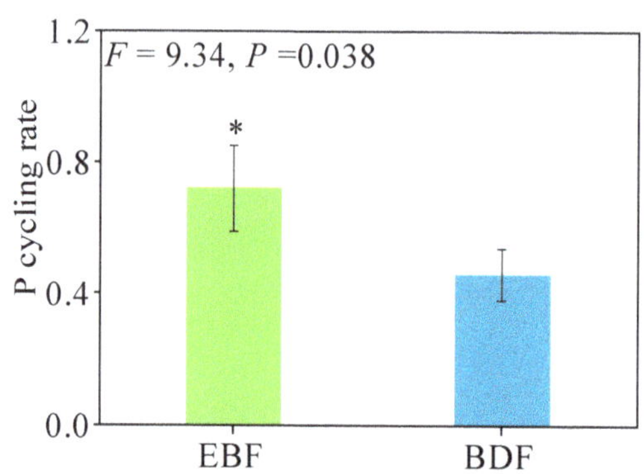

Figure 5. The ecosystem P cycling rate for bamboo-dominant forest (BDF) and secondary evergreen broadleaved forest (EBF) in Dagang Mountain National Forest Ecological Station, Jiangxi Province, China. The data are given as the mean ± SD, with SD representing the standard deviation (n = 3; * $p < 0.05$).

3.3.2. Ecosystem P-Use Efficiency

The P-use efficiency (calculated as the ratio of NPP to annual P uptake) for the EBF and the BDF was 2197 and 2861, respectively. The calculation showed that the ecosystem P-use efficiency of the BDF was significantly higher than that of the EBF (Figure 6).

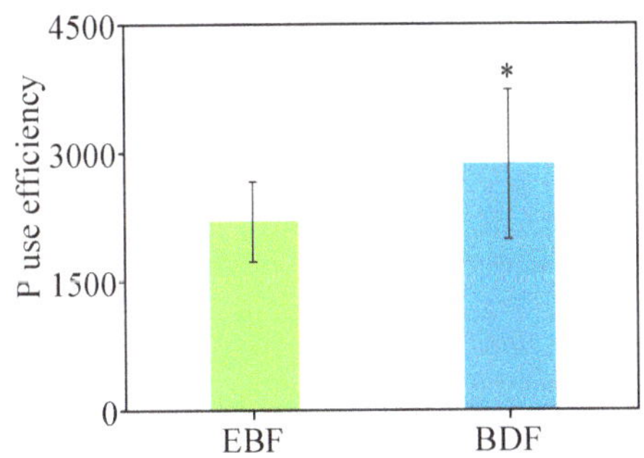

Figure 6. P-use efficiency of bamboo-dominant forest (BDF) and secondary evergreen broadleaved forest (EBF). Data are expressed as the mean ± SD; SD is the standard deviation (n = 3; * $p < 0.05$).

4. Discussion

4.1. Changed Plant P Pool after Moso Bamboo Expansion

The expansion of moso bamboo into neighboring forests is one of the most serious ecological problems in subtropical China [38], which has changed the composition and diversity of plant communities [16,17,39], thus affecting the P cycle of the ecosystem. This study suggested that moso bamboo expansion caused a series of changes in the ecosystem P cycle of EBF because of differences in ecophysiological properties in moso bamboo and broadleaved trees. The plant P storage increased after moso bamboo expansion (Figure 1b). The biomass and P content of plants are the indexes responsible for P storage. A previous study showed that there was no significant difference in the biomass between the BDF and the EBF [20]; however, the results from this study indicated that the total tissue P content of moso bamboo was higher compared to broadleaved trees (Table 1). Therefore, once most of the broadleaved trees were replaced by moso bamboo after bamboo expansion, the plant P pool of the forest ecosystem increased. Furthermore, we discovered that following the transformation of EBF into BDF, the belowground component of the plant P pool significantly increased (Figure 1a). These results have two possible explanations. First, the roots of the moso bamboo had a higher P content (Table 1). Second, the BDF showed a higher percentage of belowground biomass compared to the EBF. A previous study found that the ratio of aboveground biomass to belowground biomass shifted from 4.1 in the EBF to 2.1 in the BDF after moso bamboo expansion, indicating that the belowground biomass increased significantly [20], which led to an increase in the belowground plant P pool in bamboo stands. In summary, this is consistent with our first hypothesis that the expansion of bamboo would bring about higher P uptake and P storage in plant tissues.

4.2. Decreased Leaf Litter P Return and Soil P Pool after Moso Bamboo Expansion

Numerous studies found that moso bamboo expansion significantly altered the soil P status and dynamics. Xia et al. (2022) reported that moso bamboo expansion into broadleaved forests increased the mass fraction of soil total P and available P [40]. According to Wu et al. (2018), the expansion of moso bamboo into nearby *Cryptomeria japonica* (Thunb. ex L. f.) D. Don stands further reduced the total P content of the soil but raised the concentration of available P. [25]. In this publication, the leaf litter P pool increased (Figure 2) and the available soil P pool decreased (Figures 2 and 3) following the expansion of moso bamboo, which is consistent with our second hypothesis that the expansion of bamboo would bring about a lower soil P pool in the recipient BDF ecosystem. The soil P cycle is a sedimentary cycle, and the decomposition of leaf litter will affect the size of the available soil P pool in a short period of time [41]. The significant decrease in annual litter return combined with the increase in the litter P pool (Figure 2) suggests that the decomposition of litter in the stand is inhibited by the expansion of moso bamboo. Litter decomposition is related to litter quality. It is generally accepted that the lower the litter quality (i.e., low N and P content, high C:N and C:P ratios), the slower the decomposition rate [42]. By combining previous data [35] with the results from this paper, the litter leaf N and P content of the stand decreased by 20% and 28%, respectively, and C:N and C:P increased by 19% and 35% after the expansion of moso bamboo, respectively. This decreased litter quality may lead to slower litter decomposition, which in turn would increase the P storage in the litter layer. Furthermore, the enzymatic activity and composition of the microbial community impact the decomposition of litter material. Previous studies found that moso bamboo expansion led to a decrease in soil microbial richness and diversity [43,44] and a decrease in soil phospholipid fatty acids, soil cellulase, and xylanase activities [44–47], which may also contribute significantly to the reduced rate of litter decomposition. Thus, we hypothesize that the low quality of litter and the changes in the microbial community structure and function after moso bamboo expansion would reduce the decomposition rate of litter and reduce P input to the soil, leading to the accumulation of P in the litter pool and reduced P in the soil pool.

4.3. Reduced P Cycle Rate after Moso Bamboo Expansion

The larger the nutrient cycling flux, the faster the cycling rate. In this paper, the P cycling rate of forest ecosystems slowed down after the expansion of moso bamboo (Figure 5; 0.72 for EBF and 0.46 for BDF), which supported our third hypothesis. The P cycling rate in this paper is the ratio between P return and P uptake. It has been shown that the acquisition of P is closely related to the plant root system [48]. A previous study found that moso bamboo had a well-developed underground flagellar root system with a high fine root biomass, a long specific root size, and a fast growth rate compared to broadleaved trees [24]. It is hypothesized that after native broadleaved trees were replaced by moso bamboo, the root surface area for P uptake in the stand would increase, effectively reducing the distance of P to the root system and greatly increasing the efficiency of P uptake. Collectively, this could lead to increased P uptake in the belowground portion of the stand after moso bamboo expansion.

Nutrient return is determined by the starting nutrient content of leaf litter, which is related to the resorption efficiency of leaves [49]. In this research, the P resorption efficiency of leaves was significantly higher in moso bamboo than in broadleaved trees (i.e., 70.6% in moso bamboo and 11%~41% in trees). Therefore, we hypothesized that more P would be reabsorbed by the plant rather than returning to the soil, thus reducing the amount of P returned to the soil. In addition, the P reabsorbed by leaves may also be transferred to other plant tissues or individuals through physiological integration, and P can be recycled repeatedly within the bamboo system [50]. This not only prolongs the time that P persists in the plant P pool before it returns to the soil P pool, slowing down the ecosystem P cycling, but also reduces P consumption and improves the efficiency of P utilization by moso bamboo. The results in this study support this mechanistic interpretation (Figure 6; PUE for EBF: 2197, PUE for BDF: 2861). In conclusion, the strong nutrient uptake capacity of roots coupled with the high P resorption efficiency of leaves for the clonal moso bamboo system is responsible for decreased P cycling in forest ecosystems after moso bamboo expansion. This is consistent with our third hypothesis that the expansion of bamboo would bring about a lower forest P cycling rate and higher P-use efficiency.

5. Conclusions

This paper evaluated the biological P cycle through P pools of moso bamboo forest and invaded evergreen broadleaved forest. The obtained results demonstrated that moso bamboo expansion into broadleaved forest ecosystems leads to (1) an increase in the plant P pool, especially the belowground portion of the pool; (2) an increase in the P litter pool and a decrease in the available P in the soil pool; and (3) an increase in the forest P-use efficiency and a reduced forest P cycling rate. The reason is mainly because of the greater nutrient uptake capacity and higher nutrient use efficiency of moso bamboo compared to broadleaved trees. Moso bamboo retains more P in plant tissues rather than releasing P through leaf litter fall, resulting in the suppressed decomposition of litter and decreased soil P storage. This study is based on observations from a single site; however, factors such as the stand age and bamboo expansion intensity may bring some uncertainty to the results. Therefore, in the future, it could be possible to choose long-term monitoring and take into account those influencing factors to monitor the dynamic changes in the ecosystem P cycle coupled with moso bamboo expansion. This study not only provides an important basis for the evaluation of the ecological consequences of moso bamboo expansion and the encroachment into neighboring forests but also deepens our understanding of the important nutrient invasion mechanism of moso bamboo.

Author Contributions: S.S. conceived and designed the experiments, analyzed the data, prepared the figures and tables, and authored the initial draft of the article. L.W. took part in the data analysis and created the tables and figures. J.L. and Z.G.C. analyzed or interpreted the data, authored aspects of the manuscript, and reviewed drafts of the article. T.X., C.L., D.H., and Q.Y. assisted with the data collection and analysis. Q.S. conceived and designed the experiments, authored aspects of the manuscript, and reviewed drafts of the article. All authors have read and agreed to the published version of the manuscript.

Funding: This work was supported by the National Natural Science Foundation of China (32360801, 32060319, 42067050) and the 1000 Talents Plan award of Jiangxi province (jxsq2020101079).

Data Availability Statement: The data are available upon request.

Conflicts of Interest: The authors declare no conflicts of interest.

References

1. Santoro, V.; Schiavon, M.; Celi, L. Role of Soil Abiotic Processes on Phosphorus Availability and Plant Responses with a Focus on Strigolactones in Tomato Plants. *Plant Soil* **2024**, *494*, 1–49. [CrossRef]
2. Schipanski, M.E.; Bennett, E.M. The Phosphorus Cycle. In *Fundamentals of Ecosystem Science*, 2nd ed.; Weathers, K.C., Strayer, D.L., Likens, G.E., Eds.; Academic Press: Salt Lake City, UT, USA, 2021; Volume 9, pp. 189–213, ISBN 978-0-12-812762-9.
3. Ruttenberg, K.C. The Global Phosphorus Cycle. In *Treatise on Geochemistry*, 2nd ed.; Holland, H.D., Turekian, K.K., Eds.; Elsevier Science Ltd.: Amsterdam, The Netherlands, 2014; Volume 10, pp. 499–558, ISBN 978-0-08-043751-4.
4. Sohrt, J.; Lang, F.; Weiler, M. Quantifying Components of the Phosphorus Cycle in Temperate Forests. *Wires Water* **2017**, *4*, e1243. [CrossRef]
5. Zhu, X.Y.; Fang, X.; Wang, L.F.; Xiang, W.; Alharbi, H.A.; Lei, P.F.; Kuzyakov, Y. Regulation of Soil Phosphorus Availability and Composition during Forest Succession in Subtropics. *Forest Ecol. Manag.* **2021**, *502*, 119706. [CrossRef]
6. Luo, X.Z.; Liu, N.; Lambers, H.; Cai, H.Y.; Hou, E.Q.; Huang, Y.; Jian, S.G.; Kuang, Y.W.; Wen, D.Z.; Zhang, L.L. Plant Invasion Alters Soil Phosphorus Cycling on Tropical Coral Islands: Insights from Wollastonia Biflora and Chromolaena Odorata Invasions. *Soil Biol. Biochem.* **2024**, *193*, 109412. [CrossRef]
7. Kuehl, Y. Resources, Yield, and Volume of Bamboos. In *Bamboo: The Plant and Its Uses*; Liese, W., Köhl, M., Eds.; Springer: Cham, Switzerland, 2015; Volume 4, pp. 91–111, ISBN 978-3-319-14132-9.
8. Liu, K.W. Distribution of Bamboo Forest Resources and Species for Construction. In *Contemporary Bamboo Architecture in China*; Liu, K.W., Xu, Q.F., Wang, G., Chen, F.M., Leng, Y.B., Yang, J., Harries, K.A., Eds.; Springer: Singapore, 2022; Volume 1, pp. 1–5, ISBN 978-981-16-8309-1.
9. Wang, Y.X.; Bai, S.B.; Binkley, D.; Zhou, G.M.; Fang, F.Y. The Independence of Clonal Shoot's Growth from Light Availability Supports Moso Bamboo Invasion of Closed-Canopy Forest. *Forest Ecol. Manag.* **2016**, *368*, 105–110. [CrossRef]
10. Yang, Q.P.; Yang, G.Y.; Song, Q.N.; Shi, J.M.; Fang, A. Ecological Studies on Bamboo Expansion: Process, Consequence and Mechanism. *Chin. J. Plant Ecol.* **2015**, *39*, 110–124. (In Chinese) [CrossRef]
11. Xu, Q.F.; Liang, C.F.; Chen, J.H.; Li, Y.C.; Qin, H.; Fuhrmann, J.J. Rapid Bamboo Invasion (Expansion) and Its Effects on Biodiversity and Soil Processes+. *Glob. Ecol. Conserv.* **2020**, *21*, e00787. [CrossRef]
12. Ling, Z. Bamboo Expansion and Forest Biodiversity. In *Bamboo Expansion: Processes, Impacts, and Management*; Ling, Z., Ed.; Springer: Singapore, 2023; Volume 5, pp. 41–57, ISBN 978-981-99-4113-1.
13. Tang, C.Q. Evergreen Broad-Leaved Forests. In *The Subtropical Vegetation of Southwestern China: Plant Distribution, Diversity and Ecology*; Tang, C.Q., Ed.; Springer Netherlands: Dordrecht, The Netherlands, 2015; Volume 2, pp. 49–112, ISBN 978-94-017-9741-2.
14. Wang, X.H.; Kent, M.; Fang, X.F. Evergreen Broad-Leaved Forest in Eastern China: Its Ecology and Conservation and the Importance of Resprouting in Forest Restoration. *Forest Ecol. Manag.* **2007**, *245*, 76–87. [CrossRef]
15. Zheng, A.Y.; Lv, J.H. Spatial Patterns of Bamboo Expansion Across Scales: How does Moso Bamboo Interact with Competing Trees? *Landsc. Ecol.* **2023**, *38*, 3925–3943. [CrossRef]

16. Ouyang, M.; Yang, Q.P.; Chen, X.; Yang, G.Y.; Shi, J.M.; Fang, X.M. Effects of the Expansion of *Phyllostachys Edulis* on Species Composition, Structure and Diversity of the Secondary Evergreen Broad-Leaved Forests. *Biodivers. Sci* **2016**, *24*, 649–657. (In Chinese) [CrossRef]
17. Bai, S.B.; Zhou, G.M.; Wang, Y.X.; Liang, Q.Q.; Chen, J.; Chen, Y.Y.; Sheng, R. Plant Species Diversity and Dynamics in Forests Invaded by *Moso Bamboo* (*Phyllostachys edulis*) in Tianmu Mountain Nature Reserve. *Biodivers. Sci.* **2013**, *21*, 288–295. (In Chinese) [CrossRef]
18. Wang, L.; Ouyang, M.; Song, S.W.; Zeng, X.X.; Song, Q.N.; Liu, J.; Fang, X.; Luan, F.G.; Yang, Q.P. Response of N and P Stoichiometric Characteristics of Evergreen Broad-Leaved Forest Plant Leaf to *Phyllostachys edulis* Expansion. *Guihaia* **2023**, *43*, 1737–1746. (In Chinese) [CrossRef]
19. Song, S.W.; Ouyang, M.; Liu, J.; Song, Q.N.; Fang, X.; Luan, F.G.; Yang, Q.P. Effects of *Phyllostachys edulis* Expansion on Soil and Leaf N, P Stoichiometry in *Cunninghamia lanceolata* Forests. *Ecol. Sci.* **2024**, *43*, 1–7. (In Chinese) [CrossRef]
20. Song, Q.N.; Lu, H.; Liu, J.; Yang, J.; Yang, G.Y.; Yang, Q.P. Accessing the Impacts of Bamboo Expansion on Npp and N Cycling in Evergreen Broadleaved Forest in Subtropical China. *Sci. Rep.* **2017**, *7*, 1–9. [CrossRef]
21. Luo, W.X.; Zhang, Q.Y.; Wang, P.; Luo, J.; She, C.Y.; Guo, X.M.; Yuan, J.J.; Sun, Y.H.; Guo, R.M.; Li, Z.F.; et al. Unveiling the Impacts Moso Bamboo Invasion on Litter and Soil Properties: A Meta-Analysis. *Sci. Total Environ.* **2024**, *909*, 168532. [CrossRef] [PubMed]
22. Dong, H.Y.; Shao, S.; Liang, C.F.; Xu, Q.F.; Chen, J.H.; Qin, H. Impacts of Litter Composition on the Structure and Functional Pathways of Soil Microbial Community during *Phyllostachys edulis* Expansion. *Agronomy* **2022**, *12*, 220. [CrossRef]
23. Luan, J.W.; Li, S.Y.; Dong, W.; Liu, Y.C.; Wang, Y.; Liu, S.R. Litter Decomposition Affected by Bamboo Expansion is Modulated by Litter-Mixing and Microbial Composition. *Funct. Ecol.* **2021**, *35*, 2562–2574. [CrossRef]
24. Liu, J.; Yang, Q.P.; Song, Q.N.; Yu, D.K.; Yang, G.Y.; Qi, H.Y.; Shi, J.M. Strategy of Fine Root Expansion of *Phyllostachys pubescens* Population into Evergreen Broad-Leaved Forest. *Chin. J. Plant Ecol.* **2013**, *37*, 230–238. (In Chinese) [CrossRef]
25. Wu, C.S.; Mo, Q.F.; Wang, H.K.; Zhang, Z.J.; Huang, G.X.; Ye, Q.; Zou, Q.; Kong, F.Q.; Liu, Y.Q.; Geoff Wang, G. Moso Bamboo (*Phyllostachys Edulis* (Carriere) J. Houzeau) Invasion Affects Soil Phosphorus Dynamics in Adjacent Coniferous Forests in Subtropical China. *Ann. For. Sci.* **2018**, *75*, 24–35. [CrossRef]
26. Yang, D.; Shi, F.X.; Fang, X.M.; Zhang, R.L.; Shi, J.M.; Zhang, Y. Effect of the Moso Bamboo *Pyllostachys edulis* (Carrière) J.Houz. on Soil Phosphorus Bioavailability in a Broadleaf Forest (Jiangxi Province, China). *Forests* **2024**, *15*, 328. [CrossRef]
27. Wang, B. *Research on Forest Biodiversity in Dagangshan, Jiangxi*; China Forestry Publishing House: Beijing, China, 2005; ISBN 7503841346. (In Chinese)
28. Yang, Q.P.; Wang, B.; Guo, Q.R.; Zhao, G.D.; Fang, K.; Liu, Y.Q. Effects of *Phyllostachys Edulis* Expansion on Carbon Storage of Evergreen Broad-Leaved Forest in Dagangshan Mountain, Jiangxi. *Acta Agric. Univ. Jiangxiensis* **2011**, *33*, 529–536. (In Chinese) [CrossRef]
29. Li, R.; Werger, M.J.A.; During, H.J.; Zhong, Z.C. Carbon and Nutrient Dynamics in Relation to Growth Rhythm in the Giant Bamboo *Phyllostachys pubescens*. *Plant Soil* **1998**, *201*, 113–123. [CrossRef]
30. Killingbeck, K.T. Nutrients in Senesced Leaves: Keys to the Search for Potential Resorption and Resorption Proficiency. *Ecology* **1996**, *77*, 1716–1727. [CrossRef]
31. Hedley, M.J.; Stewart, J.W.B.; Chauhan, B.S. Changes in Inorganic and Organic Soil Phosphorus Fractions Induced by Cultivation Practices and by Laboratory Incubations. *Soil Sci. Soc. Am. J.* **1982**, *46*, 970–976. [CrossRef]
32. Li, L.; Li, S.M.; Sun, J.H.; Zhou, L.L.; Bao, X.G.; Zhang, H.G.; Zhang, F.S. Diversity Enhances Agricultural Productivity Via Rhizosphere Phosphorus Facilitation on Phosphorus-Deficient Soils. *Proc. Natl. Acad. Sci. USA* **2007**, *104*, 11192–11196. [CrossRef]
33. Lu, R.K. *Soil and Agro-Chemistry Analytical Methods*; China Agriculture Press: Beijing, China, 2000; ISBN 7-80119-925-1. (In Chinese)
34. Zhang, Y.W.; Guo, Y.; Tang, Z.; Feng, Y.; Zhu, X.; Xu, W.; Bai, Y.; Zhou, G.; Xie, Z.; Fang, J. Patterns of Nitrogen and Phosphorus Pools in Terrestrial Ecosystems in China. *Earth Syst. Sci. Data* **2021**, *13*, 5337–5351. [CrossRef]
35. Song, Q.N.; Ouyang, M.; Yang, Q.P.; Lu, H.; Yang, G.Y.; Chen, F.S.; Shi, J.M. Degradation of Litter Quality and Decline of Soil Nitrogen Mineralization after *Moso Bamboo* (*Phyllostachys pubscens*) Expansion to Neighboring Broadleaved Forest in Subtropical China. *Plant Soil* **2016**, *404*, 113–124. [CrossRef]
36. Tateno, R.; Takeda, H. Nitrogen Uptake and Nitrogen Use Efficiency above and below Ground along a Topographic Gradient of Soil Nitrogen Availability. *Oecologia* **2010**, *163*, 793–804. [CrossRef]
37. Vitousek, P. Nutrient Cycling and Nutrient Use Efficiency. *Am. Nat.* **1982**, *119*, 553–572. [CrossRef]
38. Bai, S.; Conant, R.T.; Zhou, G.; Wang, Y.; Wang, N.; Li, Y.; Zhang, K. Effects of Moso Bamboo Encroachment into Native, Broad-Leaved Forests on Soil Carbon and Nitrogen Pools. *Sci. Rep.* **2016**, *6*, 31480. [CrossRef]
39. Tong, R.; Zhou, B.Z.; Jiang, L.N.; Cao, Y.H.; Xiao, G.G.; Zhen, Y.Y. Influence of Moso Bamboo Invasion on Forest Plants and Soil: A Review. *Sheng Tai Xue Bao* **2019**, *39*, 3808–3815. (In Chinese) [CrossRef]
40. Xia, E.L.; Liu, G.L.; Nong, J.Q.; Wei, S.P.; Liu, X.Z. Changes in Soil Nutrient Characteristics in Moso Bamboo Forest Expanding into Broadleaved Forest. *Ecol. Environ.* **2022**, *31*, 1110–1117. [CrossRef]
41. Schaap, K.J.; Fuchslueger, L.; Hoosbeek, M.R.; Hofhansl, F.; Martins, N.P.; Valverde-Barrantes, O.J.; Hartley, I.P.; Lugli, L.F.; Quesada, C.A. Litter Inputs and Phosphatase Activity Affect the Temporal Variability of Organic Phosphorus in a Tropical Forest Soil in the Central Amazon. *Plant Soil* **2021**, *469*, 423–441. [CrossRef]

42. Yang, K.; Zhu, J.J.; Zhang, W.W.; Zhang, Q.; Lu, D.L.; Zhang, Y.K.; Zheng, X.; Xu, S.; Wang, G.G. Litter Decomposition and Nutrient Release from Monospecific and Mixed Litters: Comparisons of Litter Quality, Fauna and Decomposition Sites Effects. *J. Ecol.* **2022**, *110*, 1673–1686. [CrossRef]
43. Xu, Q.F.; Jiang, P.K.; Wu, J.S.; Zhou, G.M.; Shen, R.F.; Fuhrmann, J.J. Bamboo Invasion of Native Broadleaf Forest Modified Soil Microbial Communities and Diversity. *Biol. Invasions* **2015**, *17*, 433–444. [CrossRef]
44. Tian, X.K.; Wang, M.Y.; Meng, P.; Zhang, J.S.; Zhou, B.Z.; Ge, X.G.; Yu, F.H.; Li, M.H. Native Bamboo Invasions into Subtropical Forests Alter Microbial Communities in Litter and Soil. *Forests* **2020**, *11*, 314. [CrossRef]
45. Chang, E.H.; Chiu, C.Y. Changes in Soil Microbial Community Structure and Activity in a Cedar Plantation Invaded by Moso Bamboo. *Appl. Soil Ecol.* **2015**, *91*, 1–7. [CrossRef]
46. Chang, E.H.; Chen, C.P.; Tian, G.l.; Chiu, C.Y. Replacement of Natural Hardwood Forest with Planted Bamboo and Cedar in a Humid Subtropical Mountain Affects Soil Microbial Community. *Appl. Soil Ecol.* **2018**, *124*, 146–154. [CrossRef]
47. Liu, X.S.; Siemann, E.; Cui, C.; Liu, Y.Q.; Guo, X.M.; Zhang, L. Moso Bamboo (*Phyllostachys edulis*) Invasion Effects on Litter, Soil and Microbial Plfa Characteristics Depend on Sites and Invaded Forests. *Plant Soil* **2019**, *438*, 85–99. [CrossRef]
48. Wang, Y.; Wang, F.; Lu, H.; Liu, Y.; Mao, C. Phosphate Uptake and Transport in Plants: An Elaborate Regulatory System. *Plant Cell Physiol.* **2021**, *62*, 564–572. [CrossRef]
49. Shah, J.A.; Liu, W.F.; Ullah, S.; Duan, H.L.; Shen, F.F.; Liao, Y.C.; Huang, G.M.; Wu, J.P. Linkages among Leaf Nutrient Concentration, Resorption Efficiency, Litter Decomposition and Their Stoichiometry to Canopy Nitrogen Addition and Understory Removal in Subtropical Plantation. *Ecol. Process* **2024**, *13*, 27. [CrossRef]
50. Zhao, J.C.; Cai, C.J. Physiological Integration Improves Nitrogen Use Efficiency of Moso Bamboo: An Isotopic (^{15}N) Assessment. *Forest Ecol. Manag.* **2023**, *542*, 121073. [CrossRef]

forests

MDPI

Article

The Expansion of Moso Bamboo (*Phyllostachys edulis*) Forests into Diverse Types of Forests in China from 2010 to 2020

Dali Li [1,2], Juan Wei [1], Jiangzhou Wu [3], Yongde Zhong [1,4,*], Zhao Chen [5], Jianghua He [1], Shuangquan Zhang [3] and Lushan Yu [6]

[1] College of Forestry, Central South University of Forestry & Technology, Changsha 410004, China; lidali@hnou.edu.cn (D.L.); weijuan1004@163.com (J.W.); hejiangh_1983@163.com (J.H.)
[2] Academic Affairs Office, Hunan Open University, Changsha 410004, China
[3] College of National Park and Tourism, Central South University of Forestry & Technology, Changsha 410004, China; wu@csuft.edu.cn (J.W.); T20111197@csuft.edu.cn (S.Z.)
[4] National Forestry and Grassland Administration State Forestry Administration Engineering Research Center for Forest Tourism, Changsha 410004, China
[5] College of Economics and Management, Hunan Open University, Changsha 410004, China; chenchao2@hnou.edu.cn
[6] Principal's Office, Hunan Polytechnic of Environment and Biology, Hengyang 421005, China; taohualangzili@foxmail.com
* Correspondence: ade@csuft.edu.cn

Abstract: Moso bamboo (*Phyllostachys edulis*) forests, characterized by their rapid growth and clonal reproduction, have emerged as a significant threat to adjacent forest ecosystems. However, in China, the area, speed, and spatial distribution of moso bamboo forest expansion into other types of forests remains poorly understood. In this study, we present a case analysis of moso bamboo forests, employing a decade-long dataset from the forest second type inventory (FSTI) that utilizes transition matrices, neighboring ratio analysis, and spatio-temporal autocorrelation. This comprehensive investigation focuses on the spatio-temporal expansion of moso bamboo forests into diverse types of forests, with the aim of providing science-based recommendations for effective moso bamboo forest management. Our findings reveal that areas of moso bamboo forests have been expanding at an approximate annual rate of 2%, with an average expansion speed (including moso bamboo forests manually planted) of approximately 8 m per year. The length of moso bamboo–woodland ecotones (BWEs) increases as a consequence of moso bamboo forest expansion, indicating a sustained escalation in the extent of this expansion. Coniferous forests and evergreen broad-leaved forests are mainly invaded, accounting for around 58% of all invaded forests. The rate of moso bamboo forest expansion into different types of forests varies, although the rate remains fairly consistent within the same forest type. Moso bamboo forest expansion exhibits significant spatial heterogeneity. Furthermore, the area of moso bamboo forest intrusion into various types of forests in different provinces is notably influenced by the presence of moso bamboo forests and the proportional distribution of different forest types. The factors contributing to bamboo forest expansion encompass stand characteristics, soil attributes, light intensity, moso bamboo afforestation, forestry practices, and human disturbances.

Keywords: moso bamboo forests; expansion; forest types; bamboo–woodland ecotone

1. Introduction

It is generally accepted that a species within its native range is not considered "invasive". Nevertheless, some native species display behaviors akin to non-native species. Charles Elton, in his seminal 1927 book *Animal Ecology*, used the term "invasion" to refer to the spread of native species. For the same phenomenon, others have used various terminologies. In 1993, Garrott described the increasing dominance of native species in their original range as overabundance [1]. The overabundance or expansion of native species can diminish natural diversity by consolidating resources, introducing or spreading infectious diseases and parasites, altering the species composition or relative abundance of sympatric species, and even causing local extinctions, leading to significant damage similar to that caused by non-native invasive species. Although Garrott recognized the severe negative impact of overabundant native animals on other native species, there has been limited attention given to these species. Valéry (2013) reaffirmed that some native species benefiting from a change in their environment can meet both characteristic traits of a biological invasion. The rapid growth and clonal reproduction of bamboo enhance its ability to compete for space and form dense stands [2]. Griscom's study indicates that the mean percent mortality of tree seedlings was over twice as high in forest plots dominated by bamboo versus forest plots without bamboo [3]. In this way, bamboo-dominated areas significantly inhibit the development of tree seedlings, profoundly impacting the structure and dynamics of forest ecosystems [4,5].

Bamboo is one of the grass plants native to forests [6]. The expansion of bamboo is occurring worldwide, especially in Asia, owing to its rapid growth and extensive clonal reproduction [7]. In China, bamboo expansion into neighboring forests is more serious than in other Asian countries. The area of bamboo forests in China increased from 3.868 million ha to 4.6778 million ha during the ten-year period from 2008 to 2018 [8,9]. Among bamboo species, moso bamboo is one of the most widely distributed varieties, comprising over 70% of the total area of all bamboo species. Moso bamboo is a typical example of clonal reproduction, relying on its robust underground rhizomes to continually expand into neighboring communities, resulting in the continuous enlargement of moso bamboo forests [10,11]. Furthermore, in many districts of China, the management of moso bamboo forests has been extensive due to factors such as poor transportation, high costs in terms of labor and materials (land preparation, weeding, fertilization), and low economic returns. In some regions, the management of moso bamboo forests has even been abandoned [12,13]. The lack of proper management has exacerbated the unregulated growth of moso bamboo. On the other hand, conservation policies in China have led to the prohibition of timber harvesting in certain areas, as influenced by national regulations. Banning bamboo harvesting will intensify the encroachment of moso bamboo into the surrounding areas and the process of replacing other forest types [14,15]. As moso bamboo continues its invasive encroachment, it progressively influences national ecological security by modifying regional soil properties, nutrient composition [7,16,17], simplifying forest community structure and function [18,19], diminishing species diversity [20,21], and altering forest biomass [22,23]. Thus, the investigation into the characteristics of moso bamboo expansion, particularly when it encroaches upon other forest types, assumes a critical role in safeguarding national ecological security. It is essential to gain an understanding of moso bamboo expansion through dedicated research.

"Scale" as a paradigm in ecology plays an equally vital role in expansion ecology research [24]. Current studies on moso bamboo forest expansion into other forests primarily focus on plot and protected area scales [25]. At this scale, the research into moso bamboo expansion predominantly employs plot experiments to investigate the impact of bamboo forests on the local environmental conditions, such as soil elements, soil organic matter, and microbial communities. This approach further delves into the mechanisms underlying the expansion of bamboo forests into other forest types [26]. In their research, some colleagues explored the correlation between soil nitrogen mineralization and fungal decomposition capacity in broad-leaved forests invaded by moso bamboo within the Wuyi Mountain [27]. Some investigated the alteration of assembly processes and interaction patterns within the soil microbial community of subtropical broad-leaved forests due to moso bamboo expansion in Lingfeng Mountain, Anji County, Zhejiang Province, China [25]. Several studies concerning the encroachment of bamboo forests into other forested landscapes have been undertaken, primarily centered on Japan [28–31]. Japan, positioned between 31° N–46° N, displays considerable differences in climate features in contrast to China. Climatic characteristics serve as the foremost factors influencing the distribution of moso bamboo. Investigating the distribution features, alterations in the area, and the expansion rate of moso bamboo forests in China will help to understand the discrepancies in moso bamboo forest encroachment across diverse regions, thus strengthening their management.

Explorations into distinct forest regions carry strong regional imprints. The encroachment of bamboo forests is profoundly shaped by the geographical layout of forests and the prevailing regional environmental conditions [10]. Given the wide distribution of moso bamboo in China's subtropical regions, the strategies adopted by moso bamboo for infiltrating various forest types exhibit significant diversity. Furthermore, these strategies are bound by the geographical dispersion of plant landscapes and the distinct environmental features in various regions. Consequently, conducting a macro-scale analysis of moso bamboo expansion into diverse types of forests in China allows for a comprehensive understanding of the overall characteristics and trends in moso bamboo expansion, offering a comprehensive interpretation for the disparities in moso bamboo expansions.

A survey of the literature concerning factors closely linked to bamboo expansion underscores the significance of forest characteristics as the primary driving forces behind bamboo's successful spread into adjacent ecosystems [32,33]. Okutomi [5] suggests that moso bamboo has a proclivity for infiltrating forests with simpler structural compositions and lower species diversity. Moreover, Suzuki and Nakagoshi [34] have reported that the pace of bamboo expansion is intertwined with the slope and orientation of neighboring forests. Additionally, the role of natural disturbances, including wind, snowfall, and wildfires, in hastening the clonal expansion of bamboo is well recognized [35,36]. Furthermore, certain investigations indicate that soil attributes exert a notable influence on bamboo expansion. Dong [37], for instance, has observed a positive correlation between the extent of expansion and the number of new culms with soil fertility and porosity. In a complementary vein, Liang [38] have posited that adequate soil moisture acts as an impediment to rhizome growth. Consequently, it is apparent that bamboo expansion is entwined with the characteristics of aboveground forest stands and subterranean soils at the bamboo–woodland ecotone (BWE), which serves as a transition zone between bamboo forests and woodlands. The key to understanding bamboo forest expansion lies in scrutinizing the ecotone, where the forest type, soil attributes, and microbial environment dictate the pace of moso bamboo expansion. Meanwhile, fluctuations in the ecotone's movements and length imply shifts in the features of bamboo forest expansion [39]. This study endeavors to grasp the overarching trends and the velocity of bamboo forest expansion through an exploration of ecotone dynamics. Given that moso bamboo is a species with limited dispersal capabilities, natural expansion predominantly transpires at the edges of existing bamboo forests, while non-edge expansion typically results from human cultivation. An examination of anthropogenic interference is pivotal for discerning variations in artificial plantings across different regions and for curtailing the further encroachment of bamboo forests. To date, few studies have simultaneously addressed changes in the ecotones. Historically, remote sensing interpretation has served as the primary data source for macro-level research [40–42]. Nevertheless, interpreting data at this scale presents challenges, such as difficulties in distinguishing bamboo species and achieving adequate accuracy [43].

The basic data are from the forest inventory for planning and design, named the forest second type inventory (FSTI), which aims to inventory the distribution, quantity, quality, and ownership of forest and provide an objective portrayal of the forest conditions covering the whole of China, conducted by the China National Forestry and Grassland Administration [44]. The FSTI integrates plot-based field surveys and remote sensing interpretation to ensure the accuracy of forest inventory data. Remote sensing interpretation is mainly applied to delineate the boundaries of a subcompartment based on the texture and color of the image in the early stage. The forest subcompartment is the basic unit for this inventory, referred to as a patch with the same forestry characteristics inside, different in adjacent areas. Forestry characteristics encompass ownership, forest category and species, ecological public welfare forest ownership and protection level, forestry engineering category, land type, origin, dominant tree species, age class, crown closure, site type, and other factors. Typically, the area of a subcompartment does not exceed 25 hectares. Plot-based field surveys are used to verify the boundaries of a subcompartment and collect forest attributes of the subcompartment, including compartment number, subcompartment number, forest type, site attributes such as ownership and terrain, tree species, tree size, overall tree condition, etc. The inventory is conducted every 10 years, with a five-year frequency in areas or units with high management levels. FSTI field protocols are described in the national core field guide. This document describes the standards, codes, methods, and definitions for FSTI field data that are consistent and uniform across China. The current version of the Standard is GBT 26424-2010 [45], technical regulations for inventory for forest management planning and design. The outcomes constitute a database for the FSTI, including a vector polygon feature with graphics and attribution table. The smallest unit on the map is the stand patch, and the combination of patches forms a map of the FSTI. The recordings in the table correspond to a single subcompartment and each recording has all field attributions. This study mainly uses the field named "Dominant tree species" and the spatial information of the subcompartment.

Within this study, we broaden the scope of our research to include moso bamboo expansion extending from protected areas to the nation. With forest second type inventory data spanning from 2010 to 2020, we assess the area, average rate, and spatial distribution differences in moso bamboo expansion into other forests. Leveraging a decade's worth of forest second type inventory data from China, this research aims to address a set of crucial inquiries: (1) Has moso bamboo consistently encroached into different forest types, and if so, what varying expansion rates are observed? (2) What trends characterize the transformation of the critical moso bamboo-woodland ecotone during the expansion process? (3) What disparities exist in terms of expansion speed and spatial distribution as moso bamboo spreads across different forests?

2. Materials and Methods

2.1. Study Area

China is located in eastern Asia and bordered by the western coast of the Pacific Ocean to the east. Its latitude ranges from 3°51′ N to 53°33′ N, and the longitude ranges from 73°33′ E to 135°05′ E. It encompasses a diverse range of climatic zones, including tropical, subtropical, warm-temperate, temperate, cold-temperate, and plateau regions. China's landscape features plateaus, mountains, hills, and plains. China hosts an extensive distribution of plant species, horizontally extending across tropical, subtropical, and temperate zones, and vertically covering a wide array of terrains, including plains, hills, mountains, and plateaus. It boasts a rich array of forest types, with coniferous forests predominantly found in the northern and mountainous regions, while evergreen and deciduous broadleaved forests dominate the southern areas [46]. Bamboo forests are widespread in the tropical and subtropical regions of central and southern China (Figure 1). Bamboo has been cultivated in China for over 3000 years, solidifying its status as an indigenous plant. Currently, China maintains approximately 7.01 million hectares of bamboo forests, housing 39 genera and 837 species of bamboo, which constitute 51% of the world's 1642 bamboo species [47]. China holds the distinction of being the world's foremost source of bamboo resources, leading in bamboo variety, forest area, and stock volume. Notably, moso bamboo comprises approximately 70% of China's total bamboo resources. As a native invasive species, moso bamboo poses challenges to other types of forests during its expansion.

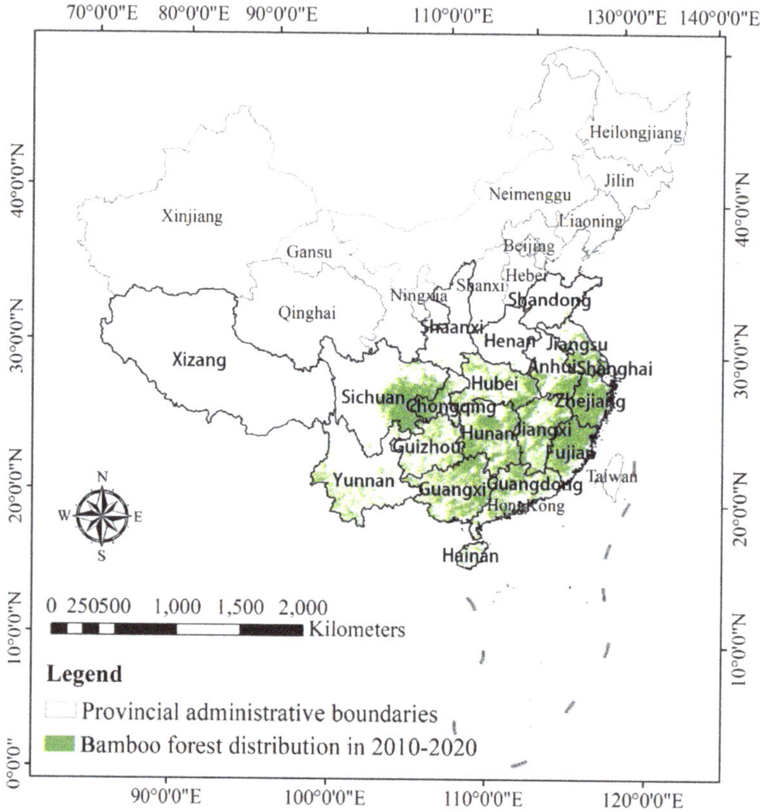

Figure 1. Distribution map of all species of bamboo stands in China [48].

2.2. Data Processing

The ArcGIS pro3.1, a geo-information system software package, was applied to select and analyze the data from the forest second type inventory database. Firstly, we obtained the vector layers of all second administration regions in 2010, 2015, and 2020 from the China National Forestry and Grassland Administration. Then, we selected the records for which the field "Dominant tree species" = "66000" to ensure that the land where moso bamboo is planted has been identified. We combined the data chosen in the three years together to obtain the region where moso bamboo forest is present from 2010 to 2020. The output layer is used to extract data from this combined region within the forest resources survey databases for these three years, resulting in the creation of a moso bamboo forest type database covering the period from 2010 to 2020. These forest stands were categorized into nine primary forest types based on the dominant tree species, including coniferous forests, evergreen broad-leaved forests, deciduous broad-leaved forests, other bamboo forests, non-forests, shrubbery, non-timber forests, other forests, and moso bamboo forests (Figure 2). Moso bamboo forests in Tibet are mainly concentrated in the Linzhi region, which, due to its small area, was virtually negligible and subsequently excluded from the datasets for further analysis.

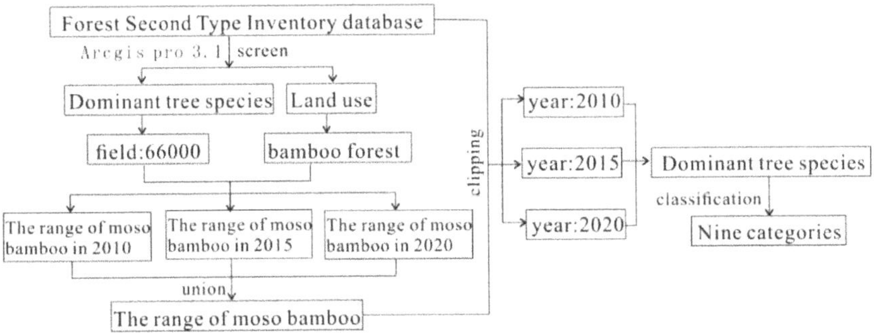

Figure 2. Diagram of data processing workflow.

2.3. *The Area Transition Matrix of Expansion Analysis*

Employing the feature to raster tool in ArcGIS Pro, the vector layers in moso bamboo forest type database were converted into raster layers with dimensions of 200 m × 200 m. The coordinate system of the resulting raster layers was then established as Albers, thereby yielding raster layers depicting forest types. Subsequent inputting of these raster into Fragstats enabled the generation of transition matrix tables depicting the alterations in bamboo forest area between the periods 2010–2015 and 2015–2020.

2.4. *Analysis of the Change in BWE*

2.4.1. Average Expansion Rate

Calculating the average expansion rate involved subtracting the total area of patches in the former year from that of the latter, divided by the length of the intersect edge in the two years (clip the previous year's boundary by the latter year's region in ArcGIS). The average expansion rate of different forest types by bamboo forest involved dividing the area invaded by bamboo forest for each forest type by the length of the expansion edge.

2.4.2. Neighboring Ratio

Using the nearest neighbor tool in ArcGIS Pro, the identification of neighboring expanding patches was carried out. Patches with a nearest neighbor distance of 0 were classified as neighboring invading patches for two consecutive years. Utilizing the reproductive characteristics of moso bamboo forests, this study proposes the concept of the neighboring ratio to gain a clearer understanding of the proportion of man-made plantations. The neighboring ratio is defined as the ratio of the area (S_n) of patches that underwent neighboring expansion compared to the preceding period, to the sum of the area of neighboring expansion (S_n) and the area of non-neighboring expansion (S_m). [49] This indicator is primarily employed to contrast the total area of expansion in proximity with that of non-neighboring expansion. A higher ratio indicates more pronounced neighboring expansion and, by extension, a reduced extent of non-neighboring areas, implying a lesser extent of human-planted moso bamboo forests.

$$NR = \frac{S_n}{S_n + S_m} \times 100\% \quad (1)$$

Patches with a nearest neighbor distance of 0 were classified as neighboring invading patches, while others were non-neighboring patches.

2.5. *Spatial Heterogeneity Analysis*

Local indicators of spatial association (LISA) is used to measure whether the distribution of spatial variables has agglomeration, including global spatial autocorrelation and local spatial autocorrelation [50]. Spatial autocorrelation analysis is an important tool for assessing the extent of spatial aggregation for variables of interest. It can reveal correlations among the attributes of study units, at either a global or local scale [51]. Here, GeoDa 1.6.7 was used for the spatial analysis of the independent variables. Spatial autocorrelation was assessed using global Moran's I and local Moran's I. The degree of spatial association was visualized in a scatter plot, and local indicators of spatial association (LISA) were calculated and visualized on the study area map [52].

3. Results

3.1. Dynamic Encroachment Process of Moso Bamboo

Overall, from 2010 to 2020, there was a remarkable increase in the extent of moso bamboo forests in China, expanding from 44,238.32 km^2 to 53,549.96 km^2 (Figure 3), at an average annual growth area rate of 2.1%. During the 2010–2015 period, the net area of moso bamboo forest encroachment into other forest types (defined as the area that moso bamboo forests invade other forest landscapes minus the area that other forest types encroach on moso bamboo forests) amounted to 3365.63 km^2, with an average annual growth area rate of 1.6%. Subsequently, during the 2015–2020 period, the net area of moso bamboo forests encroaching on other forest types was 5946.01 km^2, with an average annual growth rate of 2.4%. During the 2015–2020 period, the area of moso bamboo forests increased by 0.77 time compared to the 2010–2015 period.

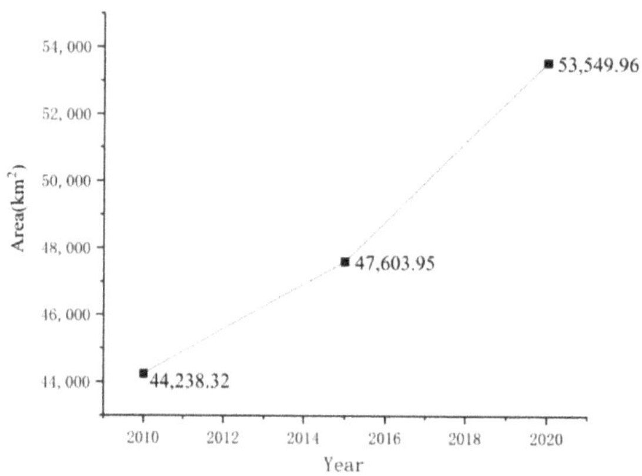

Figure 3. The area of moso Bamboo forests from 2010 to 2020.

As shown in Figure 3, during the 2010–2015 period, the predominant forest types affected by moso bamboo forest expansion were coniferous forests, evergreen broadleaved forests, non-forests, and deciduous broadleaved forests, with expansion areas measuring 7882.34 km^2, 3612.02 km^2, 2253.38 km^2, and 1638.25 km^2, respectively, collectively constituting 83.72% of the total invaded area by bamboo forests. The net area of moso bamboo forest encroachment on other forest ranked in the order coniferous forests > deciduous broad-leaved forests > shrubbery > evergreen broad-leaved forests > other bamboo forests > other forests> non-timber forests > non-forests, with transfer areas of 2366.79 km^2, 675.83 km^2, 425.93 km^2, and 367.84 km^2, respectively. Furthermore, non-forested areas encroached upon a portion of moso bamboo forests, accounting for 34.56% of the net expansion area by bamboo forests.

In the 2015–2020 period, the forest types affected by the expansion of moso bamboo forests, ranked by area in descending order, included coniferous forests, non-forests, evergreen broad-leaved forests, non-timber forests, other bamboo forests, deciduous broad-leaved forests, shrubbery, and other forests. The primary forest types impacted were coniferous forests, non-forests, evergreen broad-leaved forests, and non-timber forests, with expansion areas of 7936.27 km^2, 4908.01 km^2, 4905.03 km^2, and 1656.81 km^2, respectively, collectively representing 85.86% of the total invaded area by bamboo forests. The net area of moso bamboo forest expansion into other forest ranked as non-forests > coniferous forests > non-timber forests > other bamboo forests > deciduous broadleaved forests > evergreen broadleaved forests > other forests > shrubbery. The primary types impacted were non-forests, coniferous forests, non-timber forests, and other bamboo forests, with transfer areas of 2473.50 km^2, 1779.23 km^2, 610.44 km^2, and 530.44 km^2, respectively, cumulatively exceeding 90.71% of the net expansion area by bamboo forests.

As shown in Figure 4, when examining the expansion data for moso bamboo forests between the 2010–2015 and 2015–2020 periods, it is clear that coniferous forests, evergreen broad-leaved forests, and non-forests represent the largest area, collectively accounting for over 75% of all invaded forests. Coniferous forests and evergreen broad-leaved forests are the primary targets of moso bamboo forest expansion, with their combined expansion ratio exceeding 55% of all types. In terms of the net expansion data spanning these two stages, coniferous forests appear as the most frequently invaded forest type, with a nearly equivalent net expansion area in both periods. The most significant difference lies in non-forests. During the 2010–2015 period, moso bamboo forests encroached upon non-forested areas by 1163.04 km^2, whereas in the 2015–2020 period, moso bamboo forests invaded non-forested areas covering 2473.50 km^2. Despite a relatively large invaded area in evergreen broadleaved forests, they exhibit comparatively greater stability, resulting in a lower overall expansion ratio. The proportional areas of forest types invaded by moso bamboo forests in these two stages remain relatively consistent, with coniferous forests, evergreen broad-leaved forests, and non-forests collectively representing approximately 80% of the invaded moso bamboo forest.

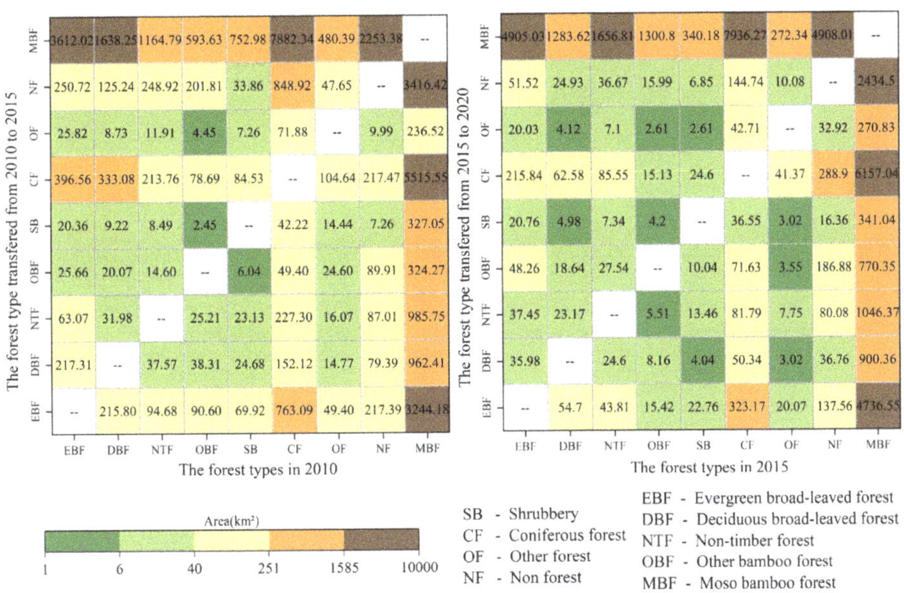

Figure 4. The transfer area of moso bamboo forests into other forests during the 2010–2015 and 2015–2020 periods.

3.2. *Dynamic Characteristics of Moso Bamboo Expansion in Bordering Other Forests*

Over the 2010–2015 period, moso bamboo forests expanded in adjacent areas by 10,921.55 km^2, while non-adjacent areas witnessed an expansion of 7458.93 km^2. During the 2015–2020 period, the expansion in adjacent areas covered 12,467.30 km^2, with non-adjacent expansion encompassing 10,139.09 km^2. In the 2010–2015 period, the average expansion extended by 8.02 m/year, while in the 2015–2020 period, the expansion rate was 7.62 m/year.

The total length of the invaded border in 2010 was 26,681.36 km, which increased to 32,065.88 km in 2015 and further to 35,391.36 km in 2020. Moso bamboo forests exhibited varying expansion rates across different forests, with the fastest expansion observed in other bamboo forests and the slowest in non-forested areas. In the 2010–2015 period, the expansion rate of moso bamboo forests (Figure 5) into other forests was as follows: other bamboo forests > shrubbery> coniferous forests > non-timber forests > other forests > deciduous broad-leaved forests > evergreen broad-leaved forests > non-forests. In the 2015–2020 period, the expansion rate of moso bamboo forests was as follows: other bamboo forests > coniferous forests > non-timber forests > evergreen broad-leaved forests > shrubbery > deciduous broad-leaved forests > other forests > non-forests.

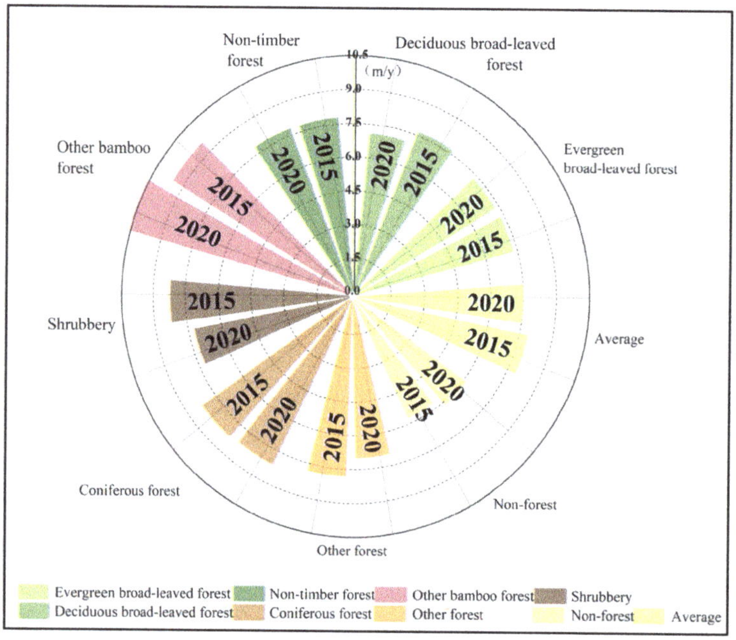

Figure 5. The average rate of bamboo expansion across different forest types and the overall average rate of expansion.

The neighboring rate of moso bamboo forests is 0.8 across the entire China. As shown in Figure 6, in the 2010–2015 period, provinces with higher bamboo forest expansion NR included Guangdong, Fujian, and Jiangxi, while in the 2015–2020 period, Anhui, Hubei, and Hainan exhibited larger NR. The main provinces where the NR increased were Hainan, Yunnan, and Anhui Provinces, while Guangdong, Chongqing, and Hubei Provinces experienced decreases in it.

3.3. Spatial Heterogeneity of Moso Bamboo Expansion into Different Types of Forests

From 2010 to 2020, the analysis reveals a significant positive correlation between the expansion area of moso bamboo forests in each province and the total area of moso bamboo forests in each province (Figure 7a,b). This suggests that the expansion of bamboo forests occurred within existing bamboo forest areas, consistently with the phenomenon of bamboo forests expanding outward through rhizome sprouting.

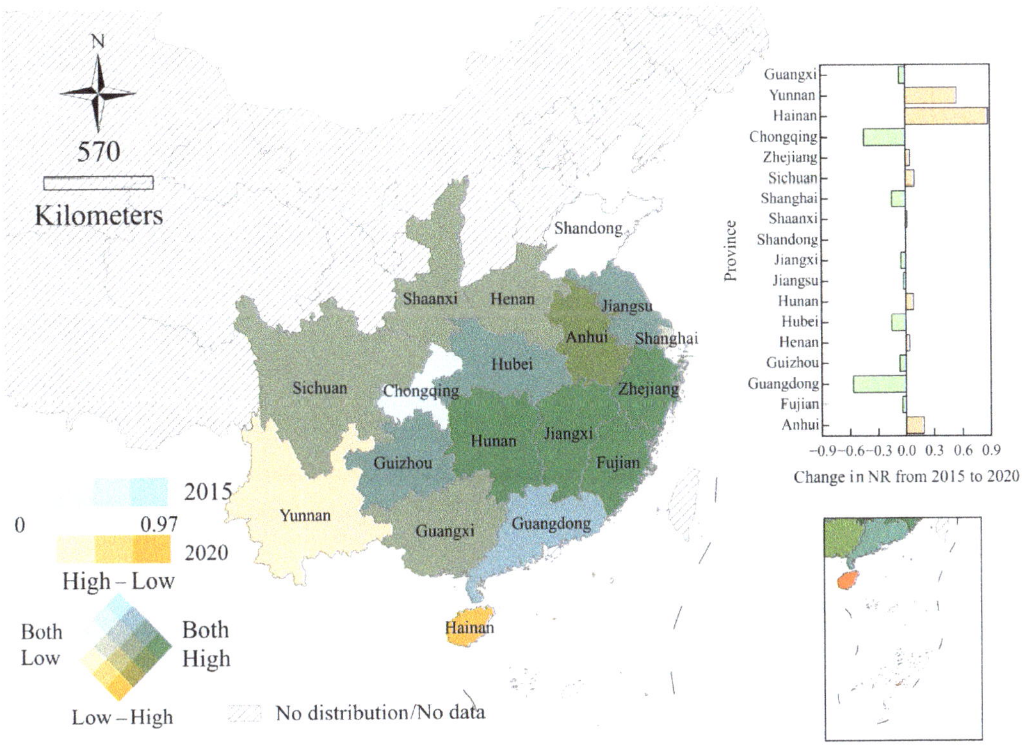

Figure 6. The neighboring rate of moso bamboo forest in every province.

A study of spatial auto-correlation in moso bamboo forests across provinces shows that while the expansion area of moso bamboo forests is positively associated with the total area of bamboo forests in each province, there are marked regional variations in the expansion area of bamboo forests. Jiangxi province and its surrounding areas are identified as core regions with high-high clustering, whereas Henan and Shandong provinces exhibit low–low clustering characteristics (Figure 7a,b).

The spatial auto-correlation analysis reveals a notable heterogeneity in bamboo forest expansion across provinces in various forest types. The expansion of moso bamboo forests in evergreen broad-leaved forests tends to cluster in specific regions, primarily centered around Jiangxi and its neighboring provinces. Significantly, during two distinct periods, 2010–2015 and 2015–2020, the regions with the most extensive moso bamboo forest expansion areas were Jiangxi and Fujian provinces. In the 2010–2015 period, Fujian accounted for the largest expansion area, comprising approximately 21.65% of the total expansion area in evergreen broad-leaved forests in 2015. In the 2015–2020 period, Jiangxi held the largest expansion area, representing about 20.88% of the total expansion area in evergreen broad-leaved forests in 2020. Moso bamboo forest expansion in deciduous broad-leaved forests primarily occurred in the middle and lower reaches of the Yangtze River.

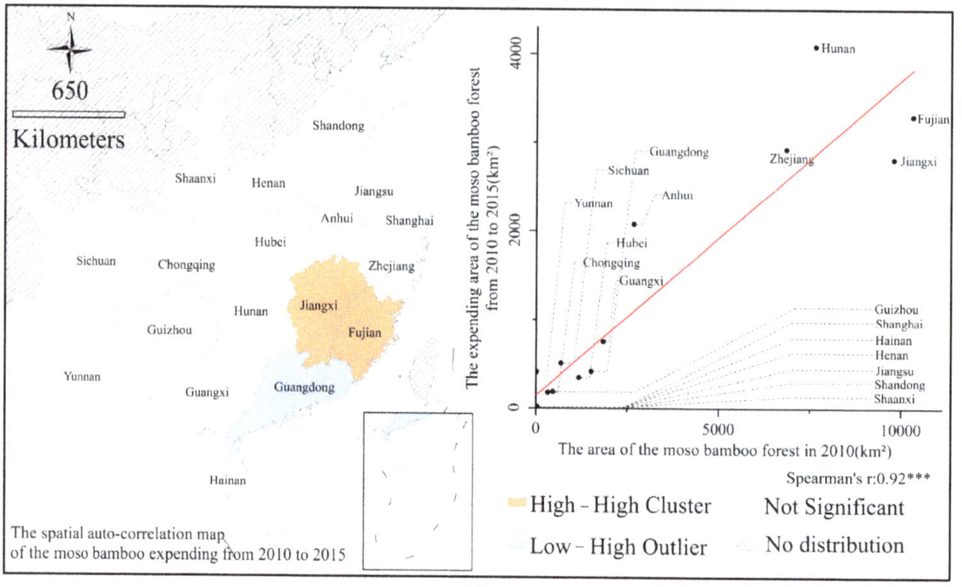

(**a**)

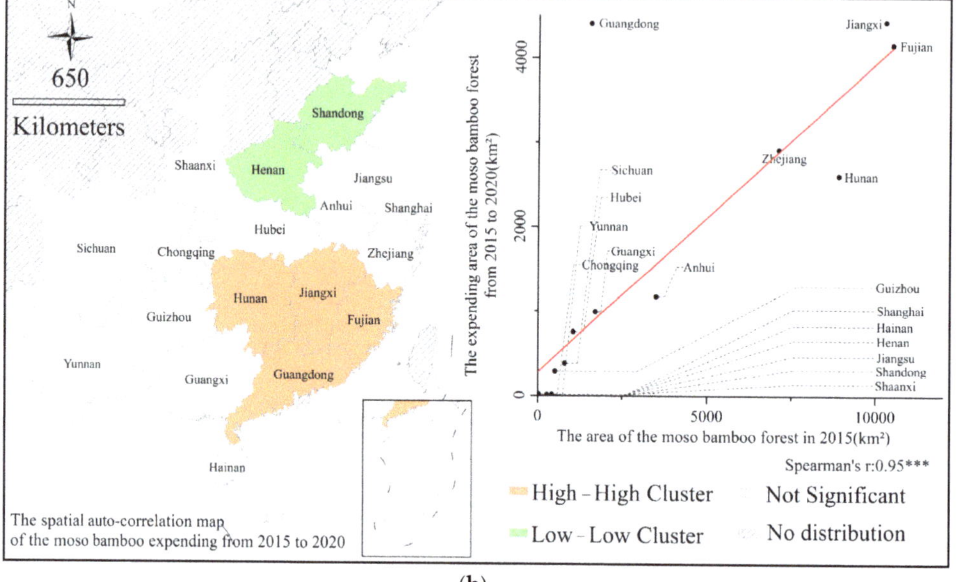

(**b**)

Figure 7. (**a**) The correlation between the area of the moso bamboo expanding and the existing extent in every province and the spatial auto-correlation map of the moso bamboo expanding in 2010–2015. (**b**) The correlation between the area of the moso bamboo expanding and the existing extent in every province and the spatial auto-correlation map of the moso bamboo expanding in 2015–2020. *** stands for the significance level >0.99.

As shown in Figure 8, during the 2010–2020 period, the provinces with significant expansion areas were Anhui and Hunan Province, collectively constituting approximately 39% of the total. In the 2010–2015 period, Zhejiang Province held the largest expansion area, covering about 33.4% of the total expansion area in deciduous broad-leaved forests. In the 2015–2020 period, Guangdong Province had the largest expansion area, accounting for approximately 31.94% of the total expansion area. The expansion of moso bamboo forests in other bamboo-rich regions appeared more dispersed, with primary occurrences in southwestern China, Guangdong Province, and Zhejiang Province. Notably, during the 2010–2015 and 2015–2020 periods, Zhejiang Province was the province with significant expansion areas. In the 2010–2015 period, Yunnan Province held the largest expansion area in other bamboo-rich regions, accounting for approximately 32.32% of the total expansion area in 2015. In the 2015–2020 period, Guangdong Province had the largest expansion area in other bamboo-rich regions, comprising approximately 79.33% of the total expansion area in 2020. Bamboo forest expansion in shrubbery was primarily concentrated in Hunan Province, with a lower proportion in the 2015–2020 period compared to the 2010–2015 period. The expansion of moso bamboo forests into coniferous forests was primarily concentrated in Jiangxi Province and its surrounding regions. Among these regions, during both periods (2010–2015 and 2015–2020), the provinces with significant expansion areas were Fujian Province and Hunan Province (Table 1). In the 2010–2015 period, Hunan Province held the largest expansion area, accounting for approximately 41.74%, while in the 2015–2020 period, Jiangxi Province had the largest expansion area, representing about 17.30%. The expansion of moso bamboo forests in non-forest areas was primarily distributed in Fujian Province, Guangdong Province, and Jiangxi Province. Notably, during both periods, the provinces with significant expansion areas were Fujian Province and Jiangxi Province. In the 2010–2015 period, Fujian Province accounted for the largest expansion area, comprising approximately 36.90%, while in the 2015–2020 period, Guangdong Province had the largest expansion area, accounting for about 22.82%. In summary, the most extensive erosion occurred in Jiangxi Province and its surrounding regions, but the distribution of erosion areas varied across different forest landscape types.

Table 1. A comparison of the major provinces in terms of invaded areas during different stages of bamboo forest expansion.

Forest Types	The Province with the Largest Invaded Area in the 2010–2015 Period	The Province with the Largest Invaded Area in the 2015–2020 Period	The Provinces Consistently Ranking within the Top Three in Invaded Areas in the Two Periods
Evergreen broad-leaved forest	Fujian Province	Jiangxi Province	Fujian Province, Jiangxi Province
Deciduous broad-leaved forest	Zhejiang Province	Guangdong Province	Anhui Province, Hunan Province
Non-timber forest	Zhejiang Province	Fujian Province	Fujian Province, Jiangxi Province
Other bamboo forest	Yunnan Province	Guangdong Province	Zhejiang Province
Shrubbery	Hunan Province	Hunan Province	Hunan Province
Coniferous forest	Hunan Province	Jiangxi Province	Fujian Province, Hunan Province
Other forest	Hunan Province	Jiangxi Province	Fujian Province, Zhejiang
Non-forest	Fujian Province	Guangdong Province	Fujian Province, Jiangxi Province

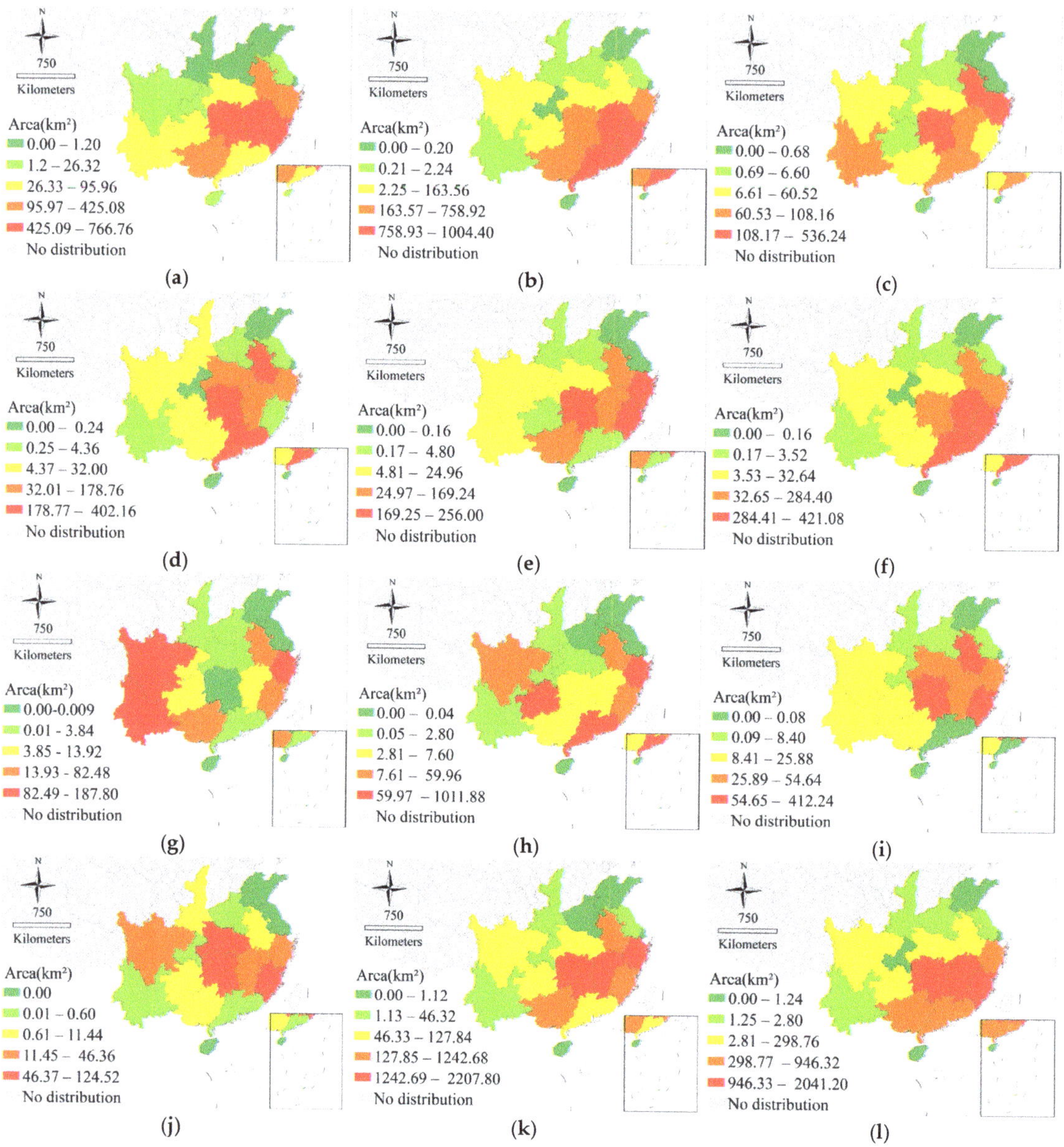

Figure 8. *Cont.*

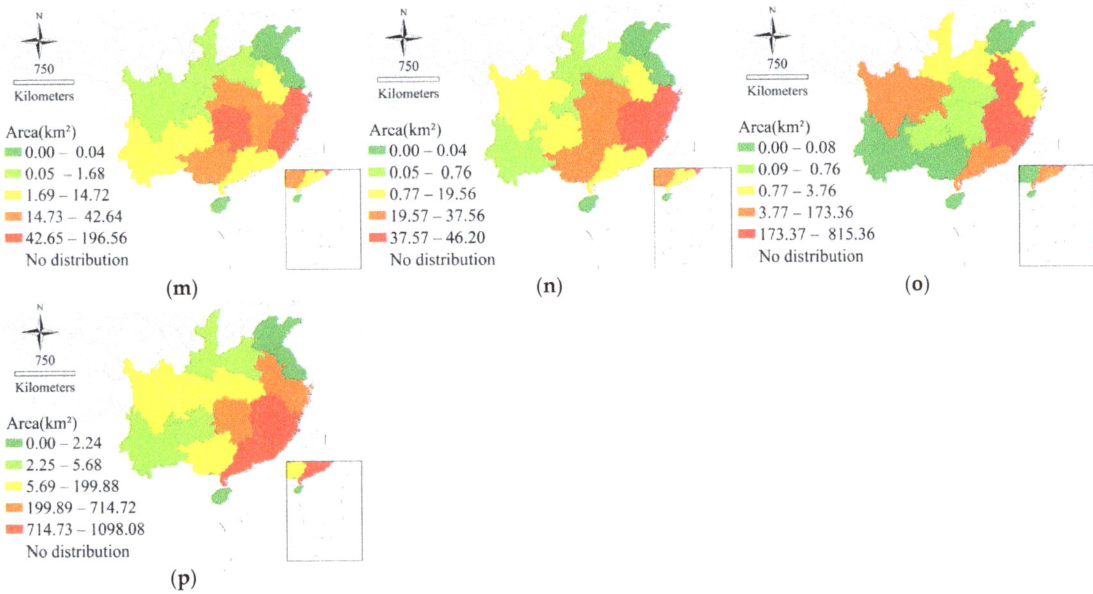

Figure 8. The area of moso bamboo forest expanding into the following areas: (**a**) evergreen broad-leaved forest (2010–2015); (**b**) evergreen broad-leaved forest (2015–2020); (**c**) deciduous broad-leaved forest (2010–2015); (**d**) deciduous broad-leaved forest (2015–2020); (**e**) non-timber forest (2010–2015); (**f**) non-timber forest (2015–2020); (**g**) other bamboo forest (2010–2015); (**h**) other bamboo forest (2015–2020); (**i**) shrubbery (2010–2015); (**j**) shrubbery (2015–2020); (**k**) coniferous forest (2010–2015); (**l**) coniferous forest (2015–2020); (**m**) other forest (2010–2015); (**n**) other forest (2015–2020); (**o**) non-forest (2010–2015); (**p**) non-forest (2015–2020).

4. Discussion

4.1. The Strength and Speed of Moso Bamboo Expansion and the Changes in BWE and NR

This study has revealed a 2.1% annual average increasing area in moso bamboo forests over a 10-year period. A study conducted by Suzuki et al. similarly indicates a 2% annual expansion rate for moso bamboo forests in Japan [34], highlighting a consistent trend in bamboo forest expansion across Asia. Nevertheless, it is crucial to recognize that the growth rate of moso bamboo is not always constant and can be subject to various influences, including non-biological factors and human activities [48]. The average speed of bamboo forest expansion into other forests is 8 m/yr, which sharply contrasts with findings in Mount Tianmu, Zhejiang Province [53], where the expansion rate is only 3 m/yr. This also differs from the review stating that "the elongation speed of bamboo whips generally reaches 4–5 m/yr" [32]. The primary reason for this disparity lies in the scale of the research. In Jiangxi Province, within nature reserves, bamboo forest expansion is a consequence of natural succession, minimally disturbed by human socio-economic activities. On a national scale, where forest types are more diverse, regions undergoing bamboo forest expansion may be influenced by human cultivation of bamboo as an economic crop. The planted bamboo patches are included in the calculation of bamboo expansion growth rates, resulting in an average growth rate higher than the annual growth rate of bamboo whips.

Over the two periods, the NR of the national moso bamboo forests averaged approximately 0.8, indicating that moso bamboo forests primarily expand naturally. Zhejiang, Fujian, Jiangxi, and Hunan Provinces exhibited relatively higher NRs in both periods. However, Jiangxi and Fujian Provinces experienced a further decrease in NR in the latter period, while Hunan and Zhejiang observed an increase. This is largely related to the bamboo forestry policies of each province. Economic incentives in Fujian and Jiangxi Provinces, driven by the bamboo industry, led to a decrease in NR due to the artificial planting of bamboo forests.

Between 2010 and 2020, the length of the critical bamboo forest ecotone has significantly increased, with the contact area between bamboo forests and other forests gradually expanding. This has led to an increase in the shape index over the three periods, signifying the growing complexity of the critical bamboo forest ecotone [44]. This suggests that moso bamboo forest expansion into other forest types will intensify in the future.

4.2. The Differences in Moso Bamboo Expansion into Various Forests

Moso bamboo forest expansion predominantly targets coniferous forests, followed by evergreen broad-leaved forests. An analysis of the expansion rates of moso bamboo forests into coniferous and evergreen broad-leaved forests confirms that the expansion rate into coniferous forests exceeds that into evergreen broad-leaved forests [54]. Bamboo forests share the same ecological niche with coniferous and evergreen broad-leaved forests, all of which are typical forest types in subtropical regions, covering extensive areas in such regions [55]. Moso bamboo, classified as a heliophile and shade-tolerant species, thrives in forests with small, sparse canopies, and low density [56]. Coniferous forests, with their narrower canopy widths, are more susceptible to moso bamboo expansion, leading to the gradual formation of mixed bamboo and fir forests, ultimately culminating in pure moso bamboo forests. Furthermore, coniferous forests generally exhibit lower biodiversity compared to evergreen broad-leaved forests. Consequently, evergreen broad-leaved forests maintain higher stability, resulting in a slower expansion rate during moso bamboo intrusion [57,58]. Over the ten-year study period, significant disparities were observed in moso bamboo forest expansion into non-forested land. In the earlier period, a substantial portion of moso bamboo forests succumbed to erosion by non-forest landscapes, while the later period witnessed an increased occurrence of moso bamboo expansion into non-forest landscapes. The variations between these two periods were primarily influenced by the bamboo industry's life cycle and adjustments in nature conservation policies. The period from 2004 to 2014 marked the mature phase of the bamboo industry's life cycle, characterized by stable market demand and a consistent increase in moso bamboo forest harvesting. This resulted in extensive felling of moso bamboo forests, converting them into farmland and barren land, leading to significant erosion between 2010 and 2015. However, after 2015, due to rising labor costs, changes in supply and demand dynamics, and enhancements in the natural conservation system, a substantial prohibition on bamboo forest logging was enforced, resulting in a distinct trend in moso bamboo forest expansion into other types of forests.

4.3. Spatial Heterogeneity in Moso Bamboo Forest Expansion in Different Provinces

The area of moso bamboo forest expansion into other forest types in each province exhibits a significant positive correlation with the existing moso bamboo forest area. This correlation aligns with the reproductive characteristics of moso bamboo (*Phyllostachys edulis*), a typical clonal plant species whose propagation and renewal primarily occur through the division of rhizomes. This growth pattern diverges from the typical "fixed growth" mode observed in many other woody plants, featuring rapid growth, sustainable strength, and a propensity for spreading.

Spatial heterogeneity in moso bamboo forest expansion is evident across different forest types in various regions. The expansion of evergreen broad-leaved and coniferous forests is prominent in Jiangxi and surrounding provinces. The expansion of deciduous broad-leaved forests mainly occurs in the middle and lower reaches of the Yangtze River. The expansion of other bamboo forests is scattered and primarily distributed in southwestern China, Guangdong, and Zhejiang Province. The expansion of shrubbery is mainly observed in Hunan province. A significant correlation is apparent between provinces experiencing moso bamboo expansion in different forest types and in the province's moso bamboo forest area, and the distribution proportion of the forest landscape in that province.

The top three provinces in terms of moso bamboo-invaded area for each landscape type belong to the top ten provinces in moso bamboo forest area ranking (percentage-wise), and the proportion of this forest type is among the highest nationwide. Although Sichuan and Hubei Province have a substantial proportion of deciduous broadleaf forests, the expansion primarily occurs in Hubei due to the smaller area of moso bamboo forests in Sichuan. Despite the larger area of shrub forests in Sichuan compared to Hunan province, moso bamboo expansion into shrub forests is most extensive in Hunan due to the significantly smaller area covered by moso bamboo forests in Sichuan.

4.4. The Possible Reasons for the Expansion of Moso Bamboo Forest

The expansion of bamboo forests can be attributed to both natural and human factors. The area of moso bamboo forest expansion due to natural factors is greater than that due to human factors. In terms of natural expansion, bamboo propagates through the outward growth of its rhizomes [11,59]. However, the rate of expansion varies significantly across different forest types, indicating that certain factors in these forests inhibit bamboo expansion. These factors differ from those influencing bamboo distribution mentioned by Yu [48], since local bamboo forests and adjacent forests share the same climate and geographic location but differ in stand characteristics, soil properties, and light intensity [60]. And various studies have shown that changes in soil physical and chemical properties (including depth, moisture content, organic carbon content, total nitrogen, total phosphorus, pH, and microorganisms), stand structure (canopy closure, canopy height, stand density, species diversity), and light intensity [56] facilitate bamboo growth. Therefore, these factors are crucial in the expansion of bamboo forests. Regarding human factors, the main causes of bamboo expansion include moso bamboo afforestation, forest management, and human disturbance [35]. The growth of bamboo forests alters soil properties and microbial composition, leading to a pattern of bamboo invasion that reduces biodiversity, changes community structure and composition, weakens ecosystem functions, and inhibits community regeneration.

4.5. Suggestions for the Management of Moso Bamboo Forests

The expansion of moso bamboo into coniferous and broad-leaved forests carries adverse repercussions for the ecosystem, encompassing decreased species diversity in both the tree and shrub layers [9], and a more homogenous forest structure. In the realm of strictly protected area management, a paramount focus should be placed on bamboo control. This can be achieved through practices such as bamboo shoot harvesting and bamboo thinning, complemented by physical barriers like stone paths and water systems. These measures are imperative to deter moso bamboo expansion [61] and to preserve the integrity of natural forest vegetation. In areas of stringent protection, clear-cutting provides a comprehensive solution to the challenge posed by bamboo expansion. As supported by the literature, clear-cutting leads to a secondary succession process characterized by increased species diversity and enhanced ecosystem stability. To curb future expansions of moso bamboo into coniferous forests, it is recommended to plant high-density native broadleaved species around the periphery of moso bamboo forests. These plantings can effectively impede the rate of expansion. To safeguard against the expansion of moso bamboo forests into non-native regions due to climate change, the establishment of a moso bamboo forest database is advised. The integration of this database into forest inspections will facilitate the continuous monitoring of moso bamboo forest introductions and natural expansion. This monitoring will prevent bamboo from spreading beyond its native areas as climate conditions evolve. Additionally, as moso bamboo is an economically significant plant, its extent is subject to considerable human influence. Governments can introduce policies such as "bamboo as a substitute for plastic initiative" to stimulate the bamboo industry, enhance the consumption of bamboo products, and help mitigate bamboo forest expansion. In future moso bamboo forest management, individual provinces can implement tailored management strategies, taking into account the varying expansion rates of moso bamboo forests and the distinctive characteristics of the bamboo industry within each province. This approach addresses the challenges posed by bamboo forest expansion from a socio-economic perspective.

4.6. Contributions and Limitations

This study broadens the research scope of moso bamboo expansion from protected areas to the national region. The choice of different research scales can impact the results of moso bamboo forest invasion studies, which will be advantageous for scale selection on moso bamboo forest invasion in future research. In contrast to the fact that the concept of BWEs is merely studied, this paper, for the first time, computes BWEs for varying temporal lengths and changes in various forest types. Delineating the trend in moso bamboo forest expansion based on length fluctuations promotes a more comprehensive exploration of changes in this forest ecotone. The conceptual formula for NR is presented, establishing the theoretical foundation and methodology for studying the expansion of local species into surrounding ecosystems. The application of the forest second type inventory data to bamboo forest expansion is employed for the first time, with the research results indicating the high reliability and validity of the forest second type inventory data.

The research at hand delves into the spatio-temporal changes in the encroachment of moso bamboo forests into other forest types between 2010 and 2020, investigating spatial characteristics, variations in tree species, and changing trends. It delivers valuable insights into the development of moso bamboo forests in China over the past decade and provides recommendations for addressing the ecological impacts of moso bamboo forest expansion. It is important to acknowledge the limitations of this study. Due to constraints in data acquisition, this study is confined to a ten-year time-frame, which is relatively brief. Subsequent research endeavors could explore longer time series and more precise data analysis by incorporating high-resolution remote sensing imagery and national forest inventory data. Moreover, human disturbances exert a significant influence on bamboo forest development in China. In comparison to natural drivers, human-induced factors are unpredictable. To study the natural development of bamboo forests, it is essential to select bamboo forests free from human interference as research subjects. Predicting the future development of bamboo forests and monitoring ecological security under conditions devoid of human interference are also crucial. Additionally, bamboo forests exhibit high sensitivity to environmental changes, making them suitable indicator plants to monitor environmental alterations in the absence of human disturbances.

5. Conclusions

This study indicates a clear positive correlation between the area invaded by moso bamboo forest and the existing moso bamboo forest area. Moso bamboo forests are expanding area at an average annual rate of 2%. As the complexity and length of bamboo–woodland ecotones (BWEs) increase, moso bamboo forests will further encroach into other forest. At different research scales, the speed of bamboo forest expansion into different forest types varies significantly. In particular, at the national scale, the average expansion rate of bamboo forests (including moso bamboo forests manually planted) is 8 m per year. The area and speed of bamboo forest expansion into different forests are not consistent. Tt is contingent upon the proportion of bamboo forest area at the location extent and the forest types being invaded. Coniferous forests and evergreen broad-leaved forests are the primary forests expanded by moso bamboo forests. However, coniferous forests and other bamboo forests are most rapidly invaded by moso bamboo. Expansion areas of bamboo forests in China exhibit pronounced spatial heterogeneity; the areas of different forest types expanded by moso bamboo forests have a substantial correlation with the bamboo forest areas in the province and the distribution proportion of the forest landscape in the province. The factors contributing to bamboo forest expansion encompass stand characteristics, soil attributes, light intensity, moso bamboo afforestation, forestry practices, and human disturbances.

Author Contributions: Conceptualization, D.L. and Y.Z.; methodology, J.W. (Juan Wei); software, D.L.; validation, J.W. (Jiangzhou Wu), Z.C. and J.H.; formal analysis, Y.Z.; data curation, S.Z.; writing—original draft preparation, D.L.; writing—review and editing, J.W. (Juan Wei) and L.Y.; visualization, D.L.; supervision, Y.Z.; project administration, Y.Z.; funding acquisition, Y.Z. All authors have read and agreed to the published version of the manuscript.

Funding: This research was funded by the Ministry of Science and Technology of The People's Republic of China, grants 2019YFD1100400.

Data Availability Statement: Dataset available on request from the authors.

Conflicts of Interest: We declare no conflict of interest.

References

1. Garrott, R.A.; White, P.J.; White, C.A.V. Overabundance: An issue for conservation biologists? *Conserv. Biol.* **1993**, *7*, 946–949. [CrossRef]
2. Valéry, L.; Fritz, H.; Lefeuvre, J.C. Another call for the end of invasion biology. *Oikos* **2013**, *122*, 1143–1146. [CrossRef]
3. Griscom, B.W.; Ashton, P.M.S. Bamboo control of forest succession: *Guadua sarcocarpa* in Southeastern Peru. *For. Ecol. Manag.* **2003**, *175*, 445–454. [CrossRef]
4. Veblen, T.T. Growth patterns of *Chusquea* bamboos in the understory of Chilean *Nothofagus* forests and their influences in forest dynamics. *Bull. Torrey Bot. Club* **1982**, *109*, 474–487. [CrossRef]
5. Okutomi, K.; Shinoda, S.; Fukuda, H. Causal analysis of the invasion of broad-leaved forest by bamboo in Japan. *J. Veg. Sci.* **1996**, *7*, 723–728. [CrossRef]
6. Peng, Z.; Lu, Y.; Li, L.; Zhao, Q.; Feng, Q.I.; Gao, Z.; Jiang, Z. The draft genome of the fast-growing non-timber forest species moso bamboo (*Phyllostachys heterocycla*). *Nat. Genet.* **2013**, *45*, 456–461. [CrossRef]
7. Ouyang, M.; Tian, D.; Pan, J.; Chen, G.; Su, H.; Yan, Z.; Fang, J. Moso bamboo (*Phyllostachys edulis*) invasion increases forest soil pH in subtropical China. *Catena* **2022**, *215*, 106339. [CrossRef]
8. He, F.; Liang, Z.; Wang, J.; Ge, Y.; Wang, H. Research Progress on the Occurrence and Prevention of Bamboo Pests in China. *Hubei For. Sci. Technol.* **2018**, *47*, 45–50.
9. Li, Y.M.; Feng, P.F. Bamboo Resources in China Based on the Ninth National Forest Inventory Data. *World Bamboo Ratt.* **2019**, *17*, 45–48.
10. Liu, X.; Siemann, E.; Cui, C.; Liu, Y.; Guo, X.; Zhang, L. Moso bamboo (*Phyllostachys edulis*) invasion effects on litter, soil and microbial PLFA characteristics depend on sites and invaded forests. *Plant Soil* **2019**, *438*, 85–99. [CrossRef]
11. Shen, R.; Bai, S.; Zhou, G.; Wang, Y.; Wang, N.; Wen, G.; Chen, J. The response of root morphological plasticity to the expansion of a population of *Phyllostachys edulis* into a mixed needle-and broad-leaved forest. *Acta Ecol. Sin.* **2015**, *36*, 326–334.
12. Jiang, L.; Mao, C.; Wu, H.; Duan, Q.; Fang, Z. Investigation and analysis of Moso forest management in songyang county, Zhejiang province. *World Bamboo Ratt.* **2015**, *13*, 28–34.
13. Wu, J.; Gu, S.; Chen, J.; Zhu, Z. Current Status and Countermeasures of *Phyllostachys pubescens* Forest Management in Huzhou, Zhejiang. *World Bamboo Ratt.* **2018**, *13*, 24–26.
14. Bai, S.; Zhou, G.; Wang, Y.; Liang, Q.; Chen, J.; Cheng, Y.; Shen, R. Plant species diversity and dynamics in forests invaded by Moso bamboo (*Phyllostachys edulis*) in Tianmu Mountain Nature Reserve. *Biodivers. Sci.* **2013**, *21*, 288–295.
15. Li, Y.; Liang, X.; Li, Y.; Wang, Q.; Chen, J.; Xu, Q. Effects of Phyllostachys edulis invasion of native broadleaf forest on soil fungal community. *Chin. J. Appl. Ecol.* **2016**, *27*, 585–592.
16. Umemura, M.; Takenaka, C. Changes in chemical characteristics of surface soils in hinoki cypress (*Chamaecyparis obtusa*) forests induced by the invasion of exotic Moso bamboo (*Phyllostachys pubescens*) in central Japan. *Plant Species Biol.* **2015**, *30*, 72–79. [CrossRef]
17. Chang, E.H.; Chiu, C.Y. Changes in soil microbial community structure and activity in a cedar plantation invaded by moso bamboo. *Appl. Soil Ecol.* **2015**, *91*, 1–7. [CrossRef]
18. Liebsch, D.; Velazco, S.J.E.; Mikich, S.B.; Marques, M.C.; Galvao, F. Effects of selective logging, fragmentation, and dominance of bamboos on the structure and diversity of Araucaria Forest fragments. *For. Ecol. Manag.* **2021**, *487*, 118961. [CrossRef]
19. Canavan, S.; Kumschick, S.; Le Roux, J.J.; Richardson, D.M.; Wilson, J.R. Does origin determine environmental impacts? Not for bamboos. *Plants People Planet* **2019**, *1*, 119–128. [CrossRef]
20. Xu, Q.F.; Jiang, P.K.; Wu, J.S.; Zhou, G.M.; Shen, R.F.; Fuhrmann, J.J. Bamboo invasion of native broadleaf forest modified soil microbial communities and diversity. *Biol. Invasions* **2015**, *17*, 433–444. [CrossRef]
21. Touyama, Y.; Yamamoto, T.; Nakagoshi, N. Myrmecofaunal change with bamboo invasion into broadleaf forests. *J. For. Res.* **1998**, *3*, 155–159. [CrossRef]
22. Chen, X.; Liu, Z.X.; Xu, J.J.; Qi, X.W.; Yu, S.Q. Restoration characteristics of community after clear-cutting of *Phyllostachys edulis* stands in Mount Tianmu. *J. Zhejiang A F Univ.* **2022**, *39*, 705–716.
23. Long, K.; Yin, R.; Kardol, P.; Wei, Q.; Li, Y.; Huang, J. Bamboo invasion alters Collembola community composition varying with life-forms. *Pest Manag. Sci.* **2023**, *79*, 2517–2526. [CrossRef] [PubMed]

24. Yihe, L.; Bojie, F. Ecological scale and scaling. *Acta Ecol. Sin.* **2001**, *21*, 2096–2105.
25. Liu, C.; Zheng, C.; Wang, L.; Zhang, J.; Wang, Q.; Shao, S.; Chen, J. Moso bamboo invasion changes the assembly process and interactive relationship of soil microbial communities in a subtropical broadleaf forest. *For. Ecol. Manag.* **2023**, *536*, 120901. [CrossRef]
26. Chen, X.; Chen, X.; Huang, S.; Fang, D. Impacts of Moso bamboo (*Phyllostachys pubescens*) invasion on species diversity and aboveground biomass of secondary coniferous and broad-leaved mixed forest. *Front. Plant Sci.* **2022**, *13*, 1001785. [CrossRef] [PubMed]
27. Chen, Z.; Li, Y.; Chang, S.X.; Xu, Q.; Li, Y.; Ma, Z.; Cai, Y. Linking enhanced soil nitrogen mineralization to increased fungal decomposition capacity with Moso bamboo invasion of broadleaf forests. *Sci. Total Environ.* **2021**, *771*, 144779. [CrossRef]
28. Suzuki, S. Chronological location analyses of giant bamboo (*Phyllostachys pubescens*) groves and their invasive expansion in a satoyama landscape area, western Japan. *Plant Species Biol.* **2015**, *30*, 63–71. [CrossRef]
29. Isagi, Y.; Torii, A. Range expansion and its mechanisms in a naturalized bamboo species, *Phyllostachys pubescens*, in Japan. *J. Sustain. For.* **1997**, *6*, 127–141. [CrossRef]
30. Takano, K.T.; Hibino, K.; Numata, A.; Oguro, M.; Aiba, M.; Shiogama, H.; Nakashizuka, T. Detecting latitudinal and altitudinal expansion of invasive bamboo *Phyllostachys edulis* and *Phyllostachys bambusoides (Poaceae)* in Japan to project potential habitats under 1.5 °C–4.0 °C global warming. *Ecol. Evol.* **2017**, *7*, 9848–9859. [CrossRef]
31. Someya, T.; Takemura, S.; Miyamoto, S.; Kamada, M. Predictions of bamboo forest distribution and associated environmental factors using natural environmental information GIS and digital national land information in Japan. *Landsc. Ecol.* **2010**, *15*, 41–54.
32. Yang, Q.P.; Yang, G.Y.; Song, Q.N.; Shi, J.M.; Ouyang, M.; Qi, H.Y.; Fang, X.M. Ecological studies on bamboo expansion: Process, consequence and mechanism. *Chin. J. Plant Ecol.* **2015**, *39*, 110–124.
33. Kobayashi, K.; Kitayama, K.; Onoda, Y. A simple method to estimate the rate of the bamboo expansion based on one-time measurement of spatial distribution of culms. *Ecol. Res.* **2018**, *33*, 1137–1143. [CrossRef]
34. Suzuki, S.; Nakagoshi, N. Expansion of bamboo forests caused by reduced bamboo-shoot harvest under different natural and artificial conditions. *Ecol. Res.* **2008**, *23*, 641–647. [CrossRef]
35. Smith, M.; Nelson, B.W. Fire favours expansion of bamboo-dominated forests in the south-west Amazon. *J. Trop. Ecol.* **2011**, *27*, 59–64. [CrossRef]
36. Tomimatsu, H.; Yamagishi, H.; Tanaka, I.; Sato, M.; Kondo, R.; Konno, Y. Consequences of forest fragmentation in an understory plant community: Extensive range expansion of native dwarf bamboo. *Plant Species Biol.* **2011**, *26*, 3–12. [CrossRef]
37. Dong, C.L. Study on growth effect on new stands of *Phyllostachys heterocycla cv. Pubescens* through rhizome expanding and growing. *J. Bamboo Res.* **2003**, *22*, 30–33.
38. Liang, C.; Ruilin, Z.; Chunfu, L.I.; Yu, D. A method to inhibit the expansion of *Phyllostachys pubescens* stands based on the analysis of underground rhizome. *J. Northeast. For. Univ.* **2003**, *31*, 68–70.
39. Tong, R.; Zhou, B.; Jiang, L.; Cao, Y.; Ge, X.; Yang, Z. Influence of Moso bamboo invasion on forest plants and soil: A review. *Acta Ecol. Sin.* **2019**, *39*, 3808–3815.
40. Li, L.; Li, N.; Lu, D.; Chen, Y. Mapping Moso bamboo forest and its on-year and off-year distribution in a subtropical region using time-series Sentinel-2 and Landsat 8 data. *Remote Sens. Environ.* **2019**, *231*, 111265. [CrossRef]
41. Ji, J.; Li, X.; Du, H.; Mao, F.; Fan, W.; Xu, Y.; Kang, F. Multiscale leaf area index assimilation for moso bamboo forest based on Sentinel-2 and MODIS data. *Int. J. Appl. Earth Obs. Geoinf.* **2021**, *104*, 102519. [CrossRef]
42. Li, S.; He, S.; Yue, T.; Du, Z.; Zhao, N.; Zhao, Y.; Yang, Y. Mapping bamboo forest and expansion intensity in China by coupling vegetation phenology and C-band SAR with Sentinel-1 and Sentinel-2 images. *Int. J. Appl. Earth Obs. Geoinf.* **2023**, *121*, 103384. [CrossRef]
43. Du, H.; Mao, F.; Li, X.; Zhou, G.; Xu, X.; Han, N.; Zhou, Y. Mapping global bamboo forest distribution using multisource remote sensing data. *IEEE J. Sel. Top. Appl. Earth Obs. Remote Sens.* **2018**, *11*, 1458–1471. [CrossRef]
44. National Forestry and Grassland Science Data Center (NFGSDC). Available online: https://www.forestdata.cn/dataDetail.html?id=CSTR:17575.11.0120211215006.010001.V1 (accessed on 15 December 2018).
45. *GBT 26424-2010*; Technical Regulations for Inventory for Forest Management Planning and Design. National Technical Committee on Forest Resources Standardization: Beijing, China, 2010.
46. Hou, X.Y. Geographic Distribution Regularities of Vegetation in China. *Acta Bot. Boreali-Occident. Sin.* **1981**, *1*, 3–13.
47. National Forestry and Grassland Administration. Extraordinary Ten Years to Build a Dream of Verdant Bamboo and Beautiful Landscapes. Available online: https://www.forestry.gov.cn/main/2614/20220930/155528310422363.html (accessed on 29 September 2022).
48. Yu, L.; Wei, J.; Li, D.; Zhong, Y.; Zhang, Z. Explaining landscape levels and drivers of Chinese Moso bamboo forests based on the plus model. *Forests* **2023**, *14*, 397. [CrossRef]
49. You, C. The puzzling explanation of structural relative numbers in the application of structural analysis. *J. Jishou Univ. (Nat. Sci. Ed.)* **2015**, *1*, 19–20+27.
50. Chen, J.Y.; Luo, H.X.; Hu, L.L.; Li, Q.H.; Jiang, W.J. Correlation between social economic development and ecosystem service value–A Case of Chongqing metropolitan area. *Res. Soil Water Conserv.* **2017**, *24*, 330–336. Available online: https://www.cnki.net/kcms/doi/10.13869/j.cnki.rswc.2017.04.052.html (accessed on 30 June 2024).

51. Yao, X.; Zeng, J.; Li, W. Spatial correlation characteristics of urbanization and land ecosystem service value in Wuhan Urban Agglomeration. *Trans. Chin. Soc. Agric. Eng.* **2015**, *31*, 249–256. [CrossRef]
52. Li, J.; He, J.; Liu, Y.; Wang, D.; Rafay, L.; Chen, C.; Lin, Y. Spatial autocorrelation analysis of multi-scale damaged vegetation in the Wenchuan earthquake-affected area, Southwest China. *Forests* **2019**, *10*, 195. [CrossRef]
53. Li, S.; Xie, D.; Ge, X.; Dong, W.; Luan, J. Altered diversity and functioning of soil and root-associated microbiomes by an invasive native plant. *Plant Soil* **2022**, *473*, 235–249. [CrossRef]
54. Bai, S.; Wang, Y.; Conant, R.T.; Zhou, G.; Xu, Y.; Wang, N.; Chen, J. Can native clonal moso bamboo encroach on adjacent natural forest without human intervention? *Sci. Rep.* **2016**, *6*, 31504. [CrossRef]
55. Wang, Z.; He, G.; Hou, Z.; Luo, Z.; Chen, S.; Lu, J.; Zhao, J. Soil C:N:P stoichiometry of typical coniferous *(Cunninghamia lanceolata)* and/or evergreen broadleaved *(Phoebe bournei)* plantations in south China. *For. Ecol. Manag.* **2021**, *486*, 118974. [CrossRef]
56. Ge, Z.; Huang, S.; Ouyang, M.; Luan, F.; Fang, X.; Yang, Q.; Song, Q. Stand Characteristics Rather than Soil Properties Contribute More to the Expansion of Moso Bamboo *(Phyllostachys edulis)* into Its Neighboring Forests in Subtropical Region. *Forests* **2022**, *13*, 2159. [CrossRef]
57. Qin, L.; Yu, S.X. Complexity of forest communities: A case study of three different forest types in Heishiding Nature Reserve, Guangdong. *Biodivers. Sci.* **2004**, *3*, 354–360.
58. Wang, X.Y.; Song, X.Z.; Pang, C.M.; Yang, H.; Zhao, M.S.; Xu, L. Effects of Structure and Site Condition of Neighbor Forests on Natural Expansion of *Phyllostachys edulis* Forest. *J. Zhejiang For. Sci. Technol.* **2019**, *39*, 19–24.
59. Cai, X.; Gao, T.; Zheng, S.; Jiang, R.; Zhang, Y.; Rong, J.; He, T.S.; Chen, L.; Zheng, Y. Effects of Enclosure Succession on the Morphological Characteristics and Nutrient Content of a Bamboo Whip System in a Moso Bamboo *(Phyllostachys edulis)* Forest on Wuyi Mountain, China. *Forests* **2023**, *14*, 2193. [CrossRef]
60. Wu, Y.; Guo, J.; Tang, Z.; Wang, T.; Li, W.; Wang, X.; Cui, H.; Hu, X.; Qi, L. Moso bamboo *(Phyllostachys edulis)* expansion enhances soil pH and alters soil nutrients and microbial communities. *Sci. Total. Environ.* **2024**, *912*, 169346. [CrossRef]
61. Ding, L.X.; Wang, Z.L.; Zhou, G.M.; Du, Q.Z. Monitoring *Phyllostachys pubescens* stands expansion in National Nature Reserve of Mount Tianmu by remote sensing. *J. Zhejiang A F Univ.* **2006**, *3*, 297–300.

forests

MDPI

Article

Modeling the Effects of Spatial Distribution on Dynamics of an Invading *Melaleuca quinquenervia* (Cav.) Blake Population

Yuanming Lu [1], Junfei Xia [2], Robert D. Holt [1] and Donald L. DeAngelis [3,*]

[1] Department of Biology, University of Florida, Gainesville, FL 32611, USA; yuanming.lu@ufl.edu (Y.L.); rdholt@ufl.edu (R.D.H.)
[2] Department of Ocean Sciences, Rosenstiel School of Marine and Atmospheric Science, University of Miami, Miami, FL 33149, USA; junfei.xia@earth.miami.edu
[3] U. S. Geological Survey, Wetland and Aquatic Research Center, Davie, FL 33314, USA
* Correspondence: don_deangelis@usgs.gov

Abstract: To predict the potential success of an invading non-native species, it is important to understand its dynamics and interactions with native species in the early stages of its invasion. In spatially implicit models, mathematical stability criteria are commonly used to predict whether an invading population grows in number in an early time period. But spatial context is important for real invasions as an invading population may first occur as a small number of individuals scatter spatially. The invasion dynamics are therefore not describable in terms of population level state variables. A better approach is spatially explicit individual-based modeling (IBM). We use an established spatially explicit IBM to predict the invasion of the non-native tree, *Melaleuca quinquenervia* (Cav.) Blake, to a native community in southern Florida. We show that the initial spatial distribution, both the spatial density of individuals and the area they cover, affects its success in growing numerically and spreading. The formation of a cluster of a sufficient number and density of individuals may be needed for the invader to locally outcompete the native species and become established. Different initial densities, identical in number and density but differing in random positions of individuals, can produce very different trajectories of the invading population through time, even affecting invasion success and failure.

Keywords: plant competition; *Melaleuca quinquenervia*; spatially explicit; agent-based model; initial condition sensitivity; spatial distributions

1. Introduction

The study of invasive species has been one of the major themes in ecology for the practical reason that invasive species pose a threat to native vegetation and to agricultural systems around the world. In attempting to estimate the risk of a non-native species invading a new environment, several factors need to be considered. One factor is how the climatic conditions of a non-native species compare with those of a potential new environment. Species distribution models, also called ecological niche models, are often used to estimate the suitability of the new environment (e.g., Elith and Leathwick 2009 [1]). However, predicting invasive success goes beyond determining suitable climate conditions and should take into consideration other traits of the potential invasive species as well as of the native community to which it has been introduced. These traits include its maximum growth rate, potential dispersal rate, competitive ability, and vulnerability to natural enemies.

Analytic models have been used to take into account some of these factors. Invasion criteria based on mathematical approaches are fundamental tools in quantitative ecology for predicting the dynamics of a species population entering an equilibrium community. Suppose the competitive interactions of populations of two species are described by the equations for biomasses of native species 1, N_1, and invasive species 2, N_2, where space can be considered one-dimensional and Lotka–Volterra competition occurs. The equations can be written as

$$\frac{dN_1}{dt} = r_1(1 - c_{11}N_1 - c_{12}N_2)N_1 + D_1\frac{\partial^2 N_1}{\partial x^2} \tag{1a}$$

$$\frac{dN_2}{dt} = r_2(1 - c_{21}N_1 - c_{22}N_2)N_2 + D_2\frac{\partial^2 N_2}{\partial x^2} \tag{1b}$$

where x is the spatial distance, r_1 and r_2 are the maximum growth rates, c_{11} and c_{22} are the intraspecific competition coefficients, c_{12} and c_{21} are the interspecific competition coefficients, and D_1 and D_2 are the dispersal rates. If the equilibrium $E(N_1^*, 0)$ exists where N_1^* is the equilibrium size of species 1 and the invader biomass N_2 is initially very small, the criterion for the invasion of species 2 can be found by letting N_2 have a small value, which is represented by n_2:

$$\frac{dn_2}{dt} = r_2(1 - c_{21}N_1^*)n_2 + D_2\frac{\partial^2 n_2}{\partial x^2} \tag{2}$$

Suppose the small biomass of invading species 2 is spread uniformly over a small spatial interval of length L. Then, analytically, it can be shown that the population of species 2 can grow if

$$r_2(1 - c_{21}N_1^*) \geq \pi^2 D_2/L^2, \tag{3}$$

based on Kierstead and Slobodkin's study (1953) [2]. More sophisticated analytic models can include such aspects as spatially heterogeneous environments [3], integro-difference equations [4,5], and the stochasticity of dispersal [6].

These analytic results are interesting as broad indicators of invasion potential. But the parameters here are only gross population-level demographic characteristics of the species and do not describe how invading individuals interact with those of the native species. Simple analytic models also do not consider detailed site-specific aspects of the new environment such as spatial heterogeneity. For these reasons, individual-based models (IBMs), also called agent-based models (ABMs), have been employed to examine the effects of factors at the individual organism level. These allow models to make use of empirical data, such as those on giant hogweed (*Heracleum mantegazzianum* Sommier & Levier) [7], Russian knapweed (*Acroptilon repens* L.) [8], reed canary grass (*Phalaris arundinacea* L)) [9], paperbark tea tree (*Melaleuca quinquenervia* (Cav.) Blake) [10,11], and monocarpic thistle (*Carduus nutans* L.) [5]. Beyond their application to specific species, IBMs also allow the investigation of general types of mechanisms that might apply to a large number of potential invasive species, e.g., [12–15].

Among the general mechanisms that are known to be important in plant competition are the effects of ecological engineering, that is, one or more species having an effect on the environment that affects other species. One such component of ecological engineering is the litter produced by plants, which can affect competitive relationships, e.g., [9,16]. Leaf litter, depending on its nutrient contents and decomposition rate, can influence soil nutrient dynamics [17]. The effects of invasive *Phalaris arundinacea* on nutrient cycling in North American wetlands was modeled by [9]. Also, especially if litter accumulation can become great due to slow decomposition rates, litter can have a negative effect on seedling survival. This can be important in competition if the accumulation of litter of one species affects the survival of seedlings of the other species. The consequences of the litter suppression of seedlings were modeled by [17] for the invasion of stiltgrass (*Microstegium vimineum* (Trin.) A. Camus) in eastern and midwestern deciduous forests in the United States. High litter accumulation depends not only on litter decomposition rates but also on the number and local spatial densities of individuals of the species producing the litter. Keeping track of local densities requires a spatially explicit IBM, which can keep track of the spatial locations of all individuals. This is important as the invasion of a new species of an existing plant community can occur in the form of various initial spatial distributions of propagules.

These facts are relevant to our modeling of the invasion of southern Florida ecosystems by the paperbark tea tree (*Melaleuca quiquenervia*) [10,11] (*Melaleuca* hereafter). Our previous modeling has shown that the success of a simulated non-native *Melaleuca* invading a native community can depend not only its relative competitive abilities but also on the way the initial invaders are scattered in the native community, as well as their dispersal rates [10,11]. However, our earlier work did not systematically investigate the impact of different initial spatial starting points. Our objective here is to study the effect of the initial spatial configuration of the invading population systematically to obtain a better understanding of the factors contributing to invasion. This approach allows us to explore how variations in the initial spatial distribution influence invasion dynamics, providing a more detailed and realistic prediction of invasion outcomes.

We hypothesize that the initial spatial configuration of the invader is an important factor in determining its ability to spread, along with the above mathematical criteria, which can be tested with modeling. Our experience also shows that random differences in the initial spatial configuration of individuals with the same mean properties can lead to very different population trajectories. Thus, we also hypothesize that the outcomes for the initial conditions with the same average properties can vary dramatically.

To study both the hypothesis that different spatial configurations of the invading individuals lead to different invasion successes and the finding that random initial differences can also lead to different outcomes, we used a model called ManHam. ManHam was developed and tested for describing the invasion of the non-native tree species *Melaleuca quinquenervia* in a native community of southern Florida [10,11]. The ManHam model was developed to address the complex challenge of invasive species management and utilizes agent-based modeling to explore effective biocontrol strategies. Our approach was informed by insights into the dynamics of the invasive species, from Australia, *Melaleuca*, which is characterized by fast growth and high seed production in the environment of southern Florida due to its virtual lack of natural enemies in its new habitat [16,18]. Another advantage of *Melaleuca* is that its leaf litter has the capacity to suppress the emergence of native seedlings, which increases its competitive ability [19,20]. *Melaleuca* initially spread rapidly through natural communities in southern Florida. For the past two decades, biological control through a number of mostly arthropod agents has been shown to be able to control *Melaleuca* [21,22]. However, it is still important to model *Melaleuca* in southern Florida's native communities to help predict what the long-term consequences of the control of the invader and the recovery of native communities will be.

2. Materials and Methods

Our modeling here uses a version of ManHam that is specially designed to study early invasion dynamics. In particular, a level of biocontrol on modeled *Melaleuca* growth rate and seedling production is imposed that makes the modeled native tree community roughly equal in competitive ability to the invading *Melaleuca*. The native community is modeled as a single 'species' to roughly represent a native hardwood hammock. Although these are somewhat artificial conditions, as biocontrol appears in reality to give an advantage to the native community, it is still useful to make this assumption to study how different initial spatial configurations of the invader affect its success. In our study, we first examine various initial spatial distributions of a relatively small number of invading pre-adult *Melaleuca* and examine how these distributions affect its early invasive dynamics. Secondly, we follow up on the results showing a high degree of variability in the simulation outcomes by determining whether this variability is simply due to the small initial number of invaders. We increase the number of invaders and examine whether this leads to more deterministic results.

In other words, the model simulates individual trees through their lifetimes from seedlings. Both age and growth in dbh are modeled. The model simulates competition between the invasive *Melaleuca* and native generic hardwood hammock species. Local competition occurs with both conspecific individuals and individuals of the other species.

The litterfall and accumulation of the invader and its suppression of native seedlings are modeled.

2.1. Model

A spatially explicit, agent-based model (ABM), ManHam, is used, as in [10,11]. The purpose of the model is to test different initial spatial configurations of immature individuals of the invasive *Melaleuca* such that the same number of individuals at different spatial configurations leads to different probabilities of success in the early phase of an invasion. The model ManHam is described in detail in [10]. Here, we just mention the important aspects of each submodel.

The model is a Monte Carlo simulator of the competition between an invasive and a native tree species on a plot of 120 × 120 m. There are two types of entities: First, there are individual trees, which are termed agents, each of which is either an invader or a native tree. Individual trees have age, diameter at breast height (dbh) and spatial location. Canopy size is allometrically related to dbh. The individual trees are distributed in continuous space; that is, they can be located at any point within the 120 × 120 m plot. The second entities are the heterogenous litter accumulations across the plot, which are kept track of at the spatial resolution of 1 × 1 m.

2.1.1. Tree Growth

Individual trees grow from seedlings to senescence. At the start of the simulation, the trees that are initiated are assigned an initial tree diameter at breast height, dbh, as well as a location in continuous space. The code iterates over all trees and computes their growth in yearly increments in a way that is similar to earlier models developed for describing woody plant growth in agent-based simulations [23]. The growth and reproduction of the *Melaleuca* in the absence of biocontrol are both higher than the rates of the model native tree species. The growth of the invader is assumed affected by biocontrol. The growth of individual invader trees is sufficiently slowed by biocontrol by 80% to be one-third of the rate of growth of native trees. Slower growth over a period of time increases the likelihood of mortality.

2.1.2. Reproduction

The possibility of reproduction is assumed to occur for each mature tree each year and, in the absence of biocontrol, is greater for the invader than for the native species. The age of maturity is set arbitrarily at 20 years. The code specifies that each tree can produce an upper limit of $N_{seedling,limit}$ viable seedlings each year, where each seedling has a probability, *brate*, to survive. Therefore, each mature tree can produce up to $N_{seedling,limit}$ seedlings each year. A low value for $N_{seedling,limit}$ is used here because of the low empirical probability of the survivorship of seedlings. The potential number of seedlings produced each year by invader trees under biocontrol is assumed to be half of that of native trees.

2.1.3. Seedling Dispersal

The distance from the parent from which an individual seedling is dispersed,

$$disper = -\ln(rand)/c_3 \tag{4}$$

is exponentially distributed, where *rand* is a number chosen randomly and uniformly on the interval (0, 1), and $1/c_3$ is the mean dispersal distance. The dispersal can be in any direction (360°) with some seedlings landing outside the plot, where they do not survive. In the version of ManHam used here, the value of the mean dispersal distance, c_3, is the same for both species. This parameter, though important, is not varied here as our main interest is the initial spatial configurations of the invading individuals.

2.1.4. Competition between Trees

Competition is described by the field of neighborhood (FON) model [24], which is used to describe the effect of close neighbors on the growth of each other. All neighbors within 5 m of each tree are determined at each point in time and put into a matrix called $Neib_i$ so that their effects on annual diameter growth through FON interactions can be determined. This is carried out using the data on neighbors within 5 m of each tree that are stored in matrix *Neib*.

2.1.5. Litterfall and Its Accumulation

The leaf litterfall and accumulation of both the invader and native trees are simulated but only that of the invader is used in the simulations. *Melaleuca* leaf litter decomposes slowly and so that it can accumulate to depths such that it suppresses the emergence of native seedlings, though with less effect on *Melaleuca*'s own seedlings. Cumulative depth is calculated through time. The potential suppression of native seedlings by litter of the invader, *Melaleuca*, is based on the amount of leaf litter $LF_{accumulation}$, in the given 1 × 1 m area in which the seedling has landed. The probability of the seedling surviving is reduced according to the following:

$$Seedling\ survival\ suppression = e^{-LF_{accumulation} * s_{litter}}, \tag{5}$$

where s_{litter} is a parameter that is constant and that measures the effect of the litter of the invader on the seedlings of the native species. The value of s_{litter} is not precisely known, and a plausible value is used here, which, in combination with biocontrol on the invader, leads to the two species being competitively similar. The parameters are estimated for *Melaleuca* based on [10].

2.1.6. Tree Mortality

There are three components of the net tree mortality. First, there is background mortality that is size-independent. Second, there is size-dependent morality that decreases with size, and third, there is density-dependent mortality. The last component is implemented by increasing the probability of mortality of a tree whose growth rate has decreased over time due to crowding. Parameters for the first two mortality sources are the same for the two species. However, based on the empirical data, it is assumed that the invader can tolerate greater crowding so that crowding-related mortality affects the native species within two meters of the stem, but it affects the invader only within one and one-half meters. Finally, all trees are assumed to have a maximum age. Details can be found in [10]. Biocontrol is assumed to affect only the invader. Insect herbivores were not simulated explicitly; a constant effect level of herbivory on each invasive tree was assumed.

2.2. *Parameter Values*

Most of the parameter values used are the same as those used for growth, reproduction, mortality, and the effects of biocontrol on growth and reproduction in [10]. Here, only the specific values noted in the model description above are given (see Table 1). In [10], the parameters of seedling dispersal, c_3, and litter suppression s_{litter} were varied for different simulations in that paper. Here, the specific values of these parameters were prescribed as those values were set such that the invasive and native trees had roughly similar competitive abilities so that the effects of spatial distribution of the invading *Melaleuca* could be examined.

The parameter values are in the ranges used by [10,11]. The non-native species was assumed to have high dispersal rates (c_3 = 0.15), and its litter had a negative suppressive effect on native seedlings (s_{litter} = 10). The advantages of the non-native species were balanced by the assumption that biocontrol reduced non-native tree growth by about two-thirds and reduced non-native seedling survival by 90%.

Table 1. Parameter values for reproduction, seedling dispersal, and litter suppression of seedlings.

Parameter	Definition	Value	Units
$brate_{invader}$	Parameter for seedling establishment	0.07	dim'less
$brate_{native}$	"	0.75	"
$N_{seedling,limit}$	Upper limit on number of viable seedlings from a single tree	2	Number
$c_{3,invader}$	Parameter related to dispersal	0.05	m^{-1}
$c_{3,native}$	"	0.05	m^{-1}
$s_{litter,invasive\ on\ invader}$	Parameter related to seedling suppression of invader	1	$1/(kg\ m^{-2})$
$s_{litter,invasive\ on\ native}$	Parameter related to seedling suppression of native	10	$1/(kg\ m^{-2})$

2.3. Simulations

Two sets of simulations are performed. In the first set of simulations, a relatively small number of invaders is established along one edge of the plot. The results of these simulations motivated a second set of simulations, in which the initial number of invaders was greatly increased.

In the first set of simulations, each simulation is started with 3500 native trees assumingly randomly distributed in the 120 × 120 m plot, with a uniform random distribution of ages between 0 and 100. A typical initial distribution is shown in Figure 1. There are an initial 200 invading *Melaleuca* saplings between 0 and 20 years. Because the age of maturity is 20, all invading individuals are initially immature.

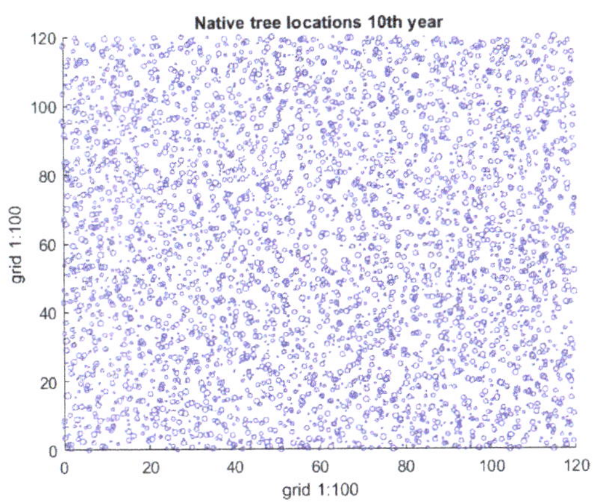

Figure 1. Typical spatial distribution of residents (native trees) on a 120 × 120 m area. The size of the circle represents initial dbh.

The invading *Melaleuca* saplings are initiated according to six different cases. The trees are randomly located in each case between 0 and 120 along the y-axis but between (case 1) 0 and 5 m, (case 2) 0 and 10 m, (case 3) 0 and 15, (case 4) 0 and 20 m, (case 5) 0 and 30 m, and (case 6) 0 and 40 m along the x-axis for the respective cases (Figure 2). That is, the invading species is assumed to be invading along the front from the left-hand side of the plot.

In the second set of simulations, the invading species is assumed to initially cover the whole left half of the plot. They are assumed to have been artificially kept from the possibility of invading the right-hand side of the plot until the start of the simulation. All simulations were performed using Matlab R(2022)a, (MathWorks).

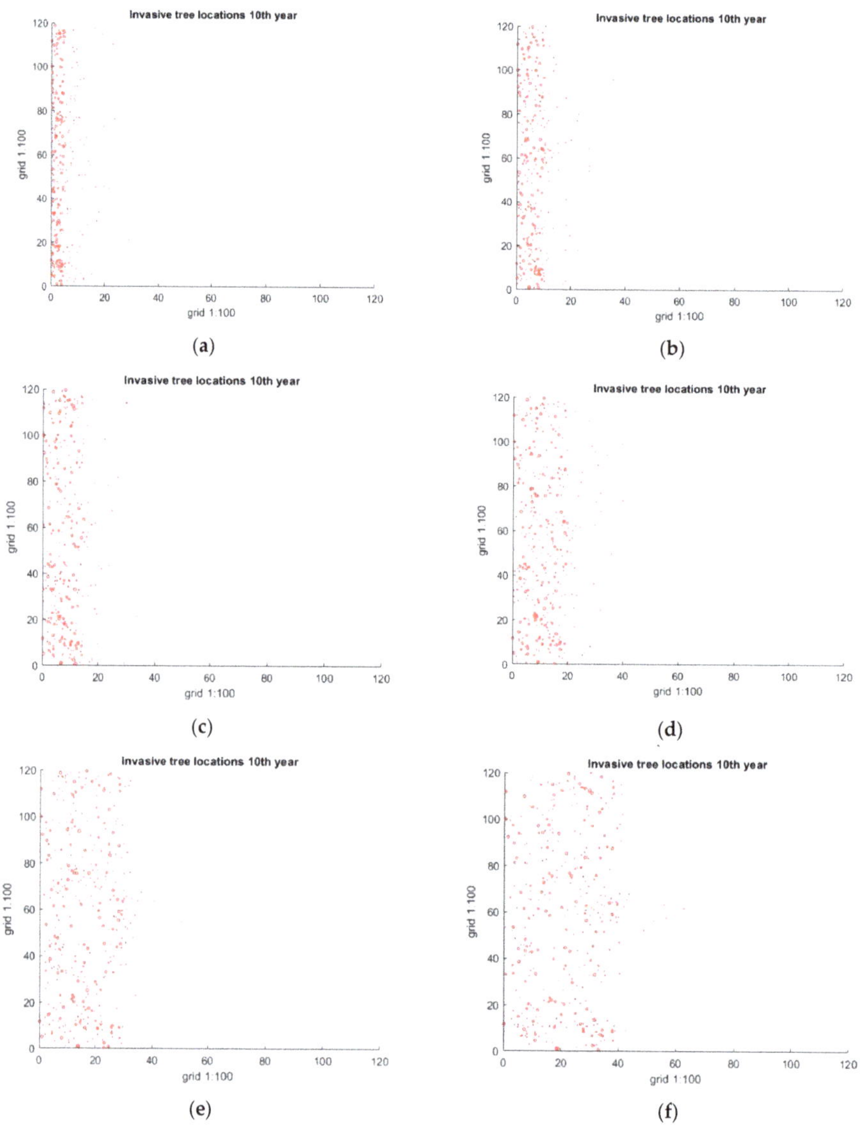

Figure 2. Typical initial distributions of 200 non-native invading saplings of ages between 0 and 15 years in six cases respectively. In the six cases, the individuals are spread within (**a**) case 1 (0–5 m), (**b**) case 2 (0–10 m), (**c**) case 3 (0–15 m), (**d**) case 4 (0–20 m), (**e**) case 5 (0–30 m), (**f**) case 6 (0–40 m) from the left-hand edge of the plot and from 0 to 120 m along the y-axis.

3. Results

3.1. Small Initial Invasions

The resident native trees have typical distributions, as shown in Figure 1. The invading *Melaleuca* trees have six initial distributions similar to those shown in Figure 2. For each of the six initial configurations, or cases, seven simulations were performed with different random number initiators so that the specific spatial locations and pre-mature ages differed. The results of Monte Carlo simulations are shown in Figure 3. Within each case, with each simulation performed with different random number initiators, the trajectories of the invading species differ. In fact, in five of the six cases, at least one extinction occurred. But there are clear differences between the six cases. In case 3 (0 to 15 m), the trajectories are all upward and tightly packed. In case 5 (0 to 30 m) and case 6 (0 to 40 m), the trajectories are more scattered, and four simulations lead to extinction in case 6. Cases 1, 2, and 4 are similar in that each has one extinction, and the trajectories show only a moderate degree of cohesion.

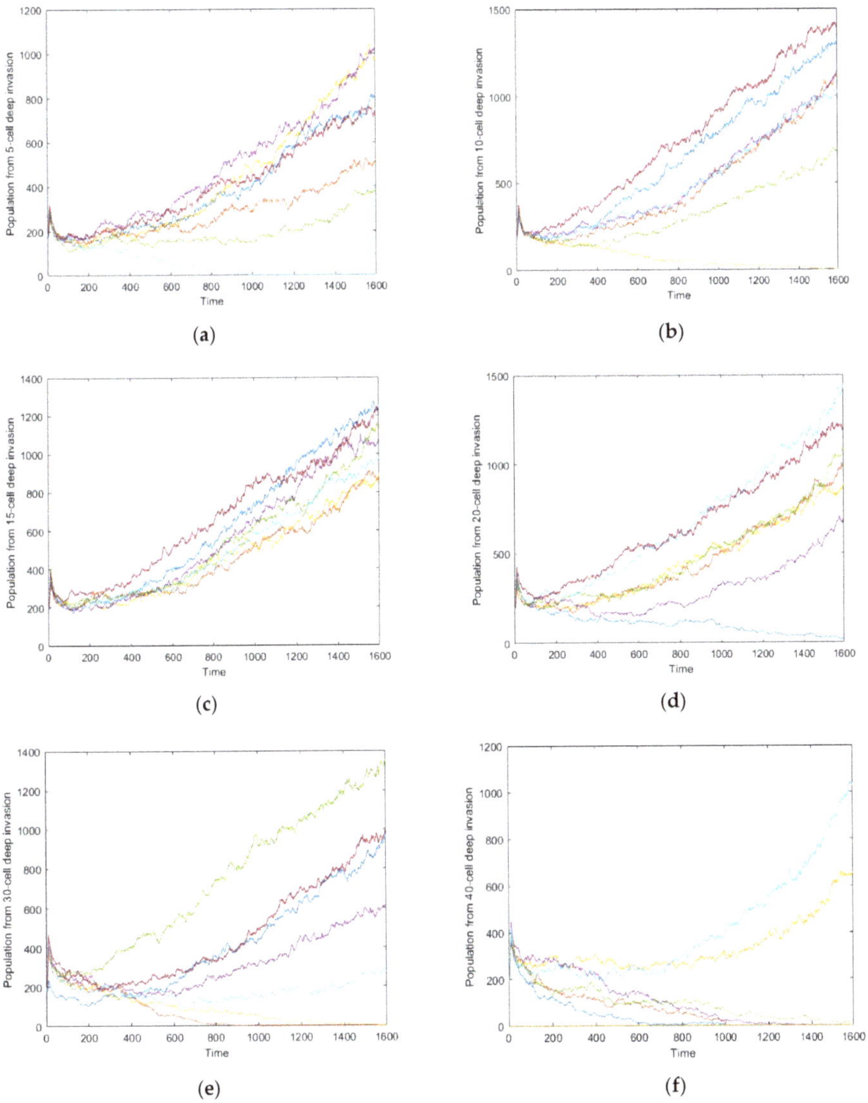

Figure 3. Results of six cases for different initial spatial invading species distribution, with six simulations each. Examples of the starting conditions are shown in Figures 1 and 2. All 200 invading (non-native) trees distributed as follows: (**a**) case 1 (0–5 m), (**b**) case 2 (0–10 m), (**c**) case 3 (0–15 m), (**d**) case 4 (0–20 m), (**e**) case 5 (0–30 m), (**f**) case 6 (0–40 m) along the left edge and across the y-axis from 0 to 120.

By examining some of the spatial patterns of individual invaders along these trajectories, one can obtain an idea of how the spatial dynamics can lead to successful invasions. Two examples are shown, one trajectory from case 1, in which the invaders were all initially within 5 m of the left edge of the plot, and one trajectory from case 6, in which all invaders were distributed within 40 m of the left edge of the plot. Ten snapshots are shown for each case, from year 10 to year 1600.

In case 1, although the individuals are first configured along a narrow line, by year 600, they start to rearrange into a tight semi-circular cluster, which then advances towards the center of the plot (Figure 4). In case 6, the initially diffused distribution of 200 individuals over $40 \times 120 = 4400$ m^2 begins to form two round clusters. These join together to form a single cluster by year 1400, which by then is rapidly advancing (Figure 5).

Each of the six panels of Figure 3 contains seven trajectories that have the same parameter values but different random number initiators. In only one of the cases (case 3 for individuals initially between 0 and 15 m) are all trajectories relatively similar, all showing clear invasion success. In the other five cases, one or more failures to invade occurred and the trajectories of the successful invasions were generally scattered. These results show that random differences in the initial configuration of individuals can lead to divergent results for the early stages of an invasive process but that a particular initial condition, case 3, leads to fairly coherent set of trajectories for different starting conditions. These results motivated the second set of simulations below.

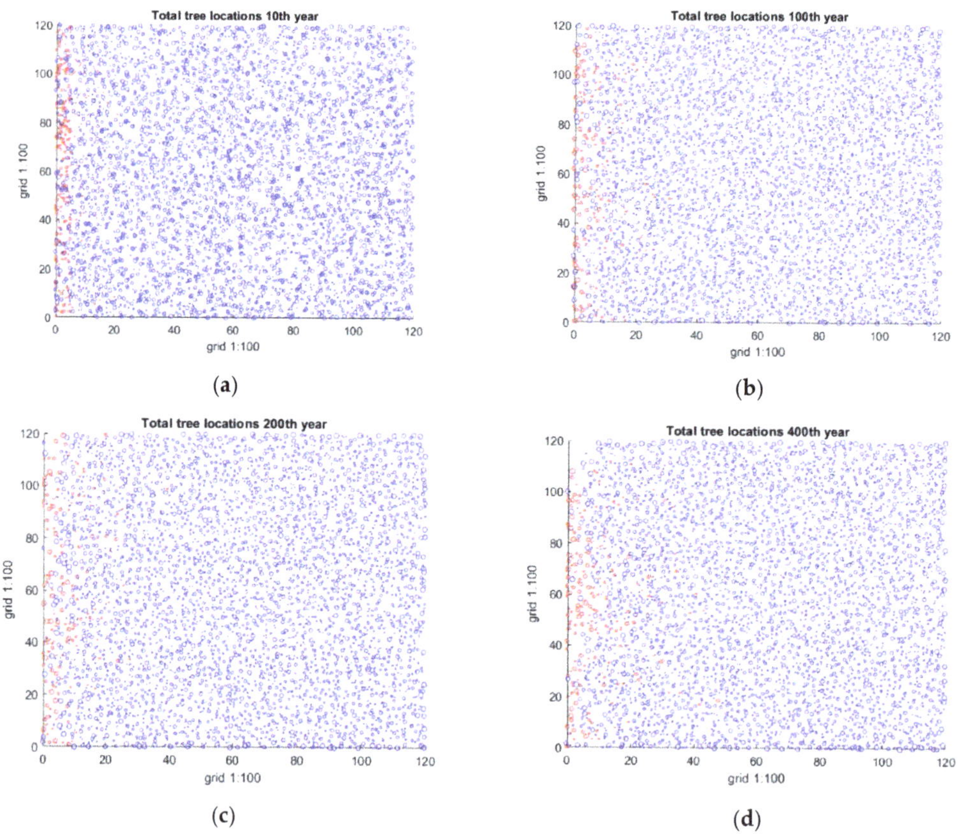

Figure 4. *Cont.*

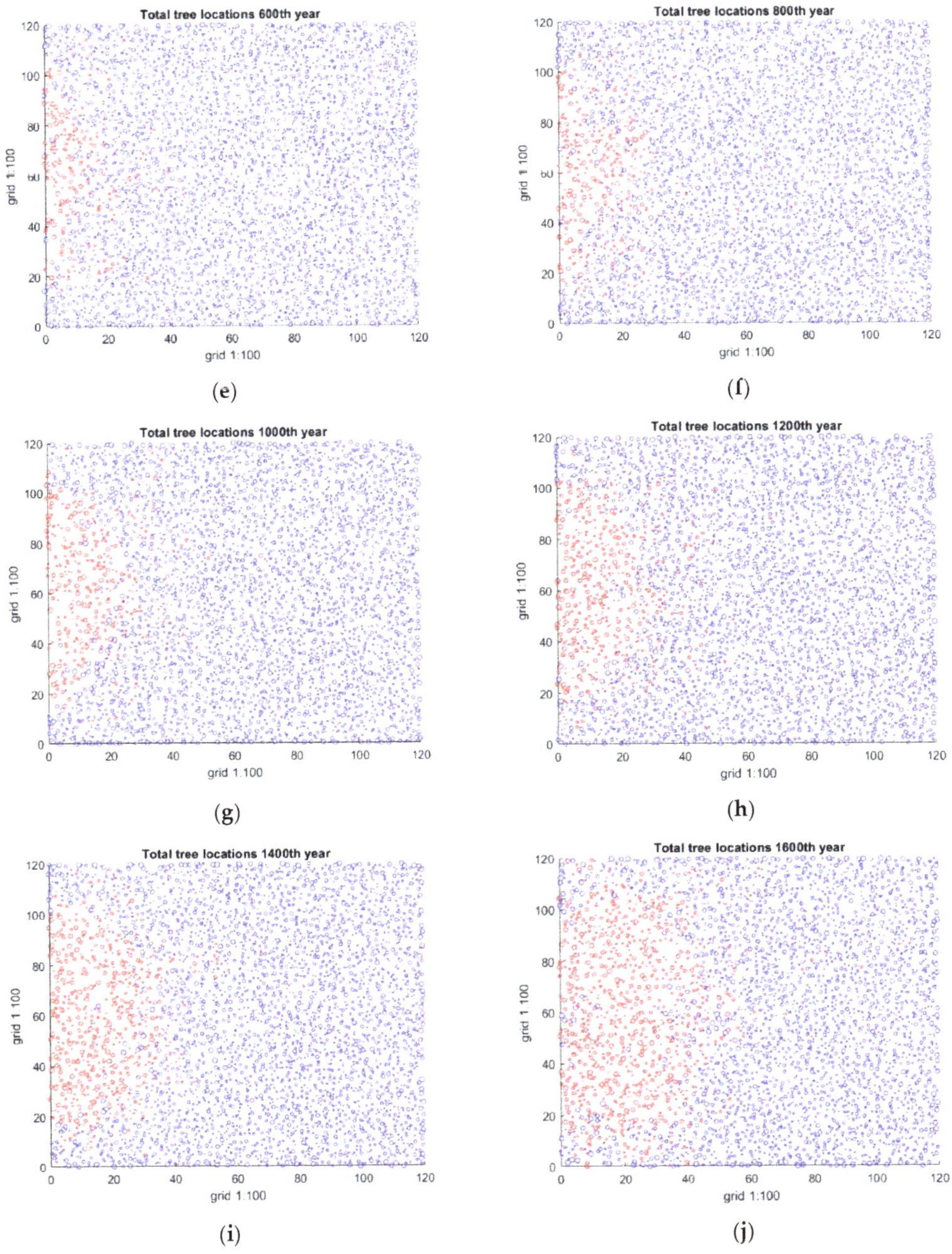

Figure 4. Simulation for case 1 (individuals configured along a line from 0 to 5 m) for a given initial random number initiator. The snapshots are at (**a**) 10 years, (**b**) 100 years, (**c**) 200 years, (**d**) 400 years, (**e**) 600 years, (**f**) 800 years, (**g**) 1000 years, (**h**) 1200 years, (**i**) 1400 years, (**j**) 1600 years. Red circles are invaders and blue circles natives. The figures show that the individuals are able to eventually form a convex cluster, within which there are almost no natives as the invaders are able to build up a thick layer of litter that suppresses native seedlings.

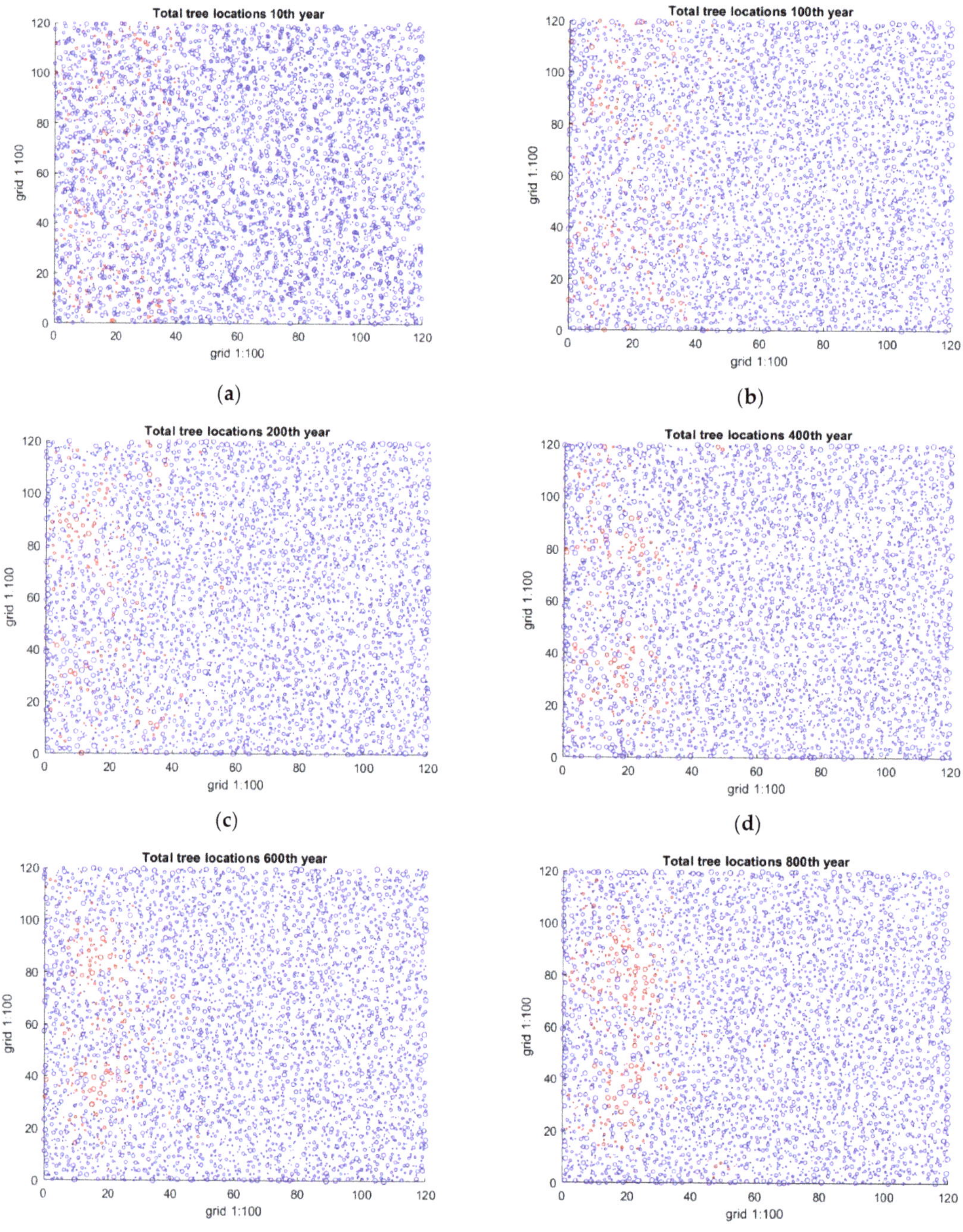

Figure 5. *Cont.*

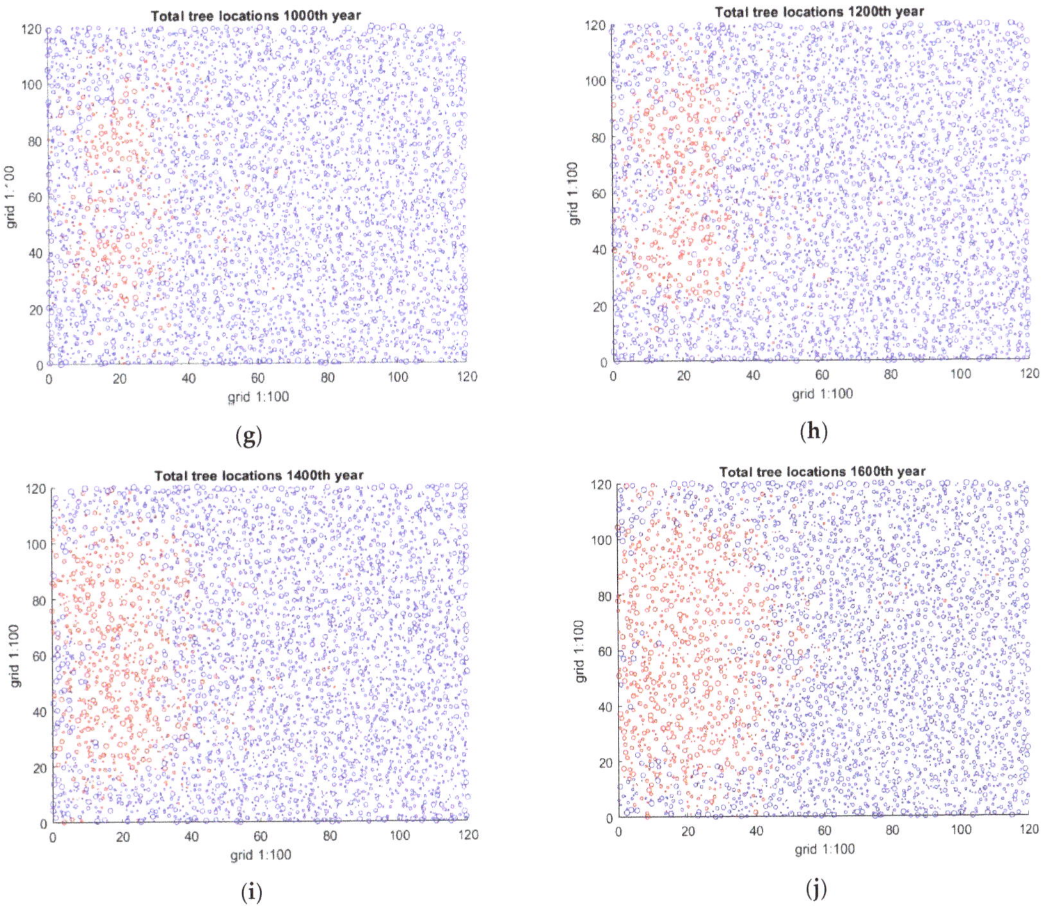

Figure 5. Simulation for case 6 (individuals configured along a line from 0 to 40 m) for a given initial random number initiator. The snapshots are at (**a**) 10 years, (**b**) 100 years, (**c**) 200 years, (**d**) 400 years, (**e**) 600 years, (**f**) 800 years, (**g**) 1000 years, (**h**) 1200 years, (**i**) 1400 years, (**j**) 1600 years. Red circles are invaders and blue circles are natives. In this case, the initially diffuse individuals are able to form into two roughly circular clusters. These are able to spread and then join to form a single large convex cluster that expands, as in case 1.

3.2. Invader Has Equal Initial Numbers to Those of the Native Population

The high degree of randomness in trajectories within each of the six cases shown above in Figure 3, which results only from the random rearrangements of individuals within the given initial areas of the invaders, is interesting as it suggests a degree of uncertainty in making predictions. However, one might argue that the small size (200 individuals) of the initial number of invaders made possible substantial differences in the initial configurations. Therefore, we repeated these invasion experiments starting with much larger numbers of invaders. Typical initial conditions are shown in Figure 6, in which 2000 individuals each of the invader and native populations initially occupy opposite sides of the plot and where the invading species is prevented from invading the right-hand side of the plot until the start of the simulation. Differences in circle size, representing different ages, are not considered in Figure 6.

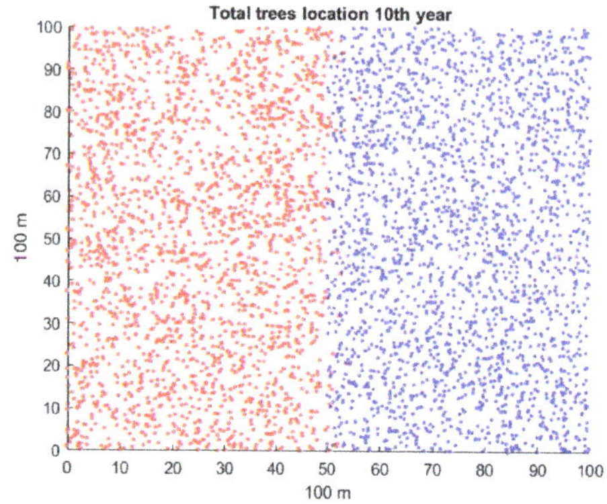

Figure 6. Starting conditions of invader (red) and native (blue). There are 2000 individuals each, spread over a range of initial ages and dbhs.

As in the above simulations for each of the six cases in Figure 3, the simulations based on the situation in Figure 6 all employ the same parameter values but start with different random number initiators for setting initial spatial locations and ages. The simulations were carried out for different periods of time, with only a few being carried out long enough to reach extinctions (Figure 7). The intention was not to show which species eventually won but to show the vast variability in trajectories in the early stages of competition. The simulations show the extreme variability of trajectories in which only the initial spatial configurations of the individuals of the two species varied.

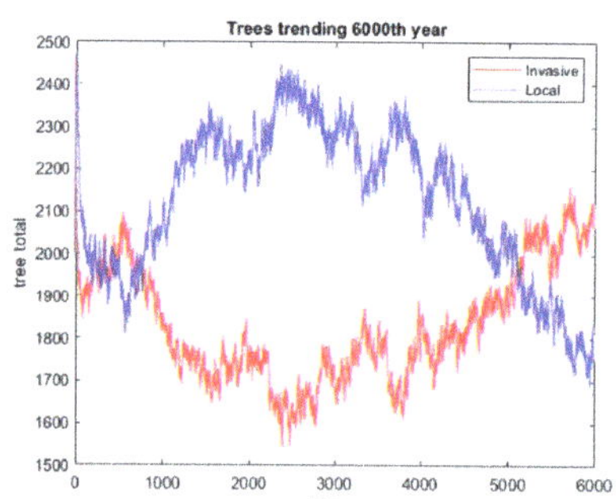

(**a**)

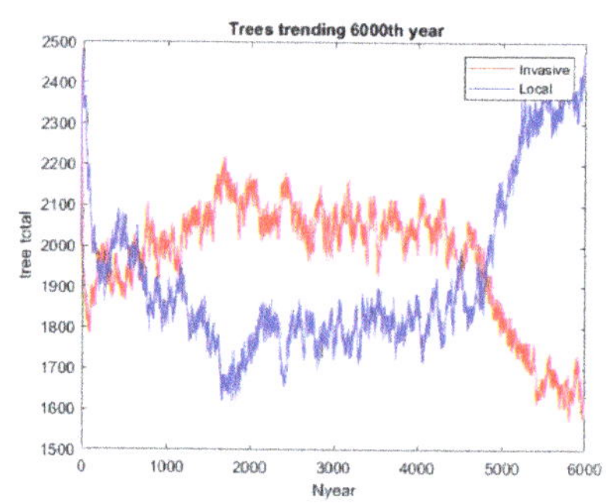

(**b**)

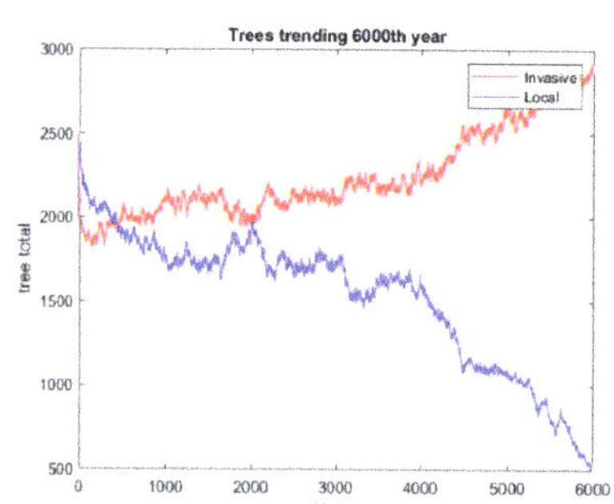

(**c**)

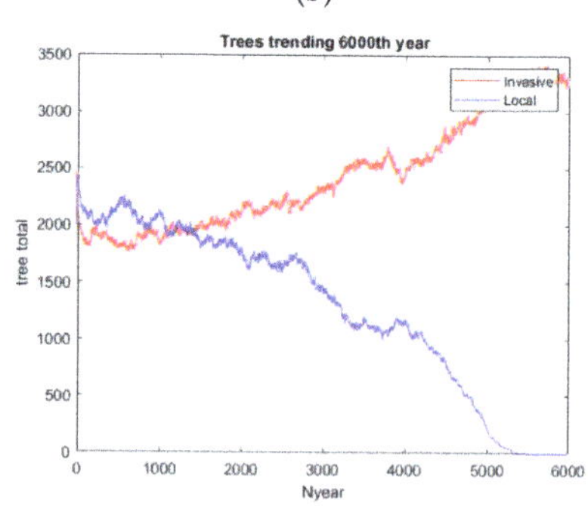

(**d**)

Figure 7. *Cont.*

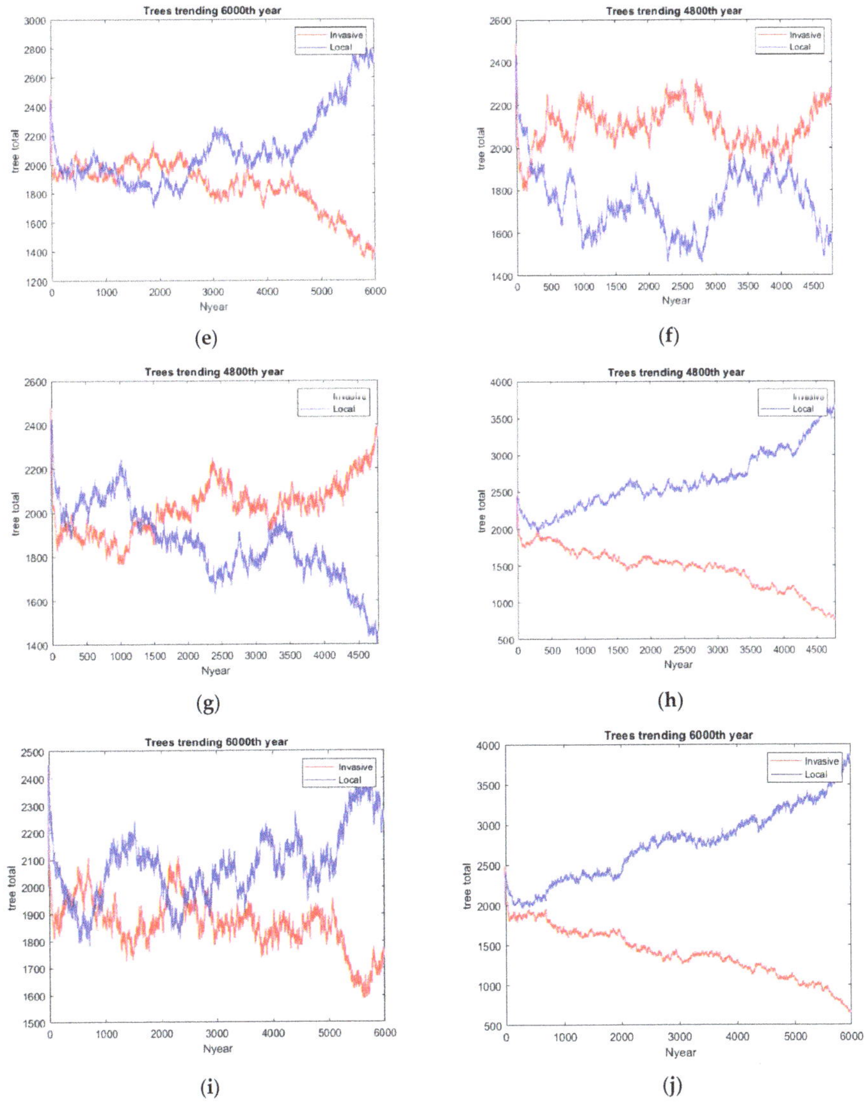

Figure 7. Simulations starting with invasive and native populations each occupying half of the plot, as in Figure 6, showing the total populations, invader (red) and native (blue). The same parameter values are used in all simulations, but different Matlab random number initiators were used in each of the simulations. (**a**) rand ('state', 36), (**b**) rand ('state', 35), (**c**) rand ('state', 34), (**d**) rand ('state', 33), (**e**) rand ('state', 32), (**f**) rand ('state', 31), (**g**) rand ('state', 30), (**h**) rand ('state', 29), (**i**) rand ('state', 37), (**j**) rand ('state', 38). The time periods of the simulations differ, as the intention was not to show the eventual winner of the competition but only the high degree of variability in population trajectories.

4. Discussion

A basic question regarding the introduction of non-native species to an environment is what the risk is that they will become invasive. Many factors have been considered in mathematical and simulation models of invasive species. Of particular importance in assessing the ultimate success of an invader is whether or not the population is able to grow in population size and spread from initially low numbers.

4.1. Invasion Success as Function of Initial Spatial Distribution of Invading Individuals

Based on the results of the simulations shown in Figure 3, an initial conclusion can be drawn that the initial configuration of individuals of an invasive tree species can influence its early success in growing and spreading. In particular, for a fixed number of immature individuals, the combination of population density and area occupied by the invading species influences the reliability of a successful invasion. For the set of starting conditions compared in Figure 3, case 3 (0 to 15 m) appears to be the best combination as a fairly tight bundle of increasing trajectories is formed. Case 6 (0 to 40) had the most (four) unsuccessful invasion attempts.

Some indication of the explanation of what leads to a successful invasion can be inferred from the examples of the case 1 simulation in Figure 4 and the case 6 simulation in Figure 5. In both cases, after some time, the invader manages to form a large dense cluster of individuals that can then grow in size at a substantial rate. The reason for this appears to be a version of the Allee effect. The Allee effect was already reflected in the classical mathematical result of [2] in that the initial size of the area occupied by the invader, L, must be large enough for inequality (3) to hold for the population to grow. In our case, the number of invading individuals is fixed, so the Allee effect depends on the initial density of individuals. Because an important mechanism for the success of invading *Melaleuca* is that it can produce litter that suppresses native seedlings, it is important that the invader exists in a high enough density that it can produce a sufficient litter layer across the space. The invader must also occupy enough space that it can gain a foothold based on its ability to produce seedlings suppressing litter. Of course, the area that is occupied by individuals and the density of individuals are tradeoffs when the number of individuals is fixed.

It appears that there is a particular tradeoff that is most effective. Case 1 (0 to 5 m) and case 2 (0 to 10 m) are generally successful but are less reliably successful than case 3 as they are slower to increase land coverage. At the opposite extreme, in case 5 (0 to 30 m) and case 6 (0 and 40 m), the initial distribution of invading *Melaleuca* was dilute. For those cases, it was more difficult for the *Melaleuca* to develop high litter densities. Nevertheless, in some simulations, even in case 6, *Melaleuca* was able to be successful in the early invasive stage (Figure 5). In these cases, although it took some time, eventually, the initially dilute density of the invaders was by chance able to form local concentrations that were sufficiently dense, occupying enough area to eventually start to grow.

4.2. Sensitivity of Invading Population Trajectories and Success to Initial Spatial Distribution

The second general feature of the simulations is that even within each of the six cases of the initial conditions shown in Figure 2, there is a large amount of variation in the trajectories. Because the initial number of invaders (200) is relatively small, to reduce the possible effect of small initial numbers on sensitivity to the initial conditions, simulations were performed with the starting conditions shown in Figure 6, that is, with 2000 individuals each of the invader and native species populations occupying opposite halves of the plot. The random number initiator was changed for each of the 10 different simulations. The trajectories show remarkable sensitivity to the initial conditions. What is most striking is the variability in the direction of population change within some of the simulations, in which the apparent dominance of one species is abruptly reversed.

It is tempting to compare the sensitivity to the initial condition exhibited here with the well-known sensitivity to the initial conditions in some deterministic mathematical models called 'deterministic chaos' (e.g., [25]). A difference, of course, is that the model here uses Monte Carlo simulations. However, the numbers in the Figure 7 simulations are large, so one might expect greater similarity in the trajectories. It is clear that a great deal remains to be learned about what drives these dynamics. Collective dynamics over spatial scales occurs within the populations. By this we mean that a locally dense cluster of invading individuals, which may occur by chance, can be self-reinforcing because they can collectively produce enough litter to exclude native seedlings. Thus, there is a small chance that the cluster can grow and that we can determine the outcome of the competitive interaction. Another factor that may help explain some of the sudden changes in the direction of population trajectories in Figure 7 is the upper limit on the lifetime of trees. Rapid die-offs of an age cohort of individuals in a large part of the landscape can lead to a change in the relative population growth rates of species populations. As demonstrated, random differences in the initial spatial configurations can produce varied results, highlighting the importance of considering spatial dynamics in invasion biology.

The third general feature of the simulations is that the results involved with some novel edge effects. The ecological dynamics at the edge of species' range limits are complex and involve multiple interacting factors, including dispersal, competition, and environmental variability. Turchin and Ellner [26] demonstrated that species at the edge of their range often exhibit different ecological behaviors compared to those in the core areas. For instance, increased dispersal ability and higher reproductive rates are often observed at range edges, which can facilitate range expansion despite suboptimal conditions [27,28]. These adaptive strategies highlight the importance of spatial dynamics and local adaptation in shaping species distributions. Edges can act as barriers or filters, affecting the dispersal of organisms and their propagules, which in turn can influence the spatial distribution of populations and community composition. For example, edges can impede the movement of pollinators, thereby reducing gene flow among plant populations and altering reproductive success and plant diversity [29].

4.3. Novelty of Results

The results here are new. Although, as discussed in the Introduction, IBMs have been used in studying invasions of specific plant species, we know of no study, systematic or otherwise, that focuses on the effects of the initial spatial distribution of invaders. It is well known that the success of invaders in highly variable. There may be many introductions of a novel species to an environment before the species becomes invasive. A consideration of not just the number but also the initial spatial configuration of individuals of the species may help explain why some invasions are successful and others are not. Also, our results, showing the high degree of variability in population trajectories, even when introduced in large numbers, are new. We showed a high degree of sensitivity to spatial configurations of 2000 invading species, which resembles the 'chaotic' behavior of some mathematical models but has a quite different origin.

4.4. Limits of Model

Our model simulates a specific invasive tree species, *Melaleuca quinquenervia*, in the specific environment of southern Florida. However, we believe that the nature of the interaction of the invasive and native species in our model has features that are general. Our spatially explicit IBM follows the general format of other models of forest dynamics, e.g., [24]. Also, the inhibitory effect of the invasive species, *Melaleuca*, on the native species through mechanisms, such as the litter suppression of native seedlings, is not uncommon. In our simulations, this allowed the invader to often be successful even though relatively strong biocontrol effects on growth and reproduction were imposed. Mechanisms of inhibition, such as through characteristics of litter or allelopathy, are common. Also, invading plant species affect the local soil microbial community, which can favor the invader [30]. Therefore, we believe that our model and results have implications for other systems.

The cases modeled here are an extremely limited slice of the vast range of initial spatial conditions that could be studied. We looked at only a few cases of invasions from one side of a plot carried out by a certain number of invaders. The parametrization was chosen such that, with a level of biocontrol imposed on the invader, the two species had very similar competitive fitness, although they differed in types of advantages. Further studies will be needed to provide stronger conclusions concerning how the initial spatial configuration influences invasive success.

5. Conclusions

We used a tested spatially explicit IBM of the competition of *Meleleuca quinquenervia*, invasive to southern Florida, with a generic native species. The model was used to explore the effects of different spatial distributions of the invasive population. The results showed that the initial density of individuals of the invader had an effect on its success. This was due to there likely being a critical threshold density for the sufficient accumulation of the invader's litter to have a strong suppressive effect on the native species. The simulations also showed extreme sensitivity of the competitor population trajectories to the initial spatial conditions, which resembled the 'chaotic' behavior seen in some mathematical models. Future work will entail studying a greater variety of initial distributions and understanding the causal mechanisms that underlie the results.

Author Contributions: Y.L. and J.X. led the development of the model. D.L.D. carried out model simulations and wrote the first draft of the manuscript. R.D.H. contributed a section in the Discussion and helped in editing. All authors have read and agreed to the published version of the manuscript.

Funding: Donald Lee DeAngelis and Yuanming Lu were supported by the U.S. Geological Survey's Greater Everglades Priority Ecosystem Science program, with Grant GR016613 given to the University of Miami and Grant G24AS00481 to the University of Florida.

Data Availability Statement: Data are contained within the article.

Acknowledgments: We appreciate the comments of Ian Pfingsten on an earlier version of this manuscript.

Conflicts of Interest: The authors declare no competing interests.

References

1. Elith, J.; Leathwick, J.R. Species distribution models: Ecological explanation and prediction across space and time. *Annu. Rev. Ecol. Evol. Syst.* **2009**, *40*, 677–697. [CrossRef]
2. Kierstead, H.; Slobodkin, L. The size of water masses containing plankton blooms. *J. Mar. Res.* **1953**, 141–147.
3. Schreiber, S.J.; Lloyd-Smith, J.O. Invasion dynamics in spatially heterogeneous environments. *Am. Nat.* **2009**, *174*, 490–505. [CrossRef]
4. Hastings, A.; Cuddington, K.; Davies, K.F.; Dugaw, C.J.; Elmendorf, S.; Freestone, A.; Harrison, S.; Holland, M.; Lambrinos, J.; Malvadkar, U.; et al. The spatial spread of invasions: New developments in theory and evidence. *Ecol. Let.* **2005**, *8*, 91–101. [CrossRef]
5. Jongejans, E.; Shea, K.; Skarpaas, O.; Kelly, D.; Sheppard, A.W.; Woodburn, T.L. Dispersal and demography contributions to population spread of *Carduus nutans* in its native and invaded ranges. *J. Ecol.* **2008**, *96*, 687–697. [CrossRef]
6. Lewis, M.A.; Pacala, S. Modeling and analysis of stochastic invasion processes. *J. Math. Biol.* **2000**, *41*, 387–429. [CrossRef]
7. Nehrbass, N.; Winkler, E. Is the Giant Hogweed still a threat? An individual-based modelling approach for local invasion dynamics of *Heracleum mantegazzianum*. *Ecol. Model.* **2007**, *201*, 377–384. [CrossRef]
8. Goslee, S.C.; Peters, D.P.; Beck, K.G. Spatial prediction of invasion success across heterogeneous landscapes using an individual-based model. *Biol. Invas.* **2006**, *8*, 193–200. [CrossRef]
9. Eppinga, M.B.; Kaproth, M.A.; Collins, A.R.; Molofsky, J. Litter feedbacks, evolutionary change and exotic plant invasion. *J. Ecol.* **2011**, *99*, 503–514. [CrossRef]
10. Lu, Y.; DeAngelis, D.L.; Xia, J.; Jiang, J. Modeling the impact of invasive species litter on conditions affecting its spread and potential regime shift. *Ecol. Model.* **2022**, *468*, 109962. [CrossRef]
11. Lu, Y.; Xia, J.; Magee, L.J.; DeAngelis, D.L. Seed dispersal and tree legacies influence spatial patterns of plant invasion dynamics. *Front. Appl. Math. Stat.* **2023**, *9*, 1086781. [CrossRef]
12. Korniss, G.; Caraco, T. Spatial dynamics of invasion: The geometry of introduced species. *J. Theor. Biol.* **2005**, *233*, 137–150. [CrossRef] [PubMed]
13. O'Malley, L.; Korniss, G.; Caraco, T. Ecological invasion, roughened fronts, and a competitor's extreme advance: Integrating stochastic spatial-growth models. *Bull. Math. Biol.* **2009**, *71*, 1160–1188. [CrossRef]
14. Radny, J.; Meyer, K.M. The role of biotic factors during plant establishment in novel communities assessed with an agent-based simulation model. *PeerJ* **2018**, *6*, e5342. [CrossRef]
15. Plank, M.J.; Simpson, M.J.; Binny, R.N. Small-scale spatial structure influences large-scale invasion rates. *Theor. Ecol.* **2020**, *13*, 277–288. [CrossRef]

16. Daufresne, T.; Hedin, L.O. Plant coexistence depends on ecosystem nutrient cycles: Extension of the resource-ratio hypothesis. *Proc. Natl. Acad. Sci. USA* **2005**, *130*, 9212–9217. [CrossRef]
17. Kortessis, N.; Kendig, A.E.; Barfield, M.; Flory, S.L.; Simon, M.W.; Holt, R.D. Litter, plant competition, and ecosystem dynamics: A theoretical perspective. *Am. Nat.* **2022**, *200*, 739–754. [CrossRef] [PubMed]
18. Rayamajhi, M.B.; Van, T.K.; Pratt, P.D.; Center, T.D.; Tipping, P.W. *Melaleuca quinquenervia* dominated forests in Florida: Analyses of natural-enemy impacts on stand dynamics. *Plant Ecol.* **2007**, *192*, 119–132. [CrossRef]
19. Rayamajhi, M.B.; Pratt, P.D.; Center, T.D.; Tipping, P.W.; Van, T.K. Aboveground biomass of an invasive tree Melaleuca (*Melaleuca quinquenervia*) before and after herbivory by adventive and introduced natural rnemies: A temporal case study in Florida. *Weed Sci.* **2008**, *56*, 451–456. [CrossRef]
20. Rayamajhi, M.B.; Pratt, P.D.; Center, T.D.; Van, T.K. Exotic tree leaf litter accumulation and mass loss dynamics compared with two sympatric native species in south Florida, USA. *Eur. J. For. Res.* **2010**, *129*, 1155–1168. [CrossRef]
21. Rayamajhi, M.B.; Pratt, P.D.; Tipping, P.W.; Center, T.D. Litter cover of the invasive tree *Melaleuca quinquenervia* influences seedling emergence and survival. *Open J. Ecol.* **2012**, *2*, 131–140. [CrossRef]
22. Center, T.; Purcell, M.; Pratt, P.; Rayamajhi, M.; Tipping, P.; Wright, S.A.; Dray, F. Biological control of *Melaleuca quinquenervia*: An Everglades invader. *BioControl* **2012**, *57*, 151–165. [CrossRef]
23. Van, T.; Rayachhetry, M.; Center, T.; Pratt, P. Litter dynamics and phenology of Melaleuca quinquenervia in south Florida. *J. Aquat. Plant Manage.* **2002**, *40*, 22–27.
24. Botkin, D.B.; Janak, J.F.; Wallis, J.R. Some ecological consequences of a computer model of forest growth. *J. Ecol.* **1972**, *60*, 849–872. [CrossRef]
25. Berger, U.; Hildenbrandt, H. A new approach to spatially explicit modelling of forest dynamics: Spacing, ageing and neighbourhood competition of mangrove trees. *Ecol. Model.* **2000**, *132*, 287–302. [CrossRef]
26. May, R.M. Simple mathematical models with very complicated dynamics. *Nature* **1976**, *261*, 459. [CrossRef] [PubMed]
27. Turchin, P.; Ellner, S.P. Living on the edge of chaos: Population dynamics of Fennoscandian voles. *Ecology* **2000**, *81*, 3099–3116. [CrossRef]
28. Levin, S.A. Dispersion and population interactions. *Am. Nat.* **1974**, *108*, 207–228. [CrossRef]
29. Aizen, M.A.; Feinsinger, P. Forest fragmentation, pollination, and plant reproduction in a Chaco dry forest, Argentina. *Ecology* **1994**, *75*, 330–335. [CrossRef]
30. Aponte, C.; García, L.V.; Marañón, T. Tree species effects on nutrient cycling and soil biota: A feedback mechanism favouring species coexistence. *For. Ecol. Manag.* **2013**, *309*, 36–46. [CrossRef]

forests

MDPI

Article

High Impacts of Invasive Weed *Lantana camara* on Plant Community and Soil Physico-Chemical Properties across Habitat Types in Central Nepal

Chandra Kumari Paudel, Achyut Tiwari *, Chitra Bahadur Baniya, Bharat Babu Shrestha and Pramod Kumar Jha

Central Department of Botany, Tribhuvan University, Kathmandu 46000, Nepal;
chandraapaudel6@gmail.com (C.K.P.); chitra.baniya@cdb.tu.edu.np (C.B.B.);
bharatbabu.shrestha@cdb.tu.edu.np (B.B.S.); pkjhaprof@gmail.com (P.K.J.)
* Correspondence: achyut.tiwari@cdb.tu.edu.np

Abstract: Although the effects of invasive alien plants on natural ecosystems are well known, the effects of specific plant species can vary across habitat types and disturbance intensity. This study was carried out to analyze the effects of *Lantana camara* on associated vegetation and soil physico-chemical properties at invaded and non-invaded sites across three different habitat types (forest edge, fallow land, and roadside) in central Nepal. We sampled 50 pairs of 5 m × 5 m (for shrub species) and 1 m × 1 m (for herbs species) plots at invaded and non-invaded sites in each habitat and recorded community variables for each species within the sampling plots for both wet (monsoon) and dry (pre-monsoon) seasons. Further, we collected soil samples from each quadrat and determined the soil physico-chemical properties. We recorded 137 species of flowering plants (119 from non-invaded and 97 from invaded plots) and classified them in accordance with life form/habit. In invaded sites, we found a significant decline in species diversity as indicated by the Simpson and Shannon diversity indices. Specifically, *L. camara* reduced the species richness, Simpson index, and Shannon diversity index by 36.84%, 11.84%, and 40.21%, respectively. Soil nutrients such as total nitrogen, soil organic carbon, and available phosphorus were significantly higher in invaded sites than non-invaded ones except for available potassium and soil pH. This study provided evidence that *Lantana L. camara* has a substantial impact on the understory plant community assemblage and the physico-chemical properties of soil. The results suggest that the protection of native plant community requires management of L. camara by implementing appropriate measures.

Keywords: biological invasion; forest edge; roadside vegetation; invasive plant; soil nutrient; species diversity; species richness

1. Introduction

Invasive alien plant species (IAPS) are the established alien species that intimidate native ecosystems and biodiversity [1]. They cause a significant threat to native ecosystem function and structure of plant communities, and are one of the major causes of biodiversity loss, that bring about changes in vegetation structure and degradation of environment [2,3]. In the changing climate, the introduction and aggressiveness of IAPS is increasing, particularly on different land uses. According to recent reports, IAPS have massive socioeconomic and environmental consequences around the world [4]. Invasive plant species affect soil properties [5], hydrology [6] and fire regimes [7] at the ecosystem level. They are known to reduce native species diversity and alter composition at the community level [8], though such impacts can be species- or site-specific. Conversely, there is also evidence of facilitation and co-occurrence of invasive and native plants because of positive interactions between plant species with similar habitat requirements and phonologies [9]. The mixed responses demonstrate that invasive plant species can have both positive and negative effects on ecosystems. As a result, more research is needed to investigate the impacts of IAPS to develop effective control strategies.

Biological invasion has been identified as one of the emerging threats to biodiversity and ecosystem services in Nepal [10], as well as one of the major underlying causes of habitat degradation in Nepal, alongside unsustainable harvesting practices, environmental pollution, overgrazing, and infrastructure development [11]. The number of IAPS in Nepal has increased over time and the extent of climatically suitable areas for most of Nepal's IAPS is likely to expand and shift upslope in the future under climate change scenarios [12]. A recent review by [13] has revealed that there are 30 species of invasive plants, and their number has been increasing at an alarming rate. Despite the significant efforts by researchers to generate new knowledge, little is known about the biology and ecology of individual IAPS [14], their diversity [15], distribution [16], and impacts [17,18]. Therefore, ecological investigations in different regions are needed to either specify the status or impact of individual IAPS on richness and diversity of native plant communities. The problems of invasive species are widespread in both developed and developing countries, but the impacts could be greater in developing countries like Nepal. The Indian subcontinent has been experiencing altered precipitation patterns and intensified monsoon seasonality in the recent decades [19], and this can play a crucial role in regulating plant productivity [20], thereby enhancing plant invasion in the region.

We studied *L. camara* as it is considered one of the world's worst invasive plant species [21]. It is the major invader of forests, pastures, and wetlands [22]. *Lanata camara* invasion in natural ecosystems has been linked to widespread loss of native plant species diversity due to native species recruitment restrictions and altered ecosystem structure and function [23]. However, there is few quantitative evidence to support these claims and more data on the patterns of species loss associated with *L. camara* invasion is urgently needed to inform management priorities. High reproductive potential, phenotypic plasticity, resilience to grazing pressure, allelopathy, and fire resistance are the factors that cause *L. camara* to become a successful invader [24]. Various studies have found that *L. camara* to be an invasive species in various regions around the world [25–31]. Despite the large number of studies on the effects of invasive species, few have investigated the effects of invasive species on species diversity and composition as well as multiple soil properties across habitat types [32]. It is crucial to investigate the effects on both species composition and soil properties because soil and plants influence each other and changes in vegetation due to invasive species would eventually result in changes in soil properties and these changes in soil properties would result in additional changes in vegetation [33]. Despite the weed's widespread occurrence in Nepal quantitative data on the impacts and scale of the problems are inadequate to inform management decisions. Therefore, data from different regions are needed both to specify their status and to analyze their impacts on diversity and richness of native plants.

In this study we assessed the impact of *Lantana camara* on understorey community composition, diversity and soil physico-chemical properties in central Nepal. We hypothesize that invasion by *L. camara* alters the community assemblage, biotic interaction among co-existing species and reduces plant diversity and soil physico-chemical properties. To test this hypothesis, we sought to (1) assess the impact of *L. camara* on species richness and diversity of understorey plant species, and (2) analyze the soil physico-chemical properties at the invaded and non-invaded sites.

2. Materials and Methods

2.1. Study Area

Study area lies in tropical and subtropical zones within four districts of central Nepal (Makawanpur, Chitwan, Tanahun, and Kaski). The elevation range of our study sites was 153 m (Chitwan) to 1490 m (Kaski) (Figure 1). The climate of the study area is a typical monsoon climate with four distinct seasons (monsoon, post-monsoon, winter, and pre-monsoon) marked by distinct dry and wet seasons, creating a seasonal variation in plant community assemblage. The average minimum and maximum temperature of study area ranges from 16 °C to 31 °C and average annual rainfall ranges from 2000 mm to 3800 mm. Forest of *Shorea robusta* Gaertn. extended throughout the flat land and it is replaced by riverine forest along river and streamside in tropical regions. *Schima-Castanopsis*, *Pinus*, and *Alnus* forests are the dominant vegetation of the subtropical regions in the study area.

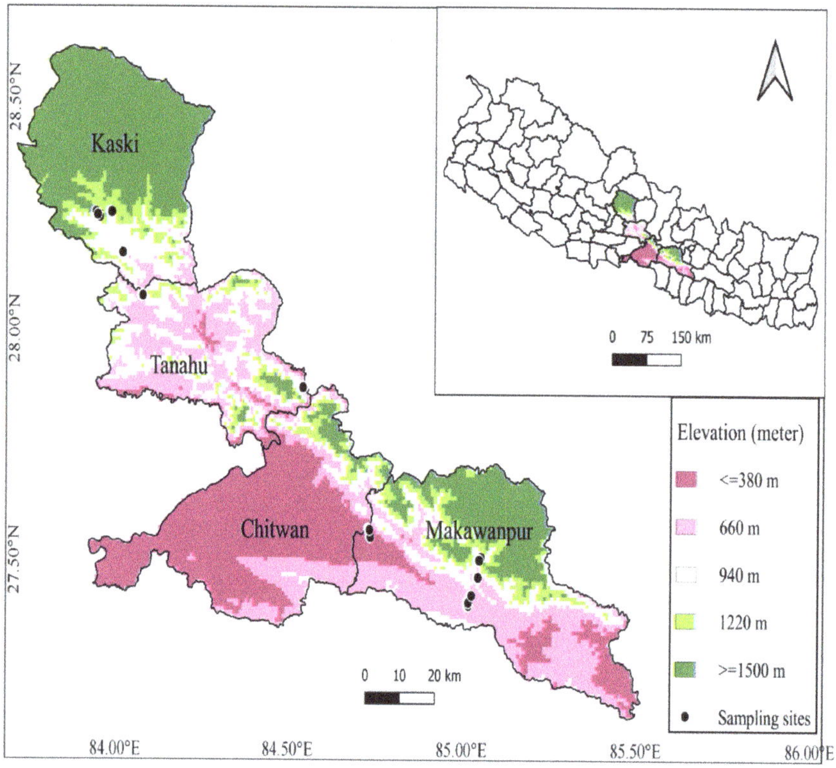

Figure 1. Map of study area showing the sampling plots location.

2.2. Field Sampling

Three different habitat types were chosen for sampling, i.e., forest-edge, fallow-land, and roadside, as such microhabitats are highly vulnerable to plant invasions and the vegetation of these sites were dominated by *L. camara* in the study area. We defined forest-edge as the transitional zone between a forested area and an adjacent landscape such as grassland, agricultural land, or urban area, and fallow-land as the agricultural land left unplanted for a substantial time and roadside as the adjacent barren land to the road. These habitat types were chosen as they appeared to be highly invaded sites by *L. camara* while conducting a reconnaissance survey prior to this study. Plot locations for each habitat were chosen arbitrarily in topographically uniforms areas with 50 m distance between invaded and non-invaded sites. The ecological sampling was carried out from 28th September to 2nd October 2020 for the wet season (monsoon) and from 16th to 20th April 2021 for the dry season (pre-monsoon) to analyze if impacts vary seasonally. A field survey was conducted at two plot types: invaded and non-invaded. We carried out paired sampling of 5 m × 5 m (for shrub species) and 1 m × 1 m (for herbs species) for each habitat [34,35]. A total of 25 pairs of plots were sampled during the monsoon and 25 pairs during pre-monsoon. Plant species were counted, and their coverage was recorded in six cover classes, i.e., 1 (0%–5%), 2 (5%–25%), 3 (25%–50%), 4 (50%–75%) 5 (75%–95%), and 6 (95%–100%) [36]

(Daubenmire, 1959), considering each plot as 100 percent based on a visual estimation method. Voucher specimens were collected except tree species (but saplings of trees were counted). Collected voucher specimens were identified at the National Herbarium and Plant Laboratories (KATH) by comparing them with identified specimens and consulting the literature [37]. The identified plants were categorized according to their life form (herbs, shrubs, and climbers).

2.3. Data Analysis

The species richness and Simpson and Shannon diversity indexes of vegetation were calculated. The species richness was defined as the number of species recorded in each sample plot.

Species diversity refers to the richness of species and their relative abundance in each region. To compare the species diversity between invaded and non-invaded sites, the Simpson and Shannon–Weiner diversity indexes were calculated.

The Shannon–Weiner Index was calculated by following equation [38]:

$$\text{Shannon Weiner Index (H)} = -\sum \text{Pi} \times \ln \text{Pi}$$

where Pi = ni/N, ni = average coverage of each individual species, N = total average coverage of all species.

Simpson Diversity Index:

$$D = 1 - \sum(\text{ni} - 1)/\text{N}\ (\text{N} - 1)$$

where D = the Simpson diversity index, ni = the coverage of the ith species, and N = total coverage of all species.

Statistical analysis was performed using the R 4.0.3 [39]. First, the normality of the data was tested. A significant value of the Shapiro–Wilk test was used to determine the normality of data ($p \geq 0.05$) and the homogeneity of variance of the data was established with Leven's test. The Simpson and Shannon diversity indexes, species richness, and soil physico-chemical properties in sites invaded and not invaded by *L. camara* during the pre-monsoon and monsoon seasons were analyzed by paired sampled *t*-test. Comparisons of the habitat types between invaded and non-invaded sites were performed using one-way ANOVA.

2.4. Soil Parameter Analysis

Soil samples were collected at a depth of 15 cm from four corners and a center, and they were mixed thoroughly to make single composite sample per plot. There were total of 100 samples (50 samples from the monsoon and 50 samples from the pre-monsoon season).

The soil samples were then left to air dry for two weeks in the shade in the lab. The soil samples were then ground and sieved through 2 mm wire mesh and a 0.5 mm mesh before being stored in airtight bags until laboratory analysis. The method outlined by [40] was used for determination of soil pH, the Walkey–Black method for soil organic carbon (SOC), and the micro-Kjeldahl method [41] for the determination of total nitrogen (TN). Available phosphorus and available potassium were estimated using a flame photometer and Olsen's bicarbonate method, respectively, as described by [42,43].

3. Results

3.1. Species Composition and Diversity

We found 137 species of flowering plants from 49 families, of which 119 were recorded in non-invaded sites and 97 in *L. camara*-invaded sites (Appendix A). Additionally, 79 plant species were present in both non-invaded and invaded sites. The number of herb species was high in non-invaded sites which decreased sharply compared to other life forms in invaded sites (Figure 2). Although all the growth forms (herb, shrub, and climber) showed changes in species composition, herbaceous species contributed the most to compositional changes. The plant family Poaceae was represented by the highest number of species (22), followed by Asteraceae (17) (Appendix A). Forty species were present only in non-invaded plots; and some of them were reported to have ethno-medicinal uses. Out of forty species, only eighteen species could successfully emerge from *L. camara*-invaded plots.

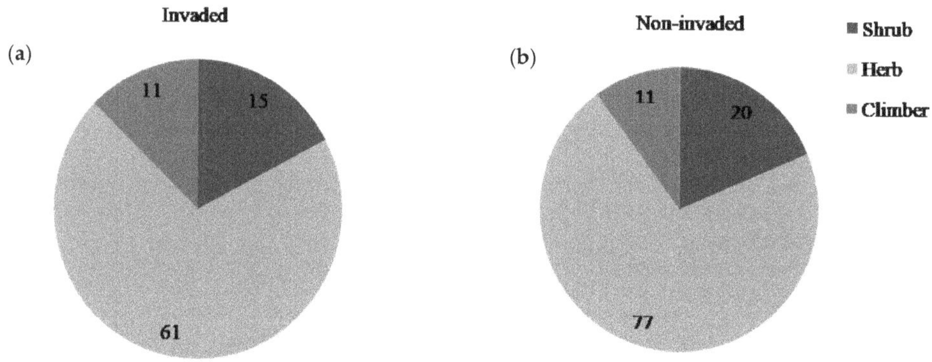

Figure 2. Life form of plant species in (**a**) invaded and (**b**) non-invaded plots.

The species richness and diversity index between invaded and non-invaded sites differed significantly (Appendix B, Figure 2). The species richness and the Simpson and Shannon diversity indices of invaded sites decreased by 36.84%, 11.84%, and 40.21%, respectively, compared to the non-invaded sites. Non-invaded sites were composed of higher richness and diversity of plant species compared to invaded sites in both the pre-monsoon and monsoon seasons. There was a significant difference in species richness across the three habitat types (Figure 3, Appendix B). In the pre-monsoon season, the forest edge exhibited the highest species richness, followed by fallow land, and roadside; however, in the monsoon season, the fallow land exhibited the highest number of species followed by the forest edge and roadside.

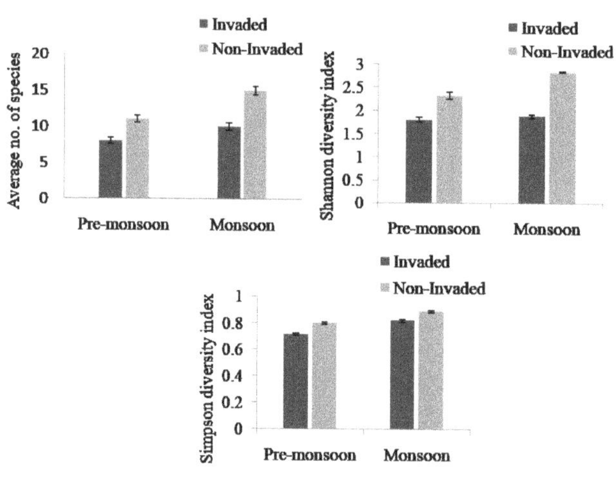

Figure 3. Species richness, Simpson diversity index, and Shannon diversity index in *Lantana camara*-invaded and non-invaded sites (paired sample *t*-test). The values shown are means ± standard error.

3.2. Soil Physico-Chemical Characteristics

The soil nutrients were higher in invaded sites compared to non-invaded sites except available Potassium and soil pH in both the pre-monsoon and monsoon seasons (Figure 4,

Appendix B). In total, the soil nutrients, like soil organic carbon, total nitrogen, and available phosphorus, of invaded sites increased by 11.61%, 28%, and 24.03%, respectively. Regarding the habitat type, total nitrogen, available phosphorus, available potassium, and soil pH were higher in fallow land followed by forest edge and roadside, while the total soil organic carbon was highest in forest edge and lowest in roadside (Figures 5 and 6, Appendix B).

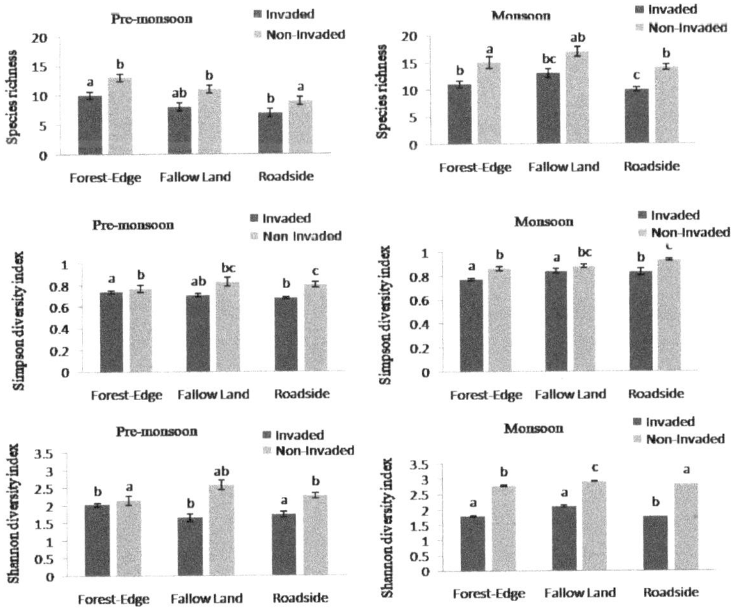

Figure 4. Species richness, Simpson diversity index, and Shannon diversity index (one-way ANOVA test) in invaded and non-invaded sites across three habitat types. The values shown are means ± standard error. Means with different superscript letters are significantly different at $p < 0.05$.

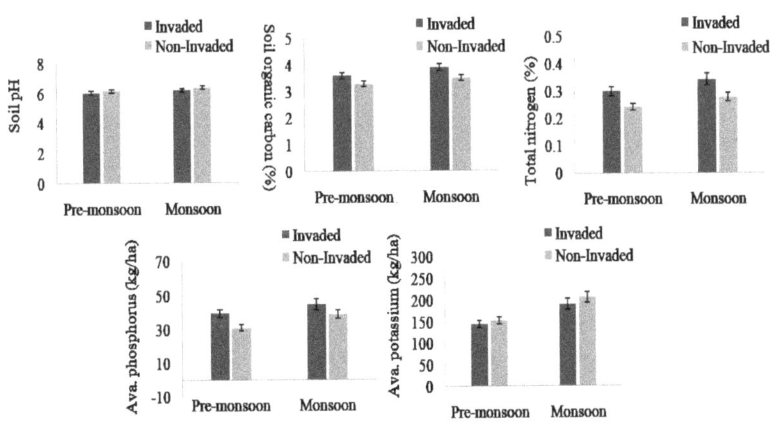

Figure 5. Soil pH, total SOC, total N, available K, and available P in invaded and non-invaded plots as obtained from paired sample t-test results, the values shown are means (±standard error), p-values indicate significant different at $p < 0.05$.

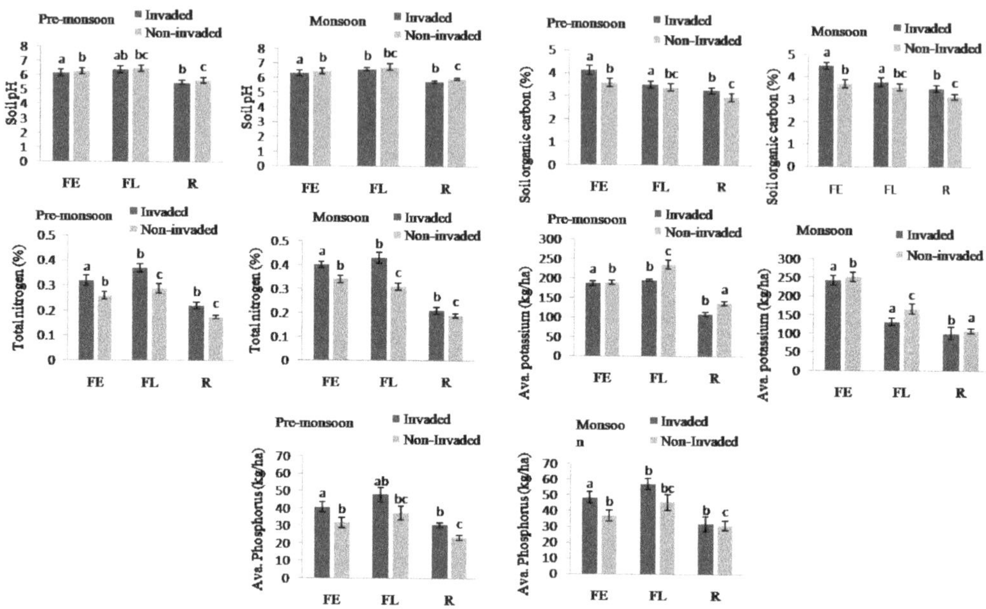

Figure 6. Soil pH, total SOC, total N, available K, and available P at invaded and non-invaded sites of three habitat types. The values shown are means (±standard error). Means with different superscript letters are significantly different at $p < 0.05$.

4. Discussion

4.1. Species Composition and Diversity

Our results indicated that plant species decreased at invaded sites, including a sharp decrease in the number of herb species (Figure 2). The extensive cover produced by *L. camara* reduced the herbaceous species probably by restricting the amount and duration of light below its canopy. This could be because of the establishment of a photosynthetically inactive light regime at ground level [44,45]. Despite the levels of other resources, light limitation below a certain threshold can prevent species from surviving [46]. Species such as *Tinospora sinensis*, *Sapium insigne*, *Calotropis gigantean*, *Solanum nigrum*, *Urena lobata*, *Woodfordia fruticosa*, *Anaphalis contorta*, *Buddleja asiatica*, *Callicarpa arborea*, *Cissampelos pareria*, *Elephantopus scaber*, and *Hydrocotyle sibthropioides* were completely lost at invaded sites (Appendix A). Soil nutrients may also play a significant role in compositional difference in plant species between *L. camara*-invaded sites and non-invaded sites. It has been reported that the changes in litter and soil chemistry under the canopy of *L. camara* facilitate its invasion and displace the native species [47].

We reported that *L. camara* invasion is associated with a decrease in species diversity and richness. The ability of *L. camara* to adapt under various ecological conditions allows it to spread quickly, which causes a reduction in diversity and the development of its monocultures. The capacity of *L. camara* to establish rapidly, its tolerance to various environmental conditions and resistance to logging, burning, and climate change reduced the diversity and richness at the invaded sites [48,49]. Vegetation at the invaded sites had declined more significantly than that of non-invaded sites [50], and very few shade tolerant species stood out against *L. camara* invasion, that can grow taller than *L. camara*. Similar trends were seen in the riparian forests along the Nayar River, where invaded sites had a nearly 30% loss in species richness. Compared to non-invaded habitats, *L. camara*-invaded sites included significantly fewer plant species [22]. In regions where *L. camara* has invaded, the richness, diversity, structure, and composition of native plant species was reduced and degraded [51]. Our results indicated that the non-invaded sites comprised a higher richness and diversity of plant species compared to invaded sites in both the pre-monsoon and monsoon seasons, and this could be due to competitive exclusion [52] or altered habitat structure [53].

Although all the habitat-types were susceptible to invasion in our study, the diversity of understorey plants species were high in fallow land and forest edge compared to roadside, which indicated that the roadside habitat was highly affected by *L. camara* invasion compared to forest edge and fallow land. The lowest invasion effect was noticed in forest edge with its high species richness. This result is consistent with those of [54], who revealed that invasive species could not get enough light under the canopy of tree, hence they cannot spread rapidly in dense forests. Forest trees and vines often provide shade for invasive plant seedlings and slow down their growth [55]. Habitats near the roads were found to be more susceptible to invasion as the invasive species are conspicuous along roadsides, and roads are more anthropogenically disturbed and provide suitable microhabitats for many IAPS [56]. Further, we observed the seasonal changes in species richness across different habitat type—forest edge, fallow land, and roadside—and these variations could be attributed to various ecological and environmental factors including changes in microclimate [57], the emergence of opportunistic species [58], and changes in water availability [59].

4.2. Soil Physico-Chemical Characteristics

Invaded sites were composed of more acidic (low pH value) soil compared to non-invaded sites (higher pH value) throughout both the pre-monsoon and monsoon seasons. This might be related to the effect of podsolization (base depletion) [60]. Soil organic carbon, nitrogen, and phosphorus contents were higher in the invaded sites compared to the non-invaded sites. The positive correlations of *L. camara* cover with soil nutrients, except soil pH and K (Appendix B), indicated that soil nutrient increases with increased *L. camara* cover. The higher nutrient content in invaded sites may be related to the higher above-ground biomass and net primary production as well as faster growth rates than those of the native species being displaced as reported by [26]. *Lantana camara*-invaded sites were associated with higher levels of organic carbon in the soil compared to the non-invaded sites [61]. The dense monoculture of *L. camara* altered the microenvironment (pH, light, and temperature) beneath it and prevented the establishment and germination (or both) of associated species as reported by [62]. The allelopathic compounds released by the root of *L. camara* prevent the development and establishment of other native plant species, which reflects the competition between them [18,63].

The high biomass (mainly leaf litter) resulting from the taller plant present in invaded sites and no consumption of *L. camara* by livestock could be two possible explanations for the observed significant increase in soil nutrient at the *L. camara*-invaded sites compared to the non-invaded sites. Due to the deep root system of *L. camara*, it can absorb and use nutrients more quickly in deeper soil layers [64], a process known as "nutrient uplift" [65], which could increase the amount of phosphorus that was accessible in the surface soil. The higher total soil organic carbon, total nitrogen, and available phosphorus could be due to the abundance of invasive species which contribute to nutrient-rich higher litterfall and biomass [66,67], which eventually increase the organic carbon and nutrient pool in the soil. In other statements, invasive species prefer nutrient-rich soil and inoculate positive plant soil feedback which encourages further invasion in such habitats [68]. However, the results of the present study showed that availability of potassium is lower in the invaded sites than the non-invaded sites. These outcomes support the observation of [69], who reported a low level of potassium in invaded sites. This might be because *L. camara* produces many seeds with a high flowering rate, which causes the highly invaded areas to have low potassium levels [70,71].

5. Conclusions

This study provides evidence that *L. camara* invasion poses a substantial impact to species richness, community assemblage, and species diversity, as well as the soil properties. The study revealed that native plant species, like *Artemisia indica*, *Woodfordia fruticosa*, *Urena lobata*, and *Achyranthes aspera*, were reduced by *L. camara* invasion. The soil nutrient levels in most of the *L. camara*-invaded sites was high and were potentially increased by its invasion, which could allow more invasive plants to enter the area and could affect the ecosystem functions. These findings are highly important to biodiversity/forest managers and farmers. Further success of the *L. camara* invasion depends on soil properties like pH, SOC, N, P and K. Hence, proper utilization of land resources, control of deforestation, and land degradation are highly recommended to contend *L. camara* invasion. Further studies on the comparative physiological traits of invasive and native species and soil microbial dynamics would help refine our knowledge on plant invasion dynamics. Long-term monitoring on permanently established plots is also needed to understand the long-term impact of invasive species on vegetation dynamics to frame appropriate policy and management strategies to control plant invasion and its impacts.

Author Contributions: Conceptualization, A.T. and P.K.J.; Validation, C.B.B.; Formal analysis, C.B.B.; Investigation, C.K.P., A.T. and B.B.S.; Data curation, C.K.P. and C.B.B.; Writing—original draft, C.K.P.; Writing—review & editing, A.T. and B.B.S.; Supervision, B.B.S. and P.K.J.; Funding acquisition, P.K.J. All authors have read and agreed to the published version of the manuscript.

Funding: This research was supported by the United States Agency for International Development (USAID) Bureau of Food Security under the Cooperative Agreement No. AID-OAA-L-15-00001 as part of the Feed the Future Innovation Lab for Integrated Pest Management.

Data Availability Statement: Data are contained within the article.

Conflicts of Interest: The authors declare no conflict of interest.

Appendix A

Name of Plants	Family	Invaded	Non-Invaded
Acacia catechu (L.f.) Wild	Fabaceae	-	+
Achyranthes aspera L.	Amaranthaceae	+	+
Acmella sp.	Asteraceae	-	+
Ageratum conyzoides L.	Asteraceae	+	+
Ageratum houstonianum Mill.	Asteraceae	+	+
Ageratina adenophora (Spreng.) R. King and H. Rob.	Asteraceae	+	+
Alternanthera phyloxeroides (Mart.) Griseb	Amaranthaceae	-	+
Anaphalis contorta (D. Don) Hook. f.	Asteraceae	-	+
Artemisia indica Willd.	Asteraceae	+	+
Arundo sp.	Poaceae	+	+
Asparagus racemosus Willd.	Asparagaceae	+	-
Axonopus Compressus (Sw.) P. Beauv.	Poaceae	-	+
Azadirachta indica A. Juss.	Meliaceae	+	-
Bidens pilosa L.	Asteraceae	+	+
Boehmeria sp.	Urticaceae	+	+
Bothriochloa ischaemum (L.) Keng	Poaceae	+	+
Bothriochloa pertusa (L.) A. Camus	Poaceae	-	+
Buddleja asiatica Lour.	Schrophulariaceae	-	+
Callicarpa arborea Roxb.	Verbenaceae	-	+
Callicarpa macrophylla Vahl	Verbenaceae	+	-
Calotropis gigantea (L.) Dryand	Apocynaceae	+	-
Cannabis sativa L.	Cannabaceae	+	+
Cardiospermum halicacabum L.	Sapindaceae	-	+
Castanopsis indica (Roxb.) Miq.	Fagaceae	-	+
Chromolaena odorata (L.) R. M. King and H. Rob.	Asteraceae	+	+
Chrysopogon aciculatus (Retz.) Trin.	Poaceae	-	+
Cissampelos pareria L.	Menispermaceae	-	+
Clerodendrum viscosum Vent.	Lamiaceae	+	+
Coccinia grandis (L.) Voigt	Cucurbitaceae	+	-

Name of Plants	Family	Invaded	Non-Invaded
Colebrookea oppositifolia sm.	Lamiaceae	+	+
Colocasia esculenta (L.) Schott	Araceae	+	+
Commelina benghalensis (L.)	Commelinaceae	+	+
Conyza bonariensis (L.) Cronquist	Asteraceae	-	+
Corchorus aestuans L.	Tiliaceae	+	+
Crassocephalum crepidiodes (Benth.) S. Moore	Asteraceae	+	+
Cucumis callosus (Rottb.) Cogn.	Cucurbitaceae	+	-
Curcuma aromatica Salisb.	Zingiberaceae	-	+
Cyanotis cristata (L.) D. Don	Commelinaceae	-	+
Cynodon dactylon (L.) Pers.	Poaceae	+	+
Cynoglossum lanceolatum Forssk.	Boraginaceae	-	+
Cyperus esculentus L.	Cyperaceae	+	+
Cyperus rotundus L.	Cyperaceae	+	+
Cyperus sp.	Cyperaceae	-	+
Dalbergia sisso DC.	Fabaceae	+	+
Desmodium heterocarpon (L.) DC.	Fabaceae	-	+
Desmodium triflorum (L.) DC.	Fabaceae	+	+
Digitaria ciliaris (Retz.) Koeler	Poaceae	+	+
Digitaria setigera Roth	Poaceae	+	-
Dioscorea bulbifera L.	Dioscoreaceae	+	+
Drepanostachyum intermedium (Munro) Keng f.	Poaceae	-	+
Drymaria diandra Blume	Caryophyllaceae	+	+
Elephantopus scaber L.	Asteraceae	-	+
Eulaliopsis binata (Retz.) C. E. Hubb	Poaceae	+	+
Euphorbia hirta L.	Euphorbiaceae	+	+
Evolvulus nummularius (L.) L.	Convolvulaceae	+	+
Ficus hispida L.	Moraceae	+	+
Ficus semicordata Buch.-Ham. ex J. E. Sm.	Moraceae	+	-
Fimbristylis dichotoma (L.) Vahl.	Cyperaceae	+	+
Flemingia macrophylla (Willd.) Merr.	Fabaceae	+	-
Galinsoga perviflora Cav.	Asteraceae	+	+
Gnaphalium affine D. Don.	Asteraceae	-	+
Gonostegia pentandra (Roxb.) Miq.	Urticaceae	+	+
Holarrhena pubescens (Buch.-Ham.) Wall. ex G. Don	Apocynaceae	+	+
Hydrocotyle sibthorpioides Lam.	Apiaceae	-	+
Hyptis suaveolens (L.) Poit.	Lamiaceae	+	+
Imperata cylindrica (L.) P. Beauv.	Poaceae	+	+
Ipomoea nil (L.) Roth	Convolvulaceae	-	+
Ipomoea purpurea (L.) Roth	Convolvulaceae	+	-
Ipomoea quamoclit L.	Convolvulaceae	+	+
Justicia adhatoda L.	Acanthaceae	+	+
Justicia simplex D. Don.	Acanthaceae	+	+
Kyllinga brevifolia Rottb.	Cyperaceae	+	+
Lantana camara L.	Verbenaceae	+	+
Leea crispa Royen ex L.	Leeaceae	+	+
Leucaena leucocephala (Lam.) de Wit	Fabaceae	+	-
Leucas cephaloates (Roth) Spreng.	Lamiaceae	-	+
Lindernia crustacea (L.) F. Muell.	Schrophulariaceae	-	+
Maesa chisia Buch.-Ham. ex D. Don	Myrsinaceae	-	+
Melastoma malabathricum L.	Melastomataceae	+	+
Mikania micrantha Kunth.	Asteraceae	+	+
Momordica charantia L.	Cucurbitaceae	+	+
Mimosa pudica L.	Fabaceae	+	+
Mimosa rubicaulis Lam.	Fabaceae	-	+
Morus alba L.	Moraceae	+	+
Murraya koenigii (L.) Spreng	Rutaceae	+	+
Oplismenus hirtellus (L.) P.Beauv.	Poaceae	+	+
Oxalis corniculata L.	Oxalidaceae	+	+
Parthenium hysterophorus L.	Asteraceae	+	+
Paspalidium flavidum (Retz.) A. Camus	Poaceae	+	+
Paspalum scrobiculatum L.	Poaceae	+	+
Paspalum sp.	Poaceae	-	+
Persicaria capitata (Buch.-Ham. ex D. Don) H. Gross	Polygonaceae	+	+
Persicaria perfoliata (L.) H. Gross	Polygonaceae	-	+
Phragmites karka (Retz.) Trin. ex Steud	Poaceae	+	+

Name of Plants	Family	Invaded	Non-Invaded
Phyllanthus urinaria L.	Euphorbiaceae	+	+
Phyllanthus virgatus G. Forst	Euphorbiaceae	+	+
Physalis angulata L.	Solanaceae	+	-
Pinus roxburghii Sarg.	Pinaceae	+	-
Piper longum L.	Piperaceae	+	-
Plumbago zeylanica L.	Plumbaginaceae	-	+
Pogonatherum paniceum (Lam.) Hack.	Poaceae	-	+
Polygonum persicaria L.	Polygonaceae	+	+
Polygonum sp.	Polygonaceae	+	+
Prunus persica (L.) Batsch.	Rosaceae	-	+
Reinwardtia indica Dumort.	Linaceae	-	+
Ricinus communis L.	Euphorbiaceae	-	+
Rubus ellipticus Sm.	Rosaceae	+	-
Rubus sp.	Rosaceae	+	+
Saccharum spontaneum L.	Poaceae	+	+
Salvia sp.	Lamiaceae	+	+
Sapium insigne (Royle) Benth. ex. Hook. f.	Euphorbiaceae	+	-
Saraca asoca (Roxb.) Willd.	Fabaceae	-	+
Schima wallichii (DC.) Korth.	Theaceae	-	+
Senna occidentalis (L.) Roxb.	Fabaceae	+	+
Senna tora (L.) Roxb.	Fabaceae	+	+
Setaria glauca (L.) P. Beauv.	Poaceae	+	+
Setaria pallidifusca (Schumach.) Stapf and C. E. Hubb.	Poaceae	+	+
Shorea robusta Gaertn.	Dipterocarpaceae	+	+
Sida acuta (L. fil.) Borss. Waalk.	Malvaceae	+	+
Sida cordifolia L.	Malvaceae	+	+
Sida rhombifolia L.	Malvaceae	+	+
Solanum aculeatissimum Jacq.	Solanaceae	+	+
Solanum nigrum L.	Solanaceae	+	-
Spermacoce alata Aubl.	Rubiaceae	+	+
Stephania glandulifera Miers	Menispermaceae	+	+
Sambucus sp.	Sambucaceae	+	-
Syzygium cumini (L.) Skeels	Myrtaceae	-	+
Tetrastigma serrulatum (Roxb.) Planch.	Vitaceae	+	+
Thysanolaena maxima (Roxb.) Kuntze	Poaceae	+	+
Tinospora sinensis (Lour.) Merr.	Menispermaceae	-	+
Tridax procumbens L.	Asteraceae	-	+
Triumfetta pilosa Roth.	Malvaceae	+	+
Urena lobata L.	Malvaceae	+	+
Urtica dioca L.	Urticaceae	-	+
Vetiveria sp.	Poaceae	+	+
Woodfordia fruticosa (L.) Kurz.	Lythraceae	+	+
Xanthium strumarium L.	Asteraceae	+	+

Appendix B

Table A1. Results of paired sample *t*-test comparing species richness and diversity between *Lantana camara*-invaded and non-invaded plots. Abbreviation: 1-D, Simpson diversity index; H′, Shannon diversity index; LI, Lantana camara-invaded; NI, non-invaded; *p*-value indicate significant different at $p < 0.05$.

Season	Simpson Index (1-D)		Shannon Index (H′)		Species Richness	
	LI	NI	LI	NI	LI	NI
Pre-monsoon	0.72 ± 0.008	0.80 ± 0.008	1.80 ± 0.547	2.32 ± 0.737	8 ± 0.469	11 ± 0.476
p-value	<0.001		<0.001		<0.001	
Monsoon	0.81 ± 0.010	0.89 ± 0.009	1.88 ± 0.035	2.84 ± 0.013	11 ± 0.404	15 ± 0.564
p-value	<0.001		<0.001		<0.001	

Table A2. Results of one-way ANOVA comparing species richness and diversity between *Lantana camara*-invaded and non-invaded plots of each habitat and plot types. The values shown are means (±standard error). Means with different superscript letters are significantly different at $p < 0.05$. Abbreviation: 1-D, Simpson diversity index; H′, Shannon diversity index; LI, Lantana camara-invaded; NI, non-invaded.

Season	Habitat	Shannon Index (H′)		Simpson Index (1-D)		Species Richness	
		LI	NI	LI	NI	LI	NI
Pre-monsoon	Forest-edge	2.02 ± 0.05^{b}	2.14 ± 0.13^{a}	0.74 ± 0.01^{a}	0.77 ± 0.03^{b}	10 ± 0.55^{a}	13 ± 0.65^{b}
	Fallow land	1.65 ± 0.10^{b}	2.57 ± 0.14^{ab}	0.71 ± 0.02^{ab}	0.83 ± 0.04^{bc}	8 ± 0.75^{ab}	11 ± 0.59^{b}
	Roadside	1.73 ± 0.07^{a}	2.26 ± 0.071^{b}	0.68 ± 0.01^{b}	0.80 ± 0.02^{c}	7 ± 0.7^{b}	9 ± 0.7^{a}
	p-value	0.011	0.048	0.007	0.008	0.009	0.003
	F-value	5.57	3.49	6.30	5.97	5.93	7.86
Monsoon	Forest-edge	1.78 ± 0.02^{a}	2.78 ± 0.02^{b}	0.77 ± 0.01^{a}	0.86 ± 0.016^{b}	11 ± 0.59^{b}	15 ± 1.07^{a}
	Fallow land	2.11 ± 0.03^{a}	2.92 ± 0.008^{c}	0.84 ± 0.02^{a}	0.88 ± 0.016^{bc}	13 ± 0.82^{bc}	17 ± 0.90^{ab}
	Roadside	1.76 ± 0.01^{b}	2.82 ± 0.009^{a}	0.83 ± 0.03^{b}	0.93 ± 0.009^{c}	10 ± 0.38^{c}	14 ± 0.52^{b}
	p-value	<0.001	<0.001	0.015	0.005	0.019	0.015
	F-value	76.72	55.33	5.13	6.82	4.76	5.10

Table A3. Paired sample *t*-test results of soil attributes of invaded and non-invaded plots. The values shown are means (±Standard error), *p*-value indicate significant different at $p < 0.05$, ns = not significant. Abbreviation: SOC, Soil organic carbon; N, Nitrogen; K, Potassium; P, Phosphorus; LI, *Lantana camara* invaded; NI, non-invaded.

Season	pH		SOC %		N %		K		P	
	LI	NI	LI	NI	LI	NI	LI	NI	LI	NI
Pre-monsoon	6.02 ± 0.139	6.15 ± 0.132	3.60 ± 0.119	3.29 ± 0.107	0.29 ± 0.016	0.24 ± 0.013	161.64 ± 8.55	185.52 ± 9.34	39.30 ± 2.21	30.62 ± 1.96
***p*-value**	ns		0.04		<0.001		<0.001		<0.001	
Monsoon	6.20 ± 0.113	6.38 ± 0.119	3.89 ± 0.128	3.43 ± 0.104	0.34 ± 0.022	0.27 ± 0.015	155.92 ± 14.45	172.78 ± 13.62	45.33 ± 3.13	37.61 ± 2.45
***p*-value**	ns		0.003		<0.001		0.026		0.05	

Table A4. One way ANOVA results of soil attributes of invaded and non-invaded plots of each habitat and plot types. The values shown are means (±standard error). Means with different superscript letters are significantly different at $p < 0.05$. Abbreviations: SOC, soil organic carbon; N, nitrogen; K, Potassium; P, Phosphorus; *Lantana camera*-invaded; NI, non-invaded

Season	Habitat	pH		SOC %		N %		K		P	
		LI	N	LI	NI	LI	NI	LI	NI	LI	NI
Pre-monsoon	Forest-edge	6.18 ± 0.24^{a}	6.31 ± 0.21^{b}	4.12 ± 0.20^{a}	3.58 ± 0.17^{b}	0.32 ± 0.02^{a}	0.26 ± 0.02^{b}	187.33 ± 6.21^{a}	189.53 ± 5.64^{b}	40.60 ± 2.93^{a}	31.76 ± 2.92^{b}
	Fallow land	6.44 ± 0.22^{ab}	6.51 ± 0.20^{bc}	3.49 ± 0.14^{a}	3.38 ± 0.16^{bc}	0.37 ± 0.02^{b}	0.29 ± 0.01^{c}	195.29 ± 2.15^{b}	235.92 ± 11.38^{c}	47.81 ± 4.27^{ab}	37.46 ± 3.88^{bc}
	Roadside	5.51 ± 0.14^{b}	5.69 ± 0.18^{c}	3.24 ± 0.14^{b}	2.96 ± 0.16^{c}	0.22 ± 0.01^{b}	0.17 ± 0.008^{c}	109.04 ± 5.35^{b}	137.16 ± 5.09^{a}	30.58 ± 1.40^{b}	23.5 ± 1.46^{c}
	p-value	0.012	0.023	0.004	0.043	<0.001	<0.001	<0.001	<0.001	0.002	0.007
	F-value	5.50	4.50	7.23	3.64	18.86	16.74	89.64	41.84	8.53	6.22
Monsoon	Forest-edge	6.31 ± 0.21^{a}	6.47 ± 0.19^{b}	4.49 ± 0.15^{a}	3.54 ± 0.18^{b}	0.40 ± 0.02^{a}	0.34 ± 0.01^{b}	242.45 ± 12.23^{a}	252.08 ± 12.58^{b}	48.47 ± 3.52^{a}	$37.0.7 \pm 3.44^{b}$
	Fallow land	6.60 ± 0.11^{b}	6.76 ± 0.23^{bc}	3.76 ± 0.19^{a}	3.70 ± 0.14^{bc}	0.43 ± 0.02^{b}	0.31 ± 0.01^{c}	131.15 ± 10.03^{a}	166.26 ± 13.92^{c}	57.16 ± 3.72^{b}	45.58 ± 5.04^{bc}
	Roadside	5.75 ± 0.09 b	5.95 ± 0.07^{c}	3.47 ± 0.15^{b}	3.11 ± 0.13^{c}	0.21 ± 0.01^{b}	0.19 ± 0.009^{c}	101.03 ± 16.43^{b}	108.11 ± 6.80^{a}	31.77 ± 4.63^{b}	30.99 ± 2.94^{c}
	p-value	0.002	0.011	<0.001	0.034	<0.001	<0.001	<0.001	<0.001	<0.001	0.042
	F-value	8.67	5.51	9.73	3.96	40.80	42.57	30.21	42.11	10.43	3.66

References

1. CBD. Alien species that threaten ecosystems, habits or species. In *Convention on Biological Diversity;* Report on Consultations Regarding International Standards; FAO: Rome, Italy, 2008.
2. Downey, P.O.; Rirchardson, D.M. Alien plant invasions and native plant extinctions: A six-threshold framework. *AoB Plants* **2016**, *8*, plw047. [CrossRef] [PubMed]
3. Bellard, C.; Cassey, P.; Blackburn, T.M. Alien species as a driver of recent extinction. *Biol. Lett.* **2016**, *12*, 20150623. [CrossRef] [PubMed]
4. IPBES. *Summary for Policymakers of the Thematic Assessment Report on Invasive Alien Species and Their Control of the Intergovernmental Science-Policy Platform on Biodiversity and Ecosystem Services;* Roy, H.E., Pauchard, A., Stoett, P., Renard Truong, T., Bacher, S., Galil, B.S., Hulme, P.E., Ikeda, T., Sankaran, K.V., McGeoch, M.A., et al., Eds.; IPBES Secretariat: Bonn, Germany, 2023. [CrossRef]
5. Gibbons, S.M.; Lekberg, Y.; Mummey, D.L.; Sangwan, N.; Ramsey, P.W.; Gilbert, J.A. Invasive plants rapidly reshape soil properties in a grassland ecosystem. *MSystems* **2017**, *2*, e00178-16. [CrossRef] [PubMed]
6. Le Maitre, D.C.; Richardson, D.M.; Chapman, R.A. Alien plant invasions in South Africa: Driving forces and the human dimension: Working for water. *S. Afr. J. Sci.* **2004**, *100*, 103–112.
7. Brooks, M.L.; D'antonio, C.M.; Richardson, D.M.; Grace, J.B.; Keeley, J.E.; DiTomaso, J.M.; Hobbs, R.J.; Pellant, M.; Pyke, D. Effects of invasive alien plants on fire regimes. *BioScience* **2004**, *54*, 677–688. [CrossRef]
8. Hejda, M.; Pyšek, P.; Jarošík, V. Impact of invasive plants on the species richness, diversity and composition of invaded communities. *J. Ecol.* **2009**, *97*, 393–403. [CrossRef]
9. Rodriguez, L.F. Can invasive species facilitate native species? Evidence of how, when, and why these impacts occur. *Biol. Invasions* **2006**, *8*, 927–939. [CrossRef]
10. Shrestha, B.B. Management of invasive alien plants in Nepal: Current practices and future prospects. In *Tropical Ecosystems: Structure, Functions and Challenges in the Face of Global Change;* Springer: Berlin/Heidelberg, Germany, 2019; pp. 45–68.
11. Chaudhary, R.P.; Uprety, Y.; Rimal, S.K. Deforestation in Nepal: Causes, consequences and responses. *Biol. Environ. Hazards Risks Disasters* **2016**, *12*, 335–372.
12. Shrestha, U.B.; Shrestha, B.B. Climate change amplifies plant invasion hotspots in Nepal. *Divers. Distrib.* **2019**, *25*, 1599–1612. [CrossRef]
13. Shrestha, B.B.; Shrestha, K.K. Invasions of alien plant species in Nepal: Patterns and process. *Invasive Alien Species Obs. Issues Around World* **2021**, *2*, 168–183. [CrossRef]
14. Maharjan, S.; Joshi, S.; Shrestha, B.B.; Devkota, A.; Jha, P.K. Life History Traits and Invasion Success of *Parthenium hysterophorus* L. in Kathmandu Valley, Nepal. *Nepal J. Sci. Technol.* **2014**, *15*, 31–38. [CrossRef]
15. Bhattarai, K.R.; Maren, I.E.; Subedi, S.C. Biodiversity and invasibility: Distribution patterns of invasive plant species in the Himalayas, Nepal. *J. Mt. Sci.* **2014**, *11*, 688–696. [CrossRef]
16. Shrestha, B.B.; Pokhrel, K.; Paudel, N.; Poudel, S.; Shabbir, A.; Adkins, S.W. Distribution of *Parthenium hysterophorus* and one of its biological control agents (Coleoptera: *Zygogramma bicolorata*) in Nepal. *Weed Res.* **2019**, *59*, 467–478. [CrossRef]
17. Bhatta, S.; Joshi, L.R.; Shrestha, B.B. Distribution and impact of invasive alien plant species in Bardia National Park, western Nepal. *Environ. Conserv.* **2020**, *47*, 197–205. [CrossRef]
18. Thapa, L.B.; Kaewchumnong, K.; Sinkkonen, A.; Sridith, K. "Soaked in rainwater" effect of *Ageratina adenophora* on seedling growth and development of native tree species in Nepal. *Flora* **2020**, *263*, 151554. [CrossRef]
19. Roxy, M.K.; Ritika, K.; Terray, P.; Murtugudde, R.; Ashok, K.; Goswami, B.N. Drying of Indian subcontinent by rapid Indian Ocean warming and a weakening land-sea thermal gradient. *Nat. Commun.* **2015**, *6*, 7423. [CrossRef]
20. Post, A.K.; Knapp, A.K. The importance of extreme rainfall events and their timing in a semi-arid grassland. *J. Ecol.* **2020**, *108*, 2431–2443. [CrossRef]
21. Lowe, S.; Browne, M.; Boudjelas, S.; De Poorter, M. *100 of the World's Worst Invasive Alien Species A Selection from the Global Invasive Species Database;* The Invasive Species Specialist Group (ISSG): Rome Italy, 2000.
22. Dobhal, P.K.; Kohli, R.K.; Batish, D.R. Evaluation of impact of *Lantana camara* L. invasion, on four major woody shrubs, along Nayar river of Pauri Garhwal in Uttarakhand Himalaya. *Int. J. Biodivers. Conserv.* **2010**, *2*, 166–172.
23. Gooden, B.; French, K.; Turner, P.J.; Downey, P.O. Impact threshold for an alien plant invader, *Lantana camara* L., on native plant communities. *Biol. Conserv.* **2009**, *142*, 2631–2641. [CrossRef]
24. Bhakat, R.K.; Maiti, P.P. Invasiveness and allelopathy as a threat to biodiversity. In Proceedings of the International Seminar on Multidisciplinary Approaches in Angiosperm Systematics, Kalyani, India, January 2012; pp. 748–751.
25. Sharma, G.P.; Raghubanshi, A.S. Lantana invasion alters soil nitrogen pools and processes in the tropical dry deciduous forest of India. *Appl. Soil Ecol.* **2009**, *42*, 134–140. [CrossRef]
26. Fan, L.; Chen, Y.; Yuan, J.; Yang, Z. The effect of *Lantana camara* Linn. invasion on soil chemical and microbiological properties and plant biomass accumulation in southern China. *Geoderma* **2010**, *154*, 370–378. [CrossRef]
27. Osunkoya, O.O.; Perrett, C. *Lantana camara* L. (Verbenaceae) invasion effects on soil physicochemical properties. *Biol. Fertil. Soils* **2011**, *47*, 349–355. [CrossRef]
28. Simba, Y.R.; Kamweya, A.M.; Mwangi, P.N.; Ochora, J.M. Impact of the invasive shrub, *Lantana camara* L. on soil properties in Nairobi National Park, Kenya. *Int. J. Biol. Divers. Conserv.* **2013**, *5*, 803–809.

29. Ruwanza, S.; Shackleton, C.M. Effects of the invasive shrub *Lantana camara* on soil properties in the Eastern Cape, South Africa. *Weed Biol. Manag.* **2016**, *16*, 67–79. [CrossRef]
30. Mahla, N.; Mlambo, D. Influence of two co-occurring invasive plant species on resident woody species and surface soil properties in Chipinge Safari Area, Zimbabwe. *Trop. Ecol.* **2019**, *60*, 129–139. [CrossRef]
31. Ruwanza, S. Effects of *Lantana camara* invasion on vegetation diversity and composition in the Vhembe Biosphere Reserve, Limpopo Province of South Africa. *Sci. Afr.* **2020**, *10*, e00610. [CrossRef]
32. Chacón, N.; Herrera, I.; Flores, S.; González, J.A.; Nassar, J.M. Chemical, physical, and biochemical soil properties and plant roots as affected by native and exotic plants in Neotropical arid zones. *Biol. Fertil. Soils* **2009**, *45*, 321–328. [CrossRef]
33. Vitousek, P.M. Biological invasions and ecosystem processes: Towards an integration of population biology and ecosystem studies. *Oikos* **1990**, *57*, 7–13. [CrossRef]
34. Kershaw, K.A. *Quantitative and Dynamic Plant Ecology*, 2nd ed.; Arnold, E., Krebs, C.J., Eds.; Ecological Methodology (No. QH541. 15. S72. K74 1999); Harper & Row: New York, NY, USA, 1973.
35. Misra, R. *Ecology Workbook*; Oxford and IBH Publishing Company: Calcutta, India, 1968.
36. Daubenmire, R. A canopy-coverage method of vegetation analysis. *Northwest Sci.* **1959**, *33*, 43–64.
37. KATH (National Herbarium & Plant Laboratories). Flora of Nepal. 2019. Available online: https://kath.gov.np/Flora_of_Nepal (accessed on 15 April 2019).
38. Magurran, A.E. *Measuring Biological Diversity*; Blackwell Publishing: Oxford, UK, 2004; 256p.
39. R Development Core Team. *R: A Language and Environment for Statistical Computing*; R Foundation for Statistical Computing: Vienna, Austria, 2020.
40. Gupta, P.K. *Methods in Environmental Analysis: Water, Soil and Air*, 1st ed.; Updesh Purohit for Agrobios, India Jodhpur Agro House: Jodhpur, India, 2004; pp. 47–48.
41. Black, C.A.; Evans, D.D.; White, J.L. Methods of soil analysis: Chemical and microbiological properties. Physical and mineralogical properties including statistics of measurement and sampling. *Science* **1965**, *151*, 982–983.
42. Olsen, S.R.; Sommers, L.E. Phosphorous. In *Methods of Soil Analysis, Part 2*, 2nd ed.; Page, A.L., Ed.; Chemical and Microbiological Properties, ASA-SSSa, Inc.: Madison, WI, USA, 1982.
43. NARC. *Methods of Soil Sample Collection and Analysis*; Soil Science Division, Council of National Agriculture Research Center: Khumaltar, Nepal, 2013; 167p.
44. Fetcher, N.; Strain, B.R.; Oberbauer, S.F. Effects of light regime on the growth, leaf morphology, and water relations of seedlings of two species of tropical trees. *Oecologia* **1983**, *58*, 314–319. [CrossRef]
45. Turton, S.M.; Duff, G.A. Light environments and floristic composition across an open forest-rainforest boundary in northeastern Queensland. *Aust. J. Ecol.* **1992**, *17*, 415–423. [CrossRef]
46. Tilman, D. *Resource Competition and Community Structure*; Princeton University Press: Princeton, NJ, USA, 1982; Volume 28, pp. 1043–1045. [CrossRef]
47. Sharma, G.P.; Raghubanshi, A.S. Tree population structure, regeneration and expected future composition at different levels of *L. camara* L. invasion in the Vindhyan tropical dry deciduous forest of India. *Lyonia* **2006**, *11*, 25–37.
48. Taylor SKumar, L.; Reid, N.; Kriticos, D.J. Climate change and the potential distribution of an invasive shrub, *Lantana camara* L. *PLoS ONE* **2012**, *7*, e35565. [CrossRef] [PubMed]
49. Zhang, Q.; Zhang, Y.; Peng, S.; Zobel, K. Climate warming may facilitate invasion of the exotic shrub *Lantana camara*. *PLoS ONE* **2014**, *9*, e105500. [CrossRef]
50. Dobhal, P.K.; Kohli, R.K.; Batish, D.R. Impact of *Lantana camara* L. invasion on riparian vegetation of Nayar region in Garhwal Himalayas (Uttarakhand, India). *J. Ecol. Nat. Environ.* **2011**, *3*, 11–22.
51. Aravindhan, V.; Rajendran, A. Impact of invasive species *Lantana camara* (L.) on the vegetation of Velliangiri Hills, the Southern Western Ghats, India. *Glob. J. Environ. Res.* **2014**, *8*, 35–40.
52. Vila, M.; Weiner, J. Are invasive plant species better competitors than native plant species?—Evidence from pair-wise experiments. *Oikos* **2004**, *105*, 229–238. [CrossRef]
53. Levine, J.M.; D'Antonio, C.M. Elton revisited: A review of evidence linking diversity and invasibility. *Oikos* **1999**, *87*, 15–26. [CrossRef]
54. Rouw, A.D. Chromolaena odorata in the farming systems of South-West Côte d'Ivoire. In *Proceedings of the Distribution, Ecology and Management of Chromolaena odorata, 1996*; ORSTOM, ICRAF and University of Guam, Mangilao, GUAM: Guam, Micronesia, 1996; pp. 76–87.
55. Tjitrosemito, S. The management of Chromolaena odorata. In Proceedings of the Third International Chromolaena Workshop on Distribution, Ecology and Management of Chromolaena odorata, 1996; Robinson, H., Ed.; ORSTOM, ICRAF and University of Guam, Mangilao, GUAM: Guam, Micronesia, 1996; pp. 135–142.
56. Adhikari, A.; Subedi, A.; Tiwari, A.; Shrestha, B.B. Impacts of road on plant invasions in the Middle Mountain region of central Nepal. *J. Mt. Sci.* **2024**, *21*, 619–632. [CrossRef]
57. Starfinger, U.; Kowarik, I.; Rode, M.; Schepker, H. From desirable ornamental plant to pest to accepted addition to the Flora?—The perception of an alien tree species through the centuries. *Biol. Invasions* **2003**, *5*, 323–335. [CrossRef]
58. Harper, K.A.; Macdonald, S.E.; Burton, P.J. Edge influence on forest structure and composition in fragmented landscapes. *Conserv. Biol.* **2005**, *19*, 768–782. [CrossRef]

59. Guo, Q.; Brown, J.H. Temporal fluctuations and experimental effects in desert plant communities. *Oecologia* **1996**, *107*, 568–577. [CrossRef] [PubMed]
60. Baskin, C.C.; Baskin, J.M. *Seeds: Ecology, Biogeography, and Evolution of Dormancy and Germination*; Academic Press: Cambridge, MA, USA, 2014.
61. Osunkoya, O.O.; Perrett, C.; Fernando, C. Population viability analysis models for *Lantana camara* L. (Verbenaceae): A weed of national significance. In Proceedings of the 17th Australasian Weeds Conference, Christchurch, New Zealand, 26–30 September 2010; Zydenbos, S.M., Ed.; New Zealand Plant Protection Society: Auckland, New Zealand, 2010; pp. 91–94.
62. Sharma, G.P.; Raghubanshi, A.S. Effect of *Lantana camara* L. cover on plant species depletion in the Vindhyan tropical dry deciduous forest of India. *Appl. Ecol. Environ. Res.* **2007**, *5*, 109–121. [CrossRef]
63. Fu, D.; Wu, X.; Huang, N.; Duan, C. Effects of the invasive herb *Ageratina adenophora* on understory plant communities and tree seedling growth in *Pinus yunnanensis* forests in Yunnan, China. *J. For. Res.* **2018**, *23*, 112–119. [CrossRef]
64. Castillo, J.M.; Leira-Doce, P.; Carrión-Tacuri, J.; Munoz-Guacho, E.; Arroyo-Solís, A.; Curado, G.; Tye, A. Contrasting strategies to cope with drought by invasive and endemic species of *Lantana* in Galapagos. *Biodivers. Conserv.* **2007**, *16*, 2123–2136. [CrossRef]
65. Dassonville, N.; Vanderhoeven, S.; Gruber, W.; Meerts, P. Invasion by *Fallopia japonica* increases topsoil mineral nutrient concentrations. *Ecoscience* **2008**, *14*, 230–240. [CrossRef]
66. Kumar, M.; Verma, A.K.; Garkoti, S.C. Lantana camara and *Ageratina adenophora* invasion alter the understory species composition and diversity of chir pine forest in central Himalaya, India. *Acta Oecologica* **2020**, *109*, 103–115. [CrossRef]
67. Kumar, M.; Garkoti, S.C. Functional traits, growth patterns, and litter dynamics of invasive alien and co-occurring native shrub species of chir pine forest in the central Himalaya, India. *Plant Ecol.* **2021**, *222*, 723–735. [CrossRef]
68. Vitti, S.; Pellegrini, E.; Casolo, V.; Trotta, G.; Boscutti, F. Contrasting responses of native and alien plant species to soil properties shed new light on the invasion of dune systems. *J. Plant Ecol.* **2020**, *13*, 667–675. [CrossRef]
69. Chatanga, P. Impacts of the Alien Species *L. camara* L. on Vegetation in Northern Gonarezhou National Park, Zimbabwe. Master's Thesis, University of Zimbabwe, Harare, Zimbabwe, 2007; pp. 1–84.
70. Thomas, S.E.; Ellison, C.A. A century of classical biological control of *Lantana camara*: Can pathogens make a significant difference. In Proceedings of the X International Symposium on Biological Control of Weeds, Bozeman, MT, USA, 4–14 July 1999; Montana State University: Bozeman, MT, USA, 2000; Volume 4, pp. 97–104.
71. ISSG. *One Hundred of the World Worst Invasive Allien Species. A Selection from the Global Invasive Database*; ISSG: Auckland, New Zealand, 2006.

forests MDPI

Article

Drivers of Plant Invasion in Stream Restoration

Douglas A. DeBerry [1,*] and Dakota M. Hunter [2,†]

1 Environment and Sustainability Program, College of William & Mary, Williamsburg, VA 23187, USA
2 Biology Department, College of William & Mary, Williamsburg, VA 23187, USA; dhunter@copperheadconsulting.com
* Correspondence: dadeberry@wm.edu
† Current address: Copperhead Environmental Consulting, Inc., Charlottesville, VA 22901, USA.

Abstract: This study characterized important environmental factors that contribute to plant invasion in the forested riparian zones surrounding stream restoration sites. We sampled vegetation and environmental variables (light availability, soil physiochemistry, and site age) across invasion gradients at multiple sites in Virginia, USA. Data analysis involved a multimetric statistical approach combining correlation and Canonical Correspondence Analysis (CCA) to arrive at a plausible model for invasion risk by species. We targeted three of the most problematic invaders in these systems: *Lespedeza cuneata* (sericea lespedeza), *Lonicera japonica* (Japanese honeysuckle), and *Microstegium vimineum* (Japanese stiltgrass). Our analysis revealed species-specific environmental drivers of invasion, with certain factors consistently important across all targeted invaders—notably, canopy cover, nitrogen availability, soil texture, and bioavailable phosphorus, as indicated by the importance of certain proxies (e.g., metal cations). The results of this research have been used to develop a suite of best practices that can be implemented at the outset of a stream restoration project to reduce the risk of invasion in the riparian forests surrounding these sites.

Keywords: stream restoration; stream mitigation; invasive plant species; invasive plant management; *Lespedeza cuneata; Lonicera japonica; Microstegium vimineum*

1. Introduction

Biological invasion has become one of the most important contemporary issues in the field of restoration ecology, with invasive vegetation representing a significant consideration in terms of post-restoration invasion risk [1]. From the first principles in invasion ecology, we know that disturbance represents a mode of introduction for invasive plants, and ecological restoration sites can be particularly susceptible because the practices used to create, restore, or enhance ecological conditions are often the same types of disturbances that leave a site vulnerable to invasion [2]. This is especially true of stream restoration sites, which have open energy cycles and are exposed to multiple vectors of plant dispersal, such as flowing water and flooding of the riparian zone [3]. Invasive plant management on these sites has increased considerably in recent decades, and in most cases, it is compulsory, i.e., it is required as a condition of an environmental permit or a banking agreement for stream restoration after invasive plant species colonize [4]. This reactive posture to management typically results in the use of non-selective herbicides and significant collateral damage to native species [5], but what is needed is a proactive approach to stream restoration with best practices aimed at inhibiting invasion from the start.

To evaluate this problem, we began by studying the "invasion gradient"—by which we mean the transition from high to low abundance of a specific plant invader—within the forested riparian corridors surrounding existing stream restoration projects in Virginia, USA. Importantly, we focused on invasion gradients where the environmental conditions were apparently the same across the gradient (i.e., similar elevation, light availability, soil conditions, etc., from the invaded to the uninvaded end of the gradient). In this manner, we could more closely inspect the conditions that "tip the scale" in favor of invasion on sites that otherwise appear similar, ultimately developing an understanding of the environmental conditions that are most consistently correlated with a higher overall cover of invasive plant species on these sites.

For the purposes of this study, an "invasive plant species" is defined as one that colonizes a new habitat, rapidly expands in space once there, and has negative consequences for the species already in the space that it enters [6]. This definition is consistent with the criteria used to develop current regional lists of invasive plant species (e.g., [7]). We use "invader" and "invasive species" interchangeably throughout and refer to the condition of invasive species dominance on a site as "invasion" or "invaded". Also, "stream restoration" means returning a degraded stream channel and its riparian corridor to a close approximation of its remaining ecological potential using the principles of natural channel design [8]. This definition accords with the one used by regulatory programs responsible for requiring stream restoration as compensatory mitigation for environmental impacts (e.g., Section 404 of the Clean Water Act and analogous state water protection laws), which is the administrative provenance for the sites in this study.

A comprehensive account of the theoretical ecology undergirding our expectations for outcomes was reviewed in a previous study by the authors (see [2]). The most important idea to highlight here is the role that the combination of stress and disturbance plays in structuring plant communities. From a plant ecology perspective, stress is defined as an abiotic or biotic factor that reduces fitness or growth (e.g., nutrient or light limitation), whereas a disturbance is a factor that results in the direct removal of biomass (e.g., flood scour, fire, or human land development; [9]). From our research, the model suggested by Alpert et al. [6] seems the most applicable to the forested ecosystems surrounding stream restoration sites. In that model, disturbance regimes of intermediate intensity and frequency combined with moderate stress regimes reduce the potential for plant invasion (Figure 1). Alternatively, sites that are recently disturbed and low in stress are prone to invasion.

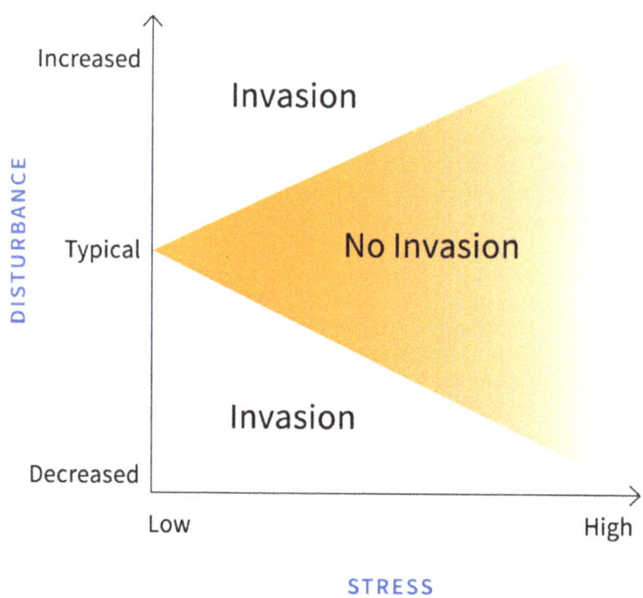

Figure 1. Conceptual model showing the relationship between stress, disturbance, and invasion (adapted from Alpert et al. [6]). Source: VHB, Inc., used with permission.

It is easy to see how a stream restoration project could result in the latter condition: the construction practices used to restore the channel and riparian corridor always involve disturbance [8], and the open nutrient and energy cycles within the floodplain of a stream system combined with potential increases in light availability from tree clearing lead to higher resource availability and, therefore, lower levels of stress [3,10]. From an invasion perspective, this condition is exacerbated by the reality that flood water is the main dispersal vector for many invasive plants [11], making stream restoration sites and their adjacent forested floodplains susceptible to recruitment by invaders originating from the upstream watershed [3].

Recognizing that disturbance is a procedural reality of stream restoration, our question is whether or not the "stress side" of the stress–disturbance dynamic can actually be manipulated during the construction phase of a project to reduce the risk of plant invasion. The purpose of this study was to answer this question by first determining which environmental factors were correlated with higher percentages of our targeted invaders in the context of the overall plant community along the invasion gradient. We anticipated that both higher levels of light and soil nutrient availability would emerge as important factors.

1.1. *Target Invaders*

We met with several stream restoration practitioners (e.g., mitigation bankers, agency representatives, environmental consultants, etc.) to create a list of the most problematic invaders within the forested riparian zones surrounding stream restoration sites. From this list, we selected three target invaders based on site-level criteria: *Lespedeza cuneata*, *Lonicera japonica*, and *Microstegium vimineum*. These species are described below.

1.1.1. *Lespedeza cuneata* (Dum.-Cours.) G. Don (Family: Fabaceae)

Lespedeza cuneata (sericea lespedeza) is a warm season perennial legume from east Asia that grows to a height of 1–2 m and can form dense populations where established [12] (Figure 2a). It was introduced to the USA in the 1890s to be tested for agricultural production and erosion control and has escaped into many natural habitats since its introduction [13]. *Lespedeza cuneata* is classified as an invasive species in most states where it has naturalized, and in Virginia, it is listed among the species with the highest risk of invasion [7]. Like most legumes, *L. cuneata* benefits from symbiosis with nitrogen-fixing bacteria, enabling it to inhabit nutrient-poor conditions and eroded soils that are typically inhospitable to other plants; however, it is shade-intolerant and does not establish well under a dense canopy [14]. With an extensive taproot, *L. cuneata* can survive extended drought conditions in the well-drained soils that it frequents, and it can also survive in a wide range of soil pH conditions from strongly acid to slightly basic [12]. Modes of invasion in this species have been studied, with general consensus that *L. cuneata* is able to modify its environment and facilitate localized dominance [15,16]. The species has a high level of tannins in its tissues, making it unpalatable to wildlife and thus resistant to herbivory [17,18]. In addition, phenolic compounds emitted into the soil through root exudates or decomposing plant residues have been shown to have allelopathic properties, increasing its competitive ability by altering nutrient uptake efficiency and decreasing germination in other species [12,16]. *Lespedeza cuneata* can reduce nutrient availability for other species via the rapid acquisition of essential nutrients (e.g., phosphorus, potassium) and slow release back to the soil via lower rates of decomposition than other plants, a phenomenon mediated by secondary compounds [17]. All of these properties contribute to localized *L. cuneata* dominance, making the species a nuisance in ecosystem restoration and a particularly problematic invader of riparian restoration projects [16,19].

Figure 2. Representative photos of study species on stream restoration sites: (a) *L. cuneata*, (b) *L. japonica*, (c) *M. vimineum*.

1.1.2. *Lonicera japonica* Thunb. (Family: Caprifoliaceae)

Lonicera japonica (Japanese honeysuckle), a perennial vine that turns woody with age, is from eastern Asia and was introduced to the USA in the early 1800s as a garden plant [13]. Its ability to escape into natural habitats was first noticed along the Potomac River near Washington, DC, in 1882, and since that time, it has expanded and become invasive in nearly every USA state east of the Mississippi River [20]. In Virginia, it is listed among the species that pose the highest risk of invasion [7]. Like most successful invaders, *L. japonica* colonization is greatly enhanced by disturbance [21], with seeds being dispersed primarily by birds that forage on the fruits [13]. Although it is a prolific seeder, it does not form persistent seedbanks, and thus, populations typically expand vegetatively [22,23]. *L. japonica* inhabits a range of habitat types, soil moistures, and pH conditions but is generally found in the well-drained, circumneutral pH soils of forests, edges, and clearings. Due to high transpiration rates, it does not tolerate prolonged drought and therefore tends to prefer mesic habitats, making riparian zones, streambanks, and floodplains particularly susceptible to invasion [20,24]. Although it can live in shady habitats, partial or full shade has been shown to inhibit *L. japonica* growth, and studies consistently highlight its preference for open environments where it can aggressively exploit canopy gaps or clearings, grow in dense thickets, and effectively smother other species [20,25] (Figure 2b). Consensus on factors contributing to invasiveness in *L. japonica* has not been reached, but all of the following functional traits have been implicated in the literature: allelopathy [26]; no known enemies and a strong compensatory response to herbivory [20]; autogenic control via the modification of environmental conditions and community structure, including nitrogen and carbon pools and the relative availability of nutrients for other species [27]; and morphological plasticity of leaf area and growth habit, which has allowed for increased resilience and expansion under disturbed conditions [28]. All of these factors make *L. japonica* a considerable threat to restoration projects, particularly in streams and riparian zones [29].

1.1.3. *Microstegium vimineum* (Trin.) A. Camus (Family: Poaceae)

Microstegium vimineum is an annual grass originating from Asia that grows in a variety of habitats and field conditions. It has been described as shade- and flood-tolerant [30–32] (Figure 2c) and is adept at seed dispersal, making it a highly invasive plant species in the region [7]. *Microstegium vimineum* has been shown to reduce native plant richness [31,33] and negatively impact native insect communities [34]. Herbicides can be effective at managing the populations within a single year [35,36], but because *M. vimineum* forms persistent seed banks, it can develop resilience to chemical treatment due to residual seed bank germination in subsequent years [37,38]. Modes of invasion seem to be facilitated by increases in light availability and soil nutrient content [32]. Riparian zones can be particularly susceptible to invasion because flowing water is a primary dispersal mechanism for *M. vimineum* seeds, so floodwaters in unidirectional lotic systems contribute greatly to its expansion and distribution in watersheds. Much attention has been paid to studying *M. vimineum* in these types of habitats [30,32,36], and its invasion potential on stream restoration sites is of particular importance in the Mid-Atlantic Region [39].

2. Materials and Methods

2.1. Study Sites

Representative field sites were chosen from a pool of 30 available sites in Virginia, USA, based on the following selection criteria: (1) the sites had to be non-tidal stream restoration projects developed under the purview of federal and/or state environmental laws, and (2) the sites had to have populations from our target list of invaders dominating at least portions of the project area. Field sites were assigned to the following age classes so that age could be evaluated as a factor: 1–2 years, 3–5 years, 6–10 years, 11–15 years, and >15 years following restoration [40]. Some of the stream restoration sites had different sections of the site that were restored at different times, correlating to the above age classes.

These sections were treated as individual sites for sampling purposes if spatial or temporal separation was deemed sufficient, and the sections fell into different age classes.

Among the stream sites screened, 21 met the suitability criteria and were selected for the study, which was conducted during the 2018 and 2019 growing seasons. Site ages ranged from 1 to 19 years post-construction and were evenly distributed across the Piedmont (10 sites) and Coastal Plain (11 sites) physiographic provinces in Virginia (Figure 3).

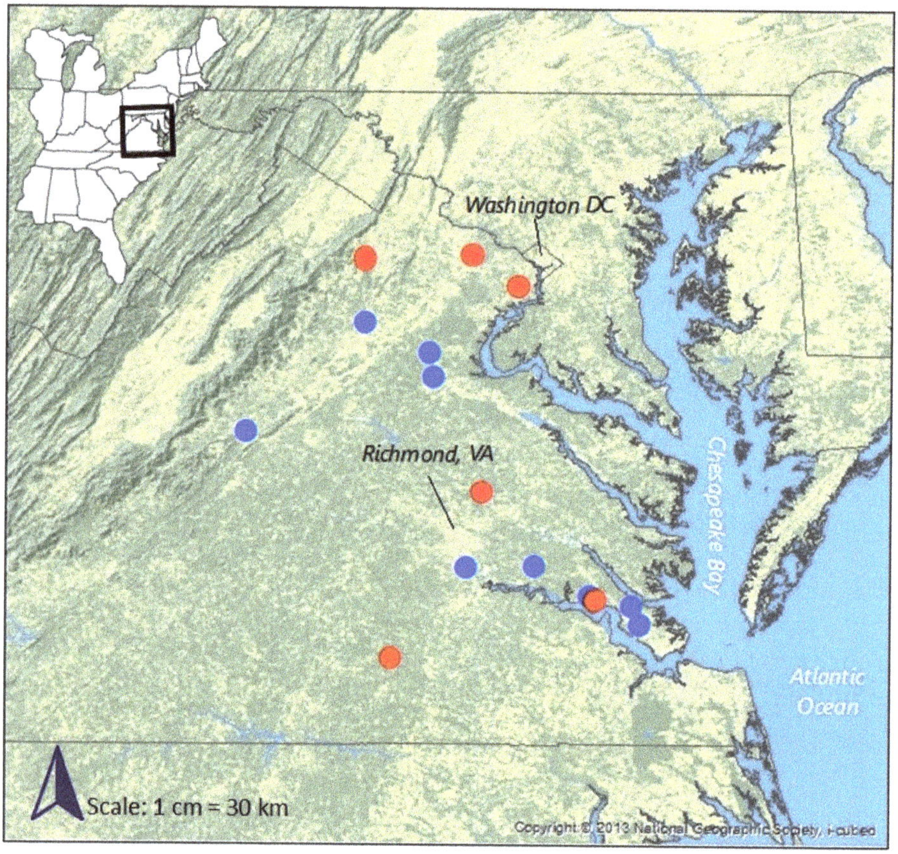

Figure 3. Stream restoration study site locations. Red symbols indicate sites in which more than one section of the site was sampled; blue symbols represent individual sites. (See text in Section 2.2 for explanation).

2.2. Sampling Methods

Hunter and DeBerry [2] described detailed methods for sampling across invasion gradients in wetland restoration sites, and these approaches were generally followed here with modifications as outlined below. The primary differences were in the transect orientation and randomization procedures used to determine plot locations.

Transect Configuration, Plot Locations, and Data Collection

Unlike wetland restoration sites, which are frequently constructed with gradual changes in relative elevation to target specific hydrologic regimes [41], stream restoration typically uses approaches that are designed to follow the geomorphology and natural contours of the surrounding landscape [8]. This means that relative elevations can change abruptly in the cross-sectional dimension of the stream valley from channel to floodplain and from floodplain to riparian buffer. To minimize the amount of apparent environmental variation across the invasion gradient, transects were aligned longitudinally with respect to the valley axis and positioned on one side of and parallel to the stream. Each of the five plots on a single transect was established at approximately the same distance away from the top-of-bank of the stream channel so that the elevation of each plot relative to the channel was approximately the same (Figure 4). The purpose of this modification was to ensure that the landscape position within the floodplain was similar for each plot along the invasion gradient. In other words, this configuration avoided the scenario of a randomly defined, straight-line transect with invaded plots down along the channel at one end of the gradient and uninvaded plots up on the ridgeline above the sidewalls of the valley (or vice versa)—an orientation that would have yielded easily detectable but unmeaningful environmental variation in the context of study objectives.

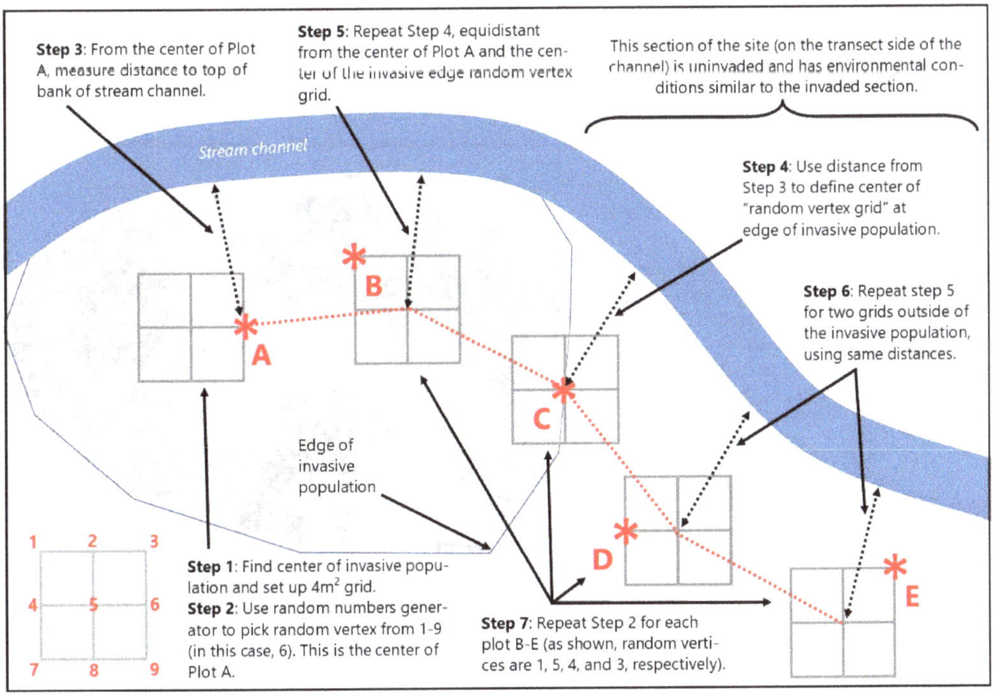

Figure 4. General layout of study design and transect configuration. Transect orientation, plot randomization, and final plot position (red asterisks) explained in text (Section 2.2).

The first plot location (Plot A, completely invaded) was determined using the 4 m^2 random vertex grid approach described by Hunter and DeBerry [2]. The distance from the center of Plot A to the top-of-bank of the stream channel was recorded, and this distance was used to define a new 4 m^2 grid centered at the edge of the invasive species population and offset the same distance from the channel bank. A third grid was established using the same offset distance from the stream bank, but this time within the invasive species population and equidistant from the center of Plot A and the center of the invasive edge grid. Using these same distances, the last two 4 m^2 grids were laid out at the uninvaded end of the transect. For each of the remaining plots (B, C, D, and E), random numbers between 1 and 9 were drawn to determine which vertex on each respective grid would be the center of each plot (Figure 4). This process ensured that all five plots along the invasion gradient were approximately the same distance from the channel (+/−1 m), with the exact location of each plot being randomized. The transects ranged in length from 30 to 100 m.

Soil sampling, canopy cover, and vegetation sampling methods all followed the same procedures described by Hunter and DeBerry [2]. There was one minor procedural difference: due to the prevalence of mature overstory canopies on the stream sites, cover estimates for tree species in plots were limited to a height of 3 m to avoid including abundance for taller trees growing outside of the transect study area. Canopy influence was assumed to be captured by canopy cover data. Also, wetland hydrology was not evaluated as an environmental factor because all stream sites were within upland portions of the forested floodplains adjacent to the restored stream channels.

2.3. Statistical Analysis

Data analysis was completed using R version 4.0.3 [42] including the packages vegan, Hmisc, and BiodiversityR [43–46]. The datasets for each invasive study species were analyzed separately due to expected variation in their relative tolerances for environmental stressors and discrepancies among growth requirements [11,13]. Correlations between environmental variables and relative invasion were calculated using the nonparametric Spearman rank-order correlation test, which is robust to deviations from normality and can detect both linear and monotonic relationships without appreciable loss of statistical power in comparison with parametric tests [47].

Canonical Correspondence Analysis (CCA) [48] was used to evaluate the overall community response to environmental variation along the invasion gradient. Rare species were removed from the abundance matrix of each dataset to neutralize their oversized effect on the X^2 distance used in CCA [49]. We used Borcard's stepwise approach based on the correspondence analysis (CA) component of CCA to evaluate the effect of progressive species removals and identify rare species accordingly [47]. Final CCA models were identified using the forward and backward model selection functionality in the ordistep() function of vegan, eventually eliminating model variables based on the significance of permutation tests in combination with the Akaike Information Criterion (AIC) [44]. We also identified and removed highly correlated variables by inspecting the variance inflation factor (VIF) [47]. The procedure described above produces a parsimonious model when all environmental variables are significant and the adjusted R^2 does not exceed that of the global model (i.e., the model with all environmental factors included) [44,50]. For CCA results, we report overall model inertia, which can be thought of as the total amount of variance in the model. In ordination approaches like CCA, "constrained" inertia represents the amount of variance explained by the environmental variables. For multivariate ecological data, values +/−20% are common [44,50]. All permutation tests were set at 1000 iterations, and all statistical analyses were evaluated at $\alpha = 0.05$.

3. Results

Two hundred eighty-six (286) species were documented in the overall stream restoration field study across 21 sites, 29 transects, and 145 plots sampled. Additional information on species composition and community properties may be found in [51]. Community modeling and environmental data are summarized below for each of the three target invasive species.

3.1. *L. cuneata*

Spearman results showed *L. cuneata* abundance significantly correlated with canopy cover ($r_s = -0.690$, $p << 0.001$). The relationship was negative, indicating that *L. cuneata* was more prevalent in areas with less canopy cover. No other environmental variables were significantly correlated with *L. cuneata* abundance in the Spearman analysis.

The CCA ordination was based on a community matrix with 11 dataset-rare species removed, leaving 148 species from the original matrix in the ordination. The final parsimonious CCA model for *L. cuneata* included five environmental variables—canopy cover, soil texture, nitrogen (N), potassium (K), and pH—which accounted for 25% of the total inertia in the ordination. All environmental factors were significant at $p \leq 0.01$ based on permutations. The ordination biplot (Figure 5) displays red arrows as vectors for environmental variables, with the vector length corresponding to the strength of correlation and vector direction indicating either a positive or negative relationship (e.g., plots aligned in the direction of and projected perpendicularly to an arrow were positively correlated with that environmental variable, and vice versa). Circles on the biplot represent plots, and circle size corresponds to the absolute abundance of *L. cuneata* within that plot (i.e., larger circles have higher abundance values). The first two ordination axes explained over 49% of the variation in the CCA model and thus were retained for the biplot. Figure 5 shows that, in general, canopy cover, pH, N, and K were negatively related to *L. cuneata* abundance in the context of the overall vegetative community. Soil texture appears positively correlated with *L. cuneata* abundance, suggesting that the invader tended to be on sites with coarser textures (i.e., texture values in the dataset were arranged on an ordinal scale from fine to coarse, so higher values corresponded to coarser-textured substrates).

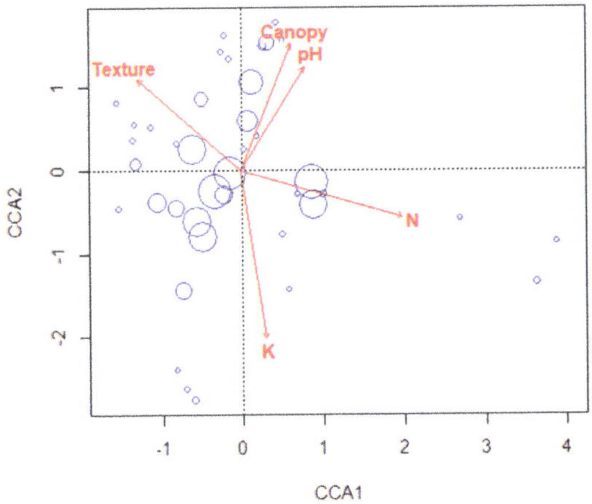

Figure 5. CCA biplot for *L. cuneata* dataset. Red arrows are vectors for environmental variables. Vector length indicates strength of correlation, and vector direction indicates a positive (pointing toward) or negative (pointing away) relationship to the plots, which are shown as circles with size corresponding to abundance of *L. cuneata* (i.e., larger circles = higher abundance). Plot relationships with environmental vectors are interpreted as perpendicular projections from blue circles to red arrows.

3.2. *L. japonica*

Spearman results indicated that *L. japonica* abundance was positively correlated with potassium (K) ($r_s = 0.298$, $p = 0.035$) and negatively correlated with canopy cover ($r_s = -0.749$, $p << 0.001$). For the CCA analysis, the *L. japonica* community matrix was reduced by 9 dataset-rare species, leaving 158 species from the original matrix in the ordination. The final parsimonious *L. japonica* CCA model included five environmental variables—canopy cover, texture, N, magnesium (Mg), and iron (Fe)—which accounted for over 20% of the total inertia in the ordination. All environmental factors were significant at $p < 0.01$ except texture ($p = 0.04$). The first two ordination axes displayed in Figure 6 explained 52% of the CCA variation. As with *L. cuneata*, canopy cover, and N showed a negative relationship with *L. japonica* abundance in the community, and texture showed a positive relationship with highly invaded sites. The other two important factors, Mg and Fe, both showed negative relationships with *L. japonica* abundance in the context of the plant community.

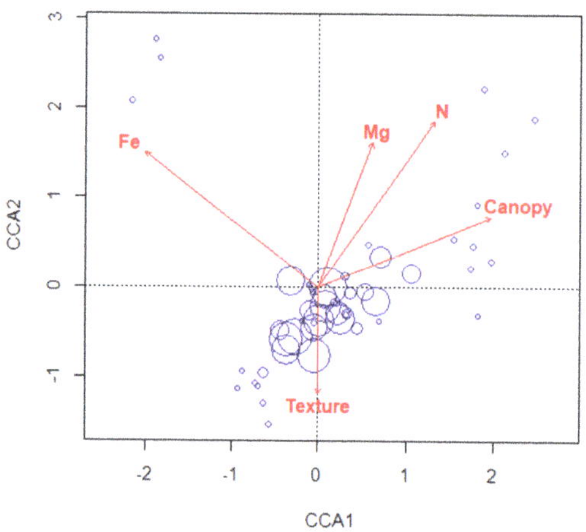

Figure 6. CCA biplot for *L. japonica* dataset. See *L. cuneata* text and Figure 5 caption for notes on interpretation.

3.3. *M. vimineum*

Spearman correlations showed that canopy cover was negatively correlated with *M. vimineum* abundance ($r_s = -0.467$, $p < 0.001$). No other environmental variables were significantly related to the *M. vimineum* invasion gradient in the correlation matrix. The CCA analysis for *M. vimineum* used a community matrix reduced by 24 dataset-rare species, leaving 167 species from the original matrix. The final parsimonious *M. vimineum* CCA model included four environmental variables—canopy cover, N, K, and manganese (Mn)—which accounted for 14% of the total inertia in the ordination. Canopy cover was significant in the model at $p < 0.001$; all other environmental factors were significant at $p < 0.05$. The first two ordination axes explained 57% of the CCA variation. As Figure 7 demonstrates, all four environmental variables showed a negative relationship with *M. vimineum* abundance in the vegetative community.

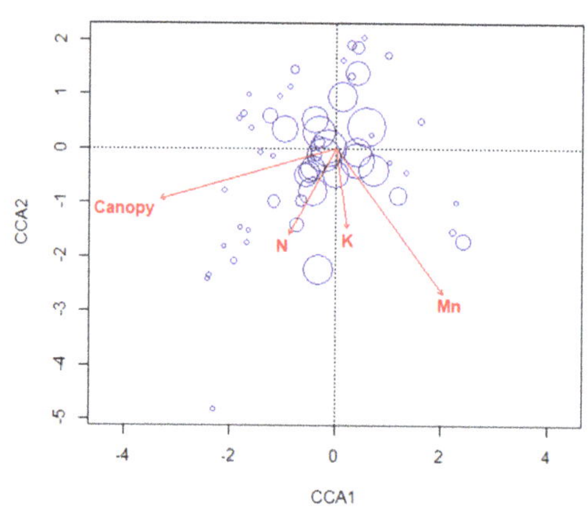

Figure 7. CCA biplot for *M. vimineum* dataset. See *L. cuneata* text and Figure 5 caption for notes on interpretation.

4. Discussion

Stream channels and their adjacent forested riparian zones are dynamic environments with open energy cycles [10]. The environmental drivers that structure plant communities in riparian zones—such as nutrient availability, sediment budgets, flood frequency and duration, and light availability—are all highly variable and subject to the pulsing forces that correspond to prevailing climate conditions and event-driven anomalies such as storms [52,53]. Floodplains are the "melting pots" of organic and inorganic matter that are shed from the upstream catchment, including the propagules of plants that use water or gravity as dispersal vectors in reproduction [54]. For stream restoration practitioners, this makes stream corridors some of the most challenging settings within which to develop planting plans and vegetation management strategies because the plant community is subject to continuous propagule pressure from upstream sources [55]. These factors are all the more challenging in view of biological invasion because, for many invasive plants, watersheds and drainage networks are the main conduits of dispersal [3].

Recognizing that most stream restoration projects will only be able to control environmental factors within project boundaries, the primary objectives of this study were directly related to the localized conditions within a stream restoration project that correlate with invasion gradients, i.e., the factors over which restoration designers and managers can have an influence. Ideally, stream restoration projects would be able to control entire watersheds, but that is neither practical due to costs nor achievable due to the intractability of acquiring land or gaining permission to access private property. Within the dynamic equilibrium (sensu [8]) ecosystem targets for a stream restoration project, there may be design and management considerations at localized scales that can help to build ecosystem resilience and resistance against future invasion [56]. With this in mind, the primary questions being addressed by this study were (1) What are the environmental drivers of plant invasion in stream restoration? (2) Based on the importance of stress and disturbance in structuring plant communities, are there best practices that can emerge from our understanding of these environmental drivers with the potential to reduce invasion risk at the outset of a restoration project?

The invasion gradients sampled in this study corresponded to light availability (canopy cover) and soil physiochemical variables, with canopy cover emerging as the most important factor across all datasets. Canopy cover was a significant variable in both the Spearman correlations and the CCA ordinations; thus, light availability exhibits a clear and distinct relationship with vegetation community development and invasive species impacts on these sites (see Figure 8). The influence of canopy cover and other environmental variables is discussed for each target invader below.

Figure 8. Representative hemispheric canopy photos taken above each plot from the invaded (**left** images) to the uninvaded (**right** images) ends of the invasion gradient. Per the methods described by Hunter and DeBerry [2], these images were analyzed to measure canopy cover. The left-to-right transition from partially open to closed canopy is apparent in all three series.

4.1. *L. cuneata*

Lespedeza cuneata is a shade-intolerant species [14], so the negative correlation between *L. cuneata* abundance and canopy cover in the stream study was not surprising. This result was also reflected in the CCA model (Figure 5), which accords with the notion that communities with *L. cuneata* as a dominant species would be more likely to have shade-intolerant associates like *Sorghastrum nutans*, *Panicum virgatum*, and *Solidago altissima*, all of which were co-dominants in the community matrix of this study. The negative relationships between *L. cuneata* abundance and soil N and K were unexpected but could have been predicted from the literature. As an "N-fixing" plant, *L. cuneata* benefits from N subsidies via a symbiotic relationship with *Rhizobium* bacteria in root nodules, which allows *L. cuneata* to inhabit nutrient-poor soils with low available N pools [16]. *Lespedeza cuneata* has been shown to reduce the availability of other macronutrients, including K, by rapid uptake and slow release back to the soil, a phenomenon that is apparently mediated by high levels of phenolic compounds (e.g., tannins) that reduce decomposition rates of its senescent tissues [17]. In dense *L. cuneata* populations on the stream sites in this study, autogenic control of nutrient availability could explain the K trends observed in the ordination.

The negative relationship between invasion and pH in the CCA model is also consistent with known ecological tolerances for *L. cuneata*, which has been documented in soils with pH as low as 4.0 [12]. On stream restoration sites, low pH could also be an indirect reflection of reduced soil nutrient status, as acidic substrates tend to be low in exchangeable bases, N, and P [57,58]. Finally, the positive relationship between *L. cuneata* dominance and coarse-textured substrates in the CCA model could signal localized disturbance in the riparian corridors of the study sites. Deposition of coarse-textured alluvium would be consistent with the type of flood-induced disturbance that could create habitat for a species like *L. cuneata* in a small stream floodplain. Anecdotally, evidence of fresh sediment accretion was noted on transects at several of the stream project sites, the consistency of which was predominantly sand and mostly occurred in close proximity to the portions of the site closest to the stream that had been more likely subjected to canopy thinning and, thus, higher light availability within the limits of disturbance during stream construction (D. DeBerry, pers. obs.). Soil like this with a low surface-to-volume particle ratio would likely also be low in ionic exchange sites for essential nutrients [58], a condition that would be consistent with the types of substrates that *L. cuneata* can exploit based on the above discussion.

All of this is consistent with *L. cuneata*'s relationship to the "high disturbance/high resource availability" model for invasion described in Section 1. With its ability to invade nutrient-poor soils, at face value, *L. cuneata* does not appear to use the "high resource" half of this strategy for dispersal and establishment. However, when the critical resource is light, *L. cuneata*'s mode of invasion does fit the paradigm because light-limited environments clearly reduce its abundance along the invasion gradient.

4.2. *L. japonica*

As with *L. cuneata*, the significant negative correlation between *L. japonica* abundance and canopy cover was anticipated and is consistent with its aggressive growth response to available light [25,27]. The positive correlation with K was not expected but is consistent with resource availability models for aggressive species [9]. Likewise, soil texture was positively related to *L. japonica* abundance in the CCA model (Figure 6), which, at the scale of the *L. japonica* transects in this study (30–80 m) most likely also signals some type of localized disturbance that is probably related to floodplain deposition.

The importance of soil N levels in the CCA model was expected, but the negative correlation between N and *L. japonica* abundance was not. Like *L. cuneata*, *L. japonica* has been shown to exhibit autogenic control over soil nutrient pools using different strategies. Morphological plasticity in *L. japonica* facilitates the opportunistic expansion of plant modules into favorable microhabitats [28], where fast-growing plants can exercise rapid uptake of mobile nutrients like N coupled to slow release back to the environment via the modification of labile carbon pools [27]. Additionally, studies have found allelopathic compounds in *L. japonica* tissues, with demonstrated effects on native species in eastern USA forests [26]. The "resource-hoarding" effects of rapid nutrient uptake and slow release could explain the lower ambient levels of soil N at the invaded end of the *L. japonica* gradient, and the "interference" effects of allelochemicals combined with morphological plasticity for micro-habitat exploitation would mean that lowering N pools would have minimal consequences for *L. japonica* itself. If this was occurring on stream sites, it would be consistent with the high disturbance/high resource concepts discussed above in that microsites exploited by *L. japonica* would have been characterized by both disturbance (habitat opening) and high resource availability (N, light), and only after *L. japonica* expansion and dominance would the relative availability of resources have been affected by autogenic controls.

The two remaining environmental variables that were important in the CCA model—soil Fe and Mg—are probably indirectly related to other factors influencing community structure along the invasion gradient. The most obvious one is the antagonistic relationship between Fe and bioavailable P, in which oxidized Fe immobilizes otherwise bioavailable P by chemical precipitation [59]. This would indirectly indicate a positive relationship between soil P and *L. japonica* abundance based on the reciprocal relationship with Fe in the model (Figure 6). The negative association between Mg and *L. japonica*-dominant sites in the CCA model could be indirectly related to soil texture. In the Mid-Atlantic Region, Mg deficiencies are often linked to coarse-textured soils from which cations are more easily leached [60]. On *L. japonica* sites, invasion was positively related to soil texture, so the Mg soil status could reflect this Mg-texture antagonism. If so, Mg would be viewed as less of a community driver and more of a secondary indicator in the *L. japonica* CCA analysis.

4.3. *M. vimineum*

As with the other two study species, the negative correlation between *M. vimineum* abundance and canopy cover is consistent with the literature. Although *M. vimineum* exhibits shade tolerance, it aggressively exploits available light and will correspond to relative light availability in forest understories [32,61,62]. The negative relationship between Mn and *M. vimineum* dominance in the CCA model (Figure 7) was unanticipated but is thought to reflect metal oxide/P dynamics. If so, the negative correlation would relate to greater P availability in *M. vimineum*-dominant sites in the same manner as described under *L. japonica* above for Fe-P complexes.

As with *L. japonica*, the influence of soil N on *M. vimineum* community dynamics was anticipated, but the direction of influence was not. Although functional traits are much different between the two species, there are similarities in their patterns of resource exploitation and habitat modification. Like *L. japonica*, *M. vimineum* rapidly acquires nutrients once it expands into a new area, can modify soil nutrient content [63], and has been shown to have allelopathic properties in competition experiments [64]. If this pattern of exploitation and resource depletion is characteristic of *M. vimineum* on stream restoration sites then it could explain the negative N relationship in the CCA model and possibly the negative K relationship as well, given the mobility of both of these elements in the soil and in plant tissues [60]. Admittedly, there is much variability in the overlap of the N and K vectors with *M. vimineum*-dominant plots in Figure 7, so it is probably prudent to view these relationships as less important than the other factors (particularly canopy cover, which was monotonically correlated with *M. vimineum* abundance).

5. Conclusions and Recommended Best Practices

By studying invasion gradients within the understory communities of forested riparian zones on stream restoration projects, we were able to identify environmental conditions that are correlated with increases in invasive plant species abundance on these sites. During the site screening phase, we chose to sample invasive populations where the apparent change in environmental conditions was negligible from the invaded end of the gradient to the uninvaded end (e.g., same relative elevations, same apparent native community structure, etc.). Through this, we were able to rephrase our environmental variation question in a more meaningful way: When there is no apparent difference in site conditions across an invasion gradient, what "tips the scale" in favor of invasion on stream restoration sites? The answer overwhelmingly pointed to canopy cover (light availability), which was directly correlated with the abundance of all three invaders. Further, based on CCA modeling, canopy cover was also important in structuring plant communities along the invasion gradients of all target species in this project. Other factors emerged in community modeling, the interpretation of which generally coincides with the literature on these species and on invasion ecology in general.

It is interesting to note that these findings were consistent with those of a related project on wetland restoration [2]. The fact that similar drivers of invasion were found in completely different ecological settings and among several different types of invaders (e.g., annual grasses, tall emergent perennials, semi-woody vines, and nitrogen-fixing forbs; [51]) suggests that there are unifying principles that can be used to assess potential risk of invasion on these types of projects. It should be noted that this study involved an extensive amount of fieldwork and analysis, and it would be neither practical nor cost-effective for stream restoration practitioners (e.g., mitigation bankers, consultants, managers, agency reviewers, etc.) to exert the same level of effort in diagnosing the relative risk of invasion and preemptively managing for it. However, our results do suggest a suite of proactive best practices that could be implemented during the construction phase of a stream restoration project to reduce or eliminate the amount of retroactive invasive species remediation required to meet performance standards in later years.

5.1. Best Practices

Recognizing that disturbance is a reality of the construction methods used to restore streams, we return to our question of whether the "stress side" of the stress-disturbance dynamic can be manipulated on stream restoration sites. We offer the following recommendations for planning and construction phase activities that could leverage the stress-disturbance relationship to preemptively inhibit plant invasions based on our research, organized by the most important environmental factors that emerged from this study.

5.1.1. Light Management

From our data, it is clear that light is the most important critical resource for plant invaders in the understories of these forested systems. This holds even for plants that are considered somewhat shade tolerant, like *M. vimineum* and *L. japonica*, two species for which any relative increase in light availability will stimulate rapid expansion [27,32]. Our thinking is that the quickest pathway to a closed forest canopy following construction is desirable, and this can be achieved in the following ways:

1. Limit tree clearing: The easiest and most effective way to promote canopy cover is to avoid removing it in the first place. Our study sites were located within riparian corridors that were already forested, so this would have been a viable alternative for the projects we studied and, based on the professional experience of both authors, most stream restoration projects within our region. Unfortunately, the techniques used to restore stream channels usually involve either relocating the channel or re-grading the existing channel to a new pattern and profile [8], all of which typically necessitate tree removal. However, with careful planning and proactive design specifications, many canopy trees can be saved, the value of which would be immeasurable in terms of proactive invasive plant management based on our results.
2. Plant larger trees: The planting plans that typically accompany stream restoration projects usually involve tree planting along with a performance standard for tree density. In Virginia, the density requirement has traditionally been 400 stems per acre (988 stems/hectare), which can be achieved by a "10-foot on-center" planting strategy [i.e., planting on a 10-foot (3.1-m) grid]. The problem with this standard is that it encourages practitioners to take the more economical route of planting

smaller, cheaper saplings to meet the density requirement. As might be expected, this will result in excess levels of sunlight within the riparian zone until the trees develop a closed canopy, which could take several years or even a few decades to evolve if the trees are grown from small saplings. One recent advancement in the area of ecological performance standards for wetland and stream mitigation is to measure "stem area at groundline" (SAG), which has a direct correlation with biomass accumulation and, therefore, ecosystem function in these systems [65]. Adopting a tree performance standard like SAG would be an important strategy to encourage restoration practitioners to plant larger trees and thereby hasten canopy closure with the disturbed portions of the stream restoration corridor.

3. Plant at a higher density: The logic behind this best practice follows from the concepts above. Although it comes at a higher cost, a high-density tree planting strategy would help to curtail understory invaders based on our research. In the cost–benefit analysis of this approach, projects would incur higher costs on the front end, but the benefits would most likely accrue over time as site managers would be spending much less money on invasive species removal over the life of a stream restoration project.
4. Plant early successional trees: Borrowed from the field of silviculture, this concept involves planting early successional trees in a "nurse crop" context, i.e., fast-growing trees with rapid canopy development that will ultimately facilitate the growth of other desirable, late-successional tree species [66]. In the eastern USA, examples of early successional trees include red maple (*Acer rubrum*), silver maple (*A. saccharinum*), river birch (*Betula nigra*), dogwood (*Cornus* spp.), sweetgum (*Liquidambar styraciflua*), sycamore (*Platanus occidentalis*), eastern cottonwood (*Populus deltoides*), and black willow (*Salix nigra*) [67–69].

5.1.2. Nutrient Management

As noted above, we believe that the most instructive model for understanding plant invasions on mitigation sites is one that combines environmental stress and disturbance to predict trends in resource acquisition and competition. Our observations over the course of this project have convinced us that the risk of invasion increases with increased disturbance and decreased stress, which is consistent with the literature on biological invasion and resource strategies of wild plants [9,56]. Stress in stream restoration is experienced by plants in various ways, but disturbance is more or less the same: sites are cleared and graded as necessary to establish appropriate elevations and desired landforms, and soil conditions are often manipulated via excavation, tilling, or the addition of soil amendments to improve conditions for target plant communities. Thus, the types of activities used in the construction sequence for a mitigation site are also the types of activities that typically "open the door" for biological invasion.

If the "disturbance" half of the stress-disturbance dynamic is an unavoidable consequence of the construction sequence, there may be alternative approaches that would allow mitigation designers and managers to manipulate the "stress" half to reduce the risk of invasion. Imposing stress on a mitigation site seems like a counterintuitive management strategy, but our research suggests that it could be used to increase native species richness and reduce the risk of invasion [6,70], particularly in the establishment phase of a mitigation site [71]. For example, a potentially low-cost stress induction method would be to use soil amendments with a high carbon-to-nitrogen ratio (e.g., sawdust, wood chips, etc.). High carbon/nitrogen materials have been shown to stimulate microbially mediated removal of nitrogen from the soil, thereby inducing a nitrogen limitation (stress) that could potentially favor native species over invaders [72,73]. Other examples include the addition of metal oxides such as aluminum and iron oxide to the soil. Metal oxides and other cationic forms with strong anion exchange capacities are known to complex with and immobilize phosphate, thus reducing its bioavailability and potentially inducing a phosphorus limitation (stress) [74,75]. Industrially manufactured forms such as alum are reasonably affordable in bulk and have been used for this same purpose in freshwater lakes [76].

5.1.3. Diversity Management

Although not specifically addressed in this study, a related project identified the potential role that species richness might play in combating invasive plants on stream restoration sites [51]. In establishing herbaceous plants on these sites, one potentially effective strategy that could preemptively exclude invaders would be to introduce a species-rich native seed mix at a higher seeding rate than typically prescribed for restoration projects [77]. This strategy, combined with the concept of limiting similarity [78], could help practitioners design planting plans that effectively target invasive species exclusion from the outset of a stream restoration project. Pairing tall, fast-growing native herbs with comparatively large woody species (see above) could allow for canopy closure throughout the growing season and subsequent reduction in light availability near the soil surface, where invasive species would benefit most from light exposure at early growth stages.

Author Contributions: Both authors contributed to the study conception and design. Material preparation, data collection, and analysis were performed by D.A.D. The majority of the manuscript was written by D.A.D., with some sections prepared and all sections reviewed by D.M.H. All authors have read and agreed to the published version of the manuscript.

Funding: This work was funded by a grant from the Resource Protection Group, Inc. (Grant No. RFP-08).

Data Availability Statement: The datasets generated during and/or analyzed during the current study are available from the authors upon request.

Acknowledgments: We would like to thank the Resource Protection Group, Inc., especially Mike Rolband and Jennifer Van Houten, for their support throughout the project. We also would like to express our gratitude to the following organizations for access to field sites and project-specific data: VHB, Stantec, WSSI, Falling Springs, and the City of Newport News. We would like to thank the following scientists for their valuable input on the sampling approach in the early phases of this project: Jim Perry, Wes Hudson, Bob Peet, Mary Kentula, Teresa Magee, Randy Chambers, Matthias Leu, and Drew Lamar.

Conflicts of Interest: The authors declare that the research was conducted in the absence of any commercial or financial relationships that could be construed as potential conflicts of interest. The funders had no role in the design of the study, the collection, analysis, or interpretation of data, the writing of the manuscript, or the decision to publish the results.

References

1. D'Antonio, C.; Meyerson, L.A. Exotic plant species as problems and solutions in ecological restoration: A synthesis. *Restor. Ecol.* **2002**, *10*, 703–713. [CrossRef]
2. Hunter, D.M.; DeBerry, D.A. Environmental drivers of plant invasion in wetland mitigation. *Wetlands* **2023**, *43*, 81. [CrossRef]
3. Richardson, D.M.; Holmes, P.M.; Esler, K.J.; Galatowitsch, S.M.; Stromberg, J.C.; Kirkman, S.P.; Pyšek, P.; Hobbs, R.J. Riparian vegetation: Degradation, alien plant invasions, and restoration prospects. *Divers. Distrib.* **2007**, *13*, 126–139. [CrossRef]
4. U.S. Army Corps of Engineers; Virginia Department of Environmental Quality. *Mitigation Banking Instrument 2018 Template*. 2018. Available online: https://www.nao.usace.army.mil/Media/Public-Notices/Article/1689864/cenao-wrr-mbi/ (accessed on 20 February 2021).
5. Kettenring, K.M.; Adams, C.R. Lessons learned from invasive plant control experiments: A systematic review and meta-analysis: Invasive plant control experiments. *J. Appl. Ecol.* **2011**, *48*, 970–979. [CrossRef]
6. Alpert, P.; Bone, E.; Holzapfel, C. Invasiveness, invasibility and the role of environmental stress in the spread of non-native plants. *Perspect. Plant Ecol. Evol. Syst.* **2000**, *3*, 52–66. [CrossRef]
7. Heffernan, K.; Engle, E.; Richardson, C. *Virginia Invasive Plant Species List*; Natural Heritage Technical Document 14–11; Virginia Department of Conservation and Recreation, Division of Natural Heritage: Richmond, VA, USA, 2014.
8. Shields, F.D.; Copeland, R.R.; Klingeman, P.C.; Doyle, M.W.; Simon, A. Design for stream restoration. *J. Hydraul. Eng.* **2003**, *129*, 575–584. [CrossRef]
9. Craine, J.M. *Resource Strategies of Wild Plants*; Princeton University Press: Princeton, NJ, USA, 2009; ISBN 978-1-4008-3064-0.
10. Naiman, R.J.; Décamps, H. The ecology of interfaces: Riparian zones. *Annu. Rev. Ecol. Syst.* **1997**, *28*, 621–658. [CrossRef]
11. Zedler, J.B.; Kercher, S. Causes and consequences of invasive plants in wetlands: Opportunities, opportunists, and outcomes. *Crit. Rev. Plant Sci.* **2004**, *23*, 431–452. [CrossRef]
12. Cummings, D.C.; Bidwell, T.G.; Medlin, C.R.; Fuhlendorf, S.D.; Elmore, R.D.; Weir, J.R. *Ecology and Management of Sericea Lespedeza*; Oklahoma Cooperative Extension Service: Durant, OK, USA, 2007.

13. Swearingen, J.; Slattery, B.; Reshetiloff, K.; Zwicker, S. *Plant Invaders of Mid-Atlantic Natural Areas*; National Park Service: Washington, DC, USA, 2010.
14. Brandon, A.L.; Gibson, D.J.; Middleton, B.A. Mechanisms for dominance in an early successional old field by the invasive non-native *Lespedeza cuneata* (Dum. Cours.) G. Don. *Biol. Invasions* **2004**, *6*, 483–493. [CrossRef]
15. Coykendall, K.E.; Houseman, G.R. *Lespedeza cuneata* invasion alters soils facilitating its own growth. *Biol. Invasions* **2014**, *16*, 1735–1742. [CrossRef]
16. Reichenborn, M.M.; Houseman, G.R.; Foster, B.L. Plant community recovery following sericea lespedeza (*Lespedeza cuneata*) removal: Testing for a soil legacy effect. *Restor. Ecol.* **2020**, *28*, 1192–1200. [CrossRef]
17. Kalburtji, K.L.; Mosjidis, J.A.; Mamolos, A.P. Litter dynamics of low and high tannin sericea lespedeza plants under field conditions. *Plant Soil* **1999**, *208*, 271–281. [CrossRef]
18. Eddy, T.A.; Davidson, J.; Obermeyer, B. Invasion dynamics and biological control prospects for sericea lespedeza in Kansas. *Great Plains Res.* **2003**, *13*, 217–230.
19. Steele, K.L.; Kabrick, J.M.; Dey, D.C.; Jensen, R.G. Restoring riparian forests in the Missouri Ozarks. *North. J. Appl. For.* **2013**, *30*, 109–117. [CrossRef]
20. Schierenbeck, K.A. Japanese honeysuckle (*Lonicera japonica*) as an invasive species: History, ecology, and context. *Crit. Rev. Plant Sci.* **2004**, *23*, 391–400. [CrossRef]
21. Surrette, S.B.; Brewer, J.S. Inferring relationships between native plant diversity and *Lonicera japonica* in upland forests in North Mississippi, USA. *Appl. Veg. Sci.* **2008**, *11*, 205–214. [CrossRef]
22. Hidayati, S.N.; Baskin, J.M.; Baskin, C.C. Dormancy-breaking and germination requirements of seeds of four *Lonicera* species (Caprifoliaceae) with underdeveloped spatulate embryos. *Seed Sci. Res.* **2000**, *10*, 459–469. [CrossRef]
23. Shelton, M.G.; Cain, M.D. Potential carry-over of seeds from 11 common shrub and vine competitors of loblolly and shortleaf pines. *Can. J. For. Res.* **2002**, *32*, 412–419. [CrossRef]
24. Miller, J.H. *Nonnative Invasive Plants of Southern Forests: A Field Guide for Identification and Control*; U.S. Department of Agriculture, Forest Service, Southern Research Station: Asheville, NC, USA, 2006; Volume 62.
25. Robertson, D.J.; Robertson, M.C.; Tague, T. Colonization dynamics of four exotic plants in a northern piedmont natural area. *Bull. Torrey Bot. Club* **1994**, *121*, 107. [CrossRef]
26. Skulman, B.W.; Mattice, J.D.; Cain, M.D.; Gbur, E.E. Evidence for allelopathic interference of Japanese honeysuckle (*Lonicera japonica*) to loblolly and shortleaf pine regeneration. *Weed Sci.* **2004**, *52*, 433–439. [CrossRef]
27. Ward, E.B.; Pregitzer, C.C.; Kuebbing, S.E.; Bradford, M.A. Invasive lianas are drivers of and passengers to altered soil nutrient availability in urban forests. *Biol. Invasions* **2020**, *22*, 935–955. [CrossRef]
28. Schweitzer, J.A.; Larson, K.C. Greater morphological plasticity of exotic honeysuckle species may make them better invaders than native species. *J. Torrey Bot. Soc.* **1999**, *126*, 15. [CrossRef]
29. Sweeney, B.W.; Czapka, S.J. Riparian forest restoration: Why each site needs an ecological prescription. *For. Ecol. Manag.* **2004**, *192*, 361–373. [CrossRef]
30. Barden, L.S. Invasion of *Microstegium vimineum* (Poaceae), an exotic, annual, shade-tolerant, C4 grass, into a North Carolina floodplain. *Am. Midl. Nat.* **1987**, *118*, 40. [CrossRef]
31. Oswalt, C.M.; Oswalt, S.N.; Clatterbuck, W.K. Effects of *Microstegium vimineum* (Trin.) A. Camus on native woody species density and diversity in a productive mixed-hardwood forest in Tennessee. *For. Ecol. Manag.* **2007**, *242*, 727–732. [CrossRef]
32. Warren, R.J.; Wright, J.P.; Bradford, M.A. The putative niche requirements and landscape dynamics of *Microstegium vimineum*: An invasive Asian grass. *Biol. Invasions* **2011**, *13*, 471–483. [CrossRef]
33. Adams, S.N.; Engelhardt, K.A.M. Diversity declines in *Microstegium vimineum* (Japanese Stiltgrass) patches. *Biol. Conserv.* **2009**, *142*, 1003–1010. [CrossRef]
34. Marshall, J.M.; Buckley, D.S. Influence of *Microstegium vimineum* presence on insect abundance in hardwood forests. *Southeast. Nat.* **2009**, *8*, 515–526. [CrossRef]
35. Judge, C.A.; Neal, J.C.; Derr, J.F. Response of Japanese stiltgrass (*Microstegium vimineum*) to application timing, rate, and frequency of postemergence herbicides 1. *Weed Technol.* **2005**, *19*, 912–917. [CrossRef]
36. Flory, S.L. Management of *Microstegium vimineum* invasions and recovery of resident plant communities. *Restor. Ecol.* **2010**, *18*, 103–112. [CrossRef]
37. Miller, N.P.; Matlack, G.R. Population expansion in an invasive grass, *Microstegium vimineum*: A test of the channelled diffusion model. *Divers. Distrib.* **2010**, *16*, 816–826. [CrossRef]
38. Ziska, L.H.; Tomecek, M.B.; Valerio, M.; Thompson, J.P. Evidence for recent evolution in an invasive species, *Microstegium vimineum*, Japanese stiltgrass. *Weed Res.* **2015**, *55*, 260–267. [CrossRef]
39. DeMeester, J.E.; de Richter, D.B. Restoring restoration: Removal of the invasive plant *Microstegium vimineum* from a North Carolina wetland. *Biol. Invasions* **2010**, *12*, 781–793. [CrossRef]
40. DeBerry, D.A.; Perry, J.E. Vegetation dynamics across a chronosequence of created wetland sites in Virginia, USA. *Wetl. Ecol Manag.* **2012**, *20*, 521–537. [CrossRef]
41. DeBerry, D.; Beisch, D.; Crayosky, T. Integrated wetland design: Sustainability within the landscape. *Geo-Strat.—Geo Inst. ASCE* **2004**, *5*, 23–25.

42. R Core Team. *R: A Language and Environment for Statistical Computing*; R Core Foundation for Statistical Computing: Vienna, Austria, 2020. Available online: https://scirp.org/reference/referencespapers?referenceid=3064798 (accessed on 20 January 2021).
43. Kindt, R.; Coe, R. *Tree Diversity Analysis: A Manual and Software for Common Statistical Methods for Ecological and Biodiversity Studies*; World Agroforestry Centre: Nairobi, Kenya, 2005.
44. Borcard, D.; Gillet, F.; Legendre, P. *Numerical Ecology with R*; Springer: New York, NY, USA, 2011; ISBN 978-1-4419-7975-9.
45. Harrell, F.E.; Dupont, C. Hmisc—Harrell Miscellaneous. R package version 4.4-2. Available online: https://CRAN.R-project.org/package=Hmisc (accessed on 20 January 2021).
46. Oksanen, J.; Blanchet, F.G.; Friendly, M.; Kindt, R.; Legendre, P.; McGlinn, D.; Minchin, P.R.; O'Hara, R.B.; Simpson, G.L.; Solymos, P.; et al. Vegan: Community Ecology Package. 2020. Available online: https://cran.r-project.org/web/packages/vegan/vegan.pdf (accessed on 20 January 2021).
47. Legendre, P.; Legendre, L. *Numerical Ecology*; Elsevier: Amsterdam, The Netherlands, 2012.
48. ter Braak, C.J.F. Canonical Correspondence Analysis: A new eigenvector technique for multivariate direct gradient analysis. *Ecology* **1986**, *67*, 1167–1179. [CrossRef]
49. Legendre, P.; Gallagher, E.D. Ecologically meaningful transformations for ordination of species data. *Oecologia* **2001**, *129*, 271–280. [CrossRef] [PubMed]
50. McCune, B.; Grace, J.B.; Urban, D.L. *Analysis of Ecological Communities*; MjM Software Design: Gleneden Beach, OR, USA, 2002; Volume 28.
51. DeBerry, D.A.; Hunter, D.M. Impacts of invasive plants on native vegetation communities in wetland and stream mitigation. *Biology* **2024**, *13*, 275. [CrossRef] [PubMed]
52. Hupp, C.R.; Osterkamp, W.R. Riparian vegetation and fluvial geomorphic processes. *Geomorphology* **1996**, *14*, 277–295. [CrossRef]
53. Hupp, C.R. Hydrology, Geomorphology and vegetation of coastal plain rivers in the Southeastern USA. *Hydrol. Process.* **2000**, *14*, 2991–3010. [CrossRef]
54. Bendix, J.; Hupp, C.R. Hydrological and geomorphological impacts on riparian plant communities. *Hydrol. Process.* **2000**, *14*, 2977–2990. [CrossRef]
55. Hughes, F.M.R.; Colston, A.; Mountford, J.O. Restoring riparian ecosystems: The challenge of accommodating variability and designing restoration trajectories. *Ecol. Soc.* **2005**, *10*, 12. [CrossRef]
56. Lockwood, J.L.; Hoopes, M.F.; Marchetti, M.P. *Invasion Ecology*; John Wiley & Sons: Hoboken, NJ, USA, 2013.
57. Mulholland, P.J. Regulation of nutrient concentrations in a temperate forest stream: Roles of upland, riparian, and instream processes. *Limnol. Oceanogr.* **1992**, *37*, 1512–1526. [CrossRef]
58. Brady, N.C.; Weil, R.R. *The Nature and Properties of Soils*; Prentice Hall International editions, 14. ed.; Prentice Hall International: London, UK, 2008; ISBN 978-0-13-243189-7.
59. Mohanty, S.; Dash, R. Chemistry of waterlogged soils. In *Wetlands Ecology and Management*; Gopal, B., Turner, R.E., Wetzel, R.G., Whigham, D.F., Eds.; National Institute of Ecology and International Scientific Publications: Jaipur, India, 1982.
60. Haering, K.C.; Abaye, A.O.; Basden, T.J.; Beegle, D.B.; Binford, G.D.; Daniels, W.L.; Duiker, S.W.; Mullins, G.L.; Taylor, R.W. *The Mid-Atlantic Nutrient Management Handbook*; Cooperative State Research, Education, and Extension Service, USDA; Virginia Cooperative Extension Service: Petersburg, VA, USA, 2015.
61. Gibson, D.J.; Spyreas, G.; Benedict, J. Life history of *Microstegium vimineum* (Poaceae), an invasive grass in Southern Illinois. *J. Torrey Bot. Soc.* **2002**, *129*, 207. [CrossRef]
62. Nord, A.N.; Mortensen, D.A.; Rauschert, E.S.J. Environmental factors influence early population growth of Japanese stiltgrass (*Microstegium vimineum*). *Invasive Plant Sci. Manag.* **2010**, *3*, 17–25. [CrossRef]
63. Ehrenfeld, J.G.; Kourtev, P.; Huang, W. Changes in soil functions following invasions of exotic understory plants in deciduous forests. *Ecol. Appl.* **2001**, *11*, 1287–1300. [CrossRef]
64. Pisula, N.L.; Meiners, S.J. Relative allelopathic potential of invasive plant species in a young disturbed woodland 1. *J. Torrey Bot. Soc.* **2010**, *137*, 81–87. [CrossRef]
65. Hudson, H.W., III; Perry, J.E. Development of Woody Ecological Performance Standards for Created/Restored Forested Wetlands; Piedmont Wetlands Research Program. 2018. Available online: https://resourceprotectiongroup.org/wp-content/uploads/2020/08/2018wssi-3addendumfinalreport.pdf (accessed on 1 March 2021).
66. McLeod, K.W.; Reed, M.R.; Nelson, E.A. Influence of a willow canopy on tree seedling establishment for wetland restoration. *Wetlands* **2001**, *21*, 395–402. [CrossRef]
67. Spencer, D.R.; Perry, J.E.; Silberhorn, G.M. Early secondary succession in bottomland hardwood forests of Southeastern Virginia. *Environ. Manag.* **2001**, *27*, 559–570. [CrossRef]
68. Diggins, T.P. A 300-year successional sequence in an Eastern United States riparian hardwood forest 1. *J. Torrey Bot. Soc.* **2013**, *140*, 65–88. [CrossRef]
69. Marks, C.O.; Yellen, B.C.; Nislow, K.H. Species distributions on successional and flooding gradients in Connecticut River floodplain forests. *Northeast. Nat.* **2021**, *28*, 577–602. [CrossRef]
70. Bryson, C.T.; Carter, R. Biology of pathways for invasive weeds 1. *Weed Technol.* **2004**, *18*, 1216–1220. [CrossRef]
71. DeBerry, D.A.; Perry, J.E. Using the Floristic Quality concept to assess created and natural wetlands: Ecological and management implications. *Ecol. Indic.* **2015**, *53*, 247–257. [CrossRef]

72. Perry, L.G.; Galatowitsch, S.M.; Rosen, C.J. Competitive control of invasive vegetation: A native wetland sedge suppresses *Phalaris arundinacea* in carbon-enriched soil. *J. Appl. Ecol.* **2004**, *41*, 151–162. [CrossRef]
73. Iannone, B.V.; Galatowitsch, S.M. Altering light and soil N to limit *Phalaris arundinacea* reinvasion in sedge meadow restorations. *Restor. Ecol.* **2008**, *16*, 689–701. [CrossRef]
74. Stauffer, A.L.; Brooks, R.P. Plant and soil responses to salvaged marsh surface and organic matter amendments at a created wetland in Central Pennsylvania. *Wetlands* **1997**, *17*, 90–105. [CrossRef]
75. Hogan, D.M.; Jordan, T.E.; Walbridge, M.R. Phosphorus retention and soil organic carbon in restored and natural freshwater wetlands. *Wetlands* **2004**, *24*, 573–585. [CrossRef]
76. Douglas, G.B.; Hamilton, D.P.; Robb, M.S.; Pan, G.; Spears, B.M.; Lurling, M. Guiding principles for the development and application of solid-phase phosphorus adsorbents for freshwater ecosystems. *Aquat. Ecol.* **2016**, *50*, 385–405. [CrossRef]
77. Yannelli, F.A.; Karrer, G.; Hall, R.; Kollmann, J.; Heger, T. Seed density is more effective than multi-trait limiting similarity in controlling grassland resistance against plant invasions in mesocosms. *Appl. Veg. Sci.* **2018**, *21*, 411–418. [CrossRef]
78. Laughlin, D.C. Applying trait-based models to achieve functional targets for theory-driven ecological restoration. *Ecol. Lett.* **2014**, *17*, 771–784. [CrossRef]

forests

MDPI

Article

The Process of Patchy Expansion for Bamboo (*Phyllostachys edulis*) at the Bamboo–Broadleaf Forest Interface: Spreading and Filling in Order

Xiaoxia Zeng [1], Huitan Luo [2], Jian Lu [2], Xianglong Zhu [2], Yaoli He [2], Chao Gong [1], Zewen Ren [1], Dongmei Huang [3], Qingni Song [1] and Qingpei Yang [1,*]

[1] College of Forestry, Jiangxi Agricultural University, Nanchang 330045, China; 18174009873@163.com (X.Z.); taitouciciyan@163.com (C.G.); renzewen2021@163.com (Z.R.); songqingni@jxau.edu.cn (Q.S.)

[2] Administration of Jiangxi Qiyunshan Nature Reserve, Ganzhou 341000, China; luohuitan2024@163.com (H.L.); lujian202403@163.com (J.L.); zhuxianglong1973@163.com (X.Z.); heyaoli2024@163.com (Y.H.)

[3] School of Humanities and Public Administration, Jiangxi Agricultural University, Nanchang 330045, China; hdmae99@163.com

* Correspondence: qingpeiyang99@163.com

Abstract: Bamboo (*Phyllostachys edulis*) expansion to native adjacent forests has become an increasingly serious problem; however, expansion patterns of bamboo are still lacking research, especially at a community scale. Quantitative research on bamboo expansion patterns plays a significant role in understanding the bamboo expansion process, as well as expansion prevention and control. We analyzed the change in expansion pattern, expansion index, and expansion rate of bamboo in the bamboo-broadleaf transition zone sample plots, specifically from 2017 to 2021 and from the bamboo forest (representing the late stage of bamboo expansion) to the bamboo expansion front (representing the early stage of bamboo expansion). We found that the expansion of bamboo is a patchy expansion, including inner filling patch, boundary expanding patch, transboundary leaping patch, expansion–infill mixed patch, and stationary patch. From the early stage of bamboo expansion (year 2017 and bamboo expansion front) to the late stage of bamboo expansion (year 2021 and bamboo forest), the type of bamboo expansion patches transitioned from boundary expanding patch to inner filling patch and boundary expansion–inner infilling mixed patch. Additionally, the expansion rate of bamboo showed a declining trend. From 2017 to 2021, the bamboo forest (position of 0–20 m) and expansion front (position of 60–80 m) declined by 0.53 m/2a and 0.47 m/2a, respectively. Our research reveals that bamboo expansion exhibits a patchy expanding process, characterized by a sequence of "first spreading outward and then filling inward", whether viewed from the type of expansion pattern or the expansion rate. This process involves continuous plaque addition, expansion, merger, and filling to complete the expansion of a bamboo population. These findings provide valuable insights into the process of bamboo expansion and have important implications for the management and control of bamboo forests.

Keywords: patchy types; expansion pattern; expansion index; expansion rate; shape index; bamboo (*Phyllostachys edulis*)

1. Introduction

Bamboo (*Phyllostachys edulis*), a giant strong woody grass, has an important ecological, economic, and cultural value [1]; therefore, bamboo has undergone extensive cultivation. However, bamboo, as a typical expansive species [2], possesses a robust rhizome system and grows rapidly [3–5]. These intrinsic strengths enable it to encroach unchecked into adjacent forests [6,7]. The 9th National Forest Inventory in China confirmed the expansion of bamboo-dominant areas, with the total bamboo area increasing from 3.00 Mhm^2 in 1972 to 6.41 Mhm^2 in 2018 [8]. Bamboo spreads uncontrollably to the surrounding forests, which seriously hinders the growth of trees and the regeneration of new saplings, resulting in a simplified community structure [9] and reduced diversity of the plant [7]. Additionally, the expansion of bamboo has increased soil pH [10], changed terrestrial water and nutrient cycles [11,12], affected soil microbial community structures [13], and so on. In recent years, considerable research has been carried out on the process of bamboo expansion [14,15] and the mechanism [14,16,17] and the expansion effects [4,18,19]. However, there is a lack of research on the bamboo expansion processes and patterns, which hinders the understanding of the bamboo expansion process and the research on expansion prevention and control.

Some studies have been carried about the process and pattern of bamboo expansion. For example, Kiyoshi et al. (1996) [20], Isagi and Torii (1997) [21], and Ding et al. (2006) [22] analyzed the expansion distance and patch number of bamboo in Kyoto, Japan and Hangzhou, China respectively. These studies made use of aerial photographs and remote sensing data to conduct the analyses. They indicate that the prevalence of bamboo expansion and the expansion pattern at the landscape scale (10 km^2), but at the community scale, bamboo primarily expands through vegetative reproduction (clonal growth) [23,24]. Specifically, mature (mother) culms provide nutrients to the rhizomes, fostering the formation of new culms (Figure 1), and via culms, the plant moves forward step by step and block by block and gradually occupies the space to realize its expansion. Therefore, to gain a better understanding of the bamboo expansion process, uncover its expansion mechanism, and develop effective control measures, we should understand the pattern and process of bamboo expansion from the community scale on the basis of individuals.

Figure 1. Diagram of vegetative reproduction (clonal growth) of bamboo. Black indicates the current stage of the mother culms and shoots, and gray represents the future formation of new culms and shoots (The distance between the shoot and the mother culm typically ranges from a few decimeters to a few meters).

The bamboo-broadleaf forest transition zone refers to the area where bamboo forests and evergreen broadleaf forests meet. The two ends of this zone are evergreen broadleaf forests or bamboo forests, and the middle is the expansion front. In zones where bamboo expansion is most active, community structure changes most rapidly, and the transition from a broadleaf forest to a bamboo forest is most significant, so it has become an important area for bamboo expansion research [25–27]. In this area, due to increased forest fragmentation caused by factors such as clonal growth of bamboo forests and community disturbance, bamboo demonstrates an epitaxial expansion process [25]. Bamboo expansion involves both spatial expansion and an increase in the number of culms [5,15,22], which includes spatial expansion scales and expansion patterns. The expansion scale is determined by the number and area of new culm patches and mature culm patches, which could be quantitatively described by the expansion index.

The expansion pattern refers to the spatial combination of the mature culm and the new culm, which is of great significance to understanding the bamboo expansion process. Isagi and Torii [21] propose from a landscape-scale perspective that new culms are consistently at the forefront of bamboo forests, expanding from the inside out. However, our perspective suggests that bamboo may first spread outward and then fill inward. Because the growth of new culms requires the mother culms to provide a lot of water and nutrients via underground rhizomes [28], this inevitably brings nutrient and energy losses to the mother bamboo plant. These losses primarily include the consumption of material transfer between new and mature culms, the consumption of new culm growth, and the consumption of rhizome growth [29]. As the distance between the new culms and the mature culms increases, the energy required to transport nutrients and maintain the growth of rhizomes increases. Therefore, when bamboo expands outward and occupies a new plot, it will fill the interior with new culms, thereby reducing the cost of nurturing new culms. Once the new culms are established, they can provide nourishment back to the mother culms and the next generation, enabling further outward expansion to occupy additional plot.

Subtropical natural reserves are mostly suitable habitats for bamboo, and the expansion, through vegetative reproduction (clonal growth), of bamboo in these areas is becoming increasingly severe [22,30–32]. Among them, the Qiyunshan National Nature Reserve (QNNR) located in Chongyi County, Jiangxi Province, China, is known as the hometown of bamboo. The area covered by bamboo forests in this reserve accounts for approximately 17% of the total protected area [33]. It serves as an ideal experimental platform for studying the process and patterns of bamboo expansion at the community scale and for quantifying its expanding rate. Therefore, we selected a representative bamboo–broadleaf transition zone in the QNNR as a research site. Our aim was to test the hypothesis that the pattern of bamboo expansion is first spread outward and then filled inward. Testing this hypothesis illuminates the process and pattern of bamboo expansion into broadleaf forests, offering crucial reference points for controlling bamboo forest expansion.

2. Material and Methods

2.1. Study Species

Bamboo (*Phyllostachys edulis*), a tree-like bamboo species of Poaceae with height of 10–20 m, is naturally distributed in warm temperate and subtropical China [34]. And bamboo has high temperature and humidity requirements for its surrounding environment and is primarily distributed in moist hilly areas between 300 and 800 m above sea level [1]. Bamboo is a type of clonal plant, composed of above-ground bamboo culms and underground rhizomes. The process of bamboo's nutritional reproduction (clonal growth) involves mature (mother) culms absorbing nutrients to support the underground rhizomes. These rhizomes elongate and give rise to shoots. When the shoots develop into new culms, the new generation of culms, in turn, absorb nutrients to supply the underground rhizomes. This process continues in an infinite cycle (Figure 1) [1]. According to the definition of clonal plants [35], the process of bamboo expansion is essentially its clonal growth, involving the increase in the number of culms and the expansion of the population's spatial distribution. Moreover, the process of shoot and rhizome generation is usually not completed within the same year but occurs alternately (in cycles of two years). In one year, there is a large-scale recruitment shoot (referred to as an "on-year"), and in the next year, there is rhizome generation (referred to as a "off-year").

2.2. Study Site

The study was conducted at the Qiyunshan National Nature Reserve (QNNR) in Jiangxi Province, China (113°54′–114°07′ E, 25°41′–25°54′ N, Figure 2), with an elevation of around 2061.3 m. It belongs to the humid subtropical monsoon climate zone, which is characterized by a warm climate and abundant rainfall [36,37]. The mean annual precipitation ranges from 1522 to 1660 mm, and the mean annual temperature is between 18.0 and ~18.4 °C [38]. QNNR's zonal vegetation is evergreen broadleaf forest, but large numbers of bamboo forests remain in some areas.

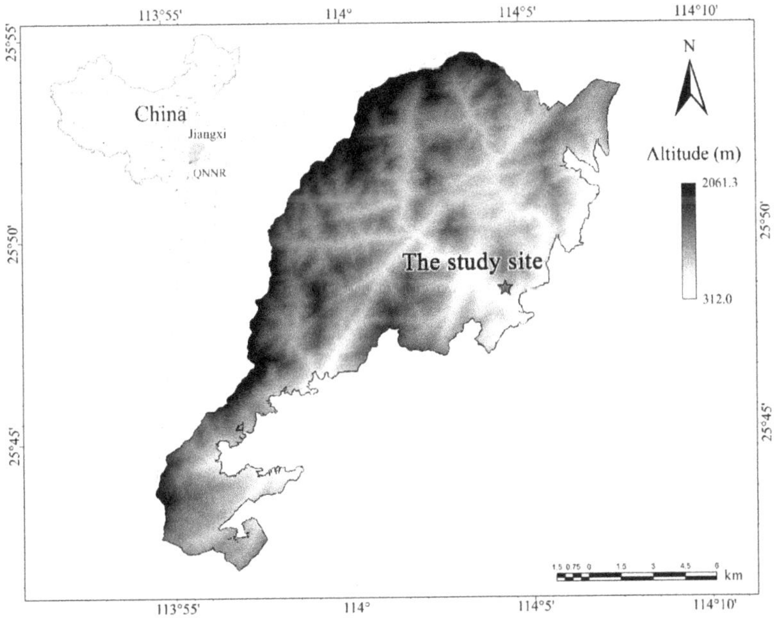

Figure 2. The study location and experimental plot. Locations of Jiangxi province (grey area) and QNNR (red point) are pictured (top left), as well as experimental plot (red star) in the QNNR area.

2.3. Survey Methods and Forest Plot Data

In April 2017, we selected a typical bamboo–broadleaf transition zone plot, which expanded along the slope, according to the CTFS (Centre for Tropical Forest Science) sample plot survey method, at QNNR (Figure 2). The size of plot was 100 m × 50 m, and the coordinate in the lower left corner was NE. Adjacent grid method was used to divide the plot into 200 sub-plots of 5 m × 5 m. We investigated the DBH, total height, and spatial coordinates (x, y) of each plant including trees and bamboo culms with a DBH greater than 1.0 cm in the plot. Following that, in August 2017, 2018, 2019, 2020, and 2021, we investigated the DBH, total height, and spatial coordinates (x and y) of each newly grown culm within the plot. There was a total of 945 culms in this study (Figure 3).

2.4. Data Processing

We selected the population information of bamboo in the major years of 2017, 2019, and 2021. Generally, the bamboo has a biennial bearing cycle, and in QNNR, the singular year is the on-year for it, and the even year is the off-year. In terms of the small number of culms recorded in the off-year, we incorporated them into the previous on-year. We calculated the number of bamboo patches, patch areas, and expansion rates through analyzing the spatial position of all bamboos in the plot. We determined the expansion mode, shape index of patch, and expansion distance based on the spatial relationship between new culms and mother culms to divide the expansion spatial pattern.

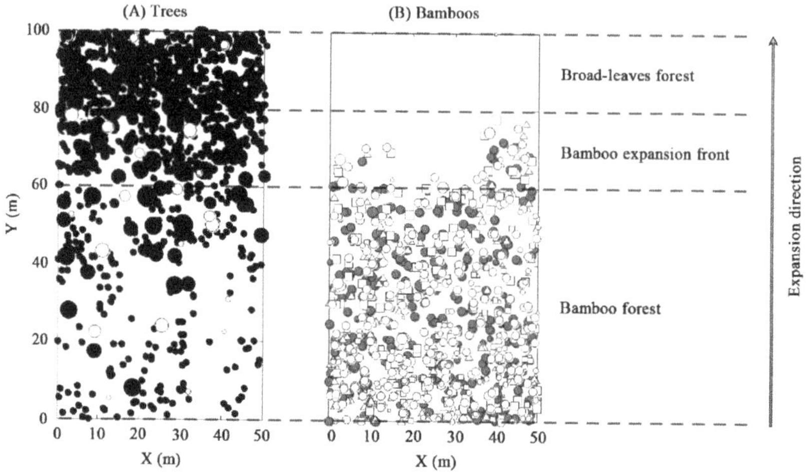

Figure 3. Spatial distribution of trees (**A**) and bamboos (**B**) in bamboo–broadleaf forest transition zone. Tree DBH is indicated by the size of the circle; solid black circles represent alive trees, and hollow circles represent dead trees (**A**). The solid gray circle represents culms established before 2017; the triangle represents culms established in 2017; the square represents culms established in 2019, and the hollow circle represents culms established in 2021 (**B**).

2.4.1. Patch Determination

According to the spatial coordinates of the bamboo culms, the *K*-means clustering analysis method was employed to partition the patches. The *K*-means clustering algorithm divides space into *K* non-overlapping clusters (referred to as patches in this study) and classifies each observation (coordinates of culms in this study) to the nearest centroid [39,40]. Due to the correlation between the number and size of patches and the distance between bamboo culms, this study used the Euclidean distance method to calculate the distance between each pair of culms in the plot. The culms with distances less than or equal to 4 m were merged to form a cluster, based on the crown radius of culms (usually ~2 m) [41].

Afterward, we used the convex hull to obtain the coordinates of the edge vertices of the convex polygon (when the number of culms within the patch is $\geq$3), determining the shape of the patch. The convex hull operates by taking a set of points X and two reference points v_i and v_j. The algorithm starts from point v_i and finds its K-nearest neighbors. From each neighbor, we searched the point v_k such that $\overrightarrow{v_i v_k}$ has the smallest angle (θ_{min}) with respect to line $\overrightarrow{v_i v_j}$. This procedure was repeated until $v_k = v_j$ [42].

$$\theta_{min} = \underset{k}{min}\left[cos^{-1}\left(\frac{\overrightarrow{v_i v_k} \cdot \overrightarrow{v_i v_j}}{\| \overrightarrow{v_i v_k} \| \cdot \| \overrightarrow{v_i v_j} \|}\right)\right] \quad (1)$$

2.4.2. Patch Characteristics

We utilized the poly area function to calculate the area of patch, and other indexes about patch including shape index (*SI*), elongation index (*G*), and fractal dimension (*FD*). The calculation formula is as follows:

Shape index of patch (*SI*):

$$SI = \frac{P_i}{2\sqrt{\pi A_i}} \quad (2)$$

Elongation index of patch (*G*):

$$G = \frac{P_i}{\sqrt{A_i}} \quad (3)$$

Fractal dimension of patch (*FD*):

$$FD = \frac{1}{N}\sum_{i=1}^{N} ln(\frac{P_i}{4lnA_i}) \tag{4}$$

where P_i (m) and A_i (m^2) in these three functions are the perimeter and area of *i*-patch, respectively. *N* represents the total number of patches.

2.4.3. Expansion Patch Types

According to the spatial position between new culms and mother culms, the expansion pattern of bamboo can be distributed as following five types: inner filling patch (IF), boundary expanding patch (BE), transboundary leaping patch (TL), expansion–infill mixed patch (BM), and stationary patch (ST, Figure 4).

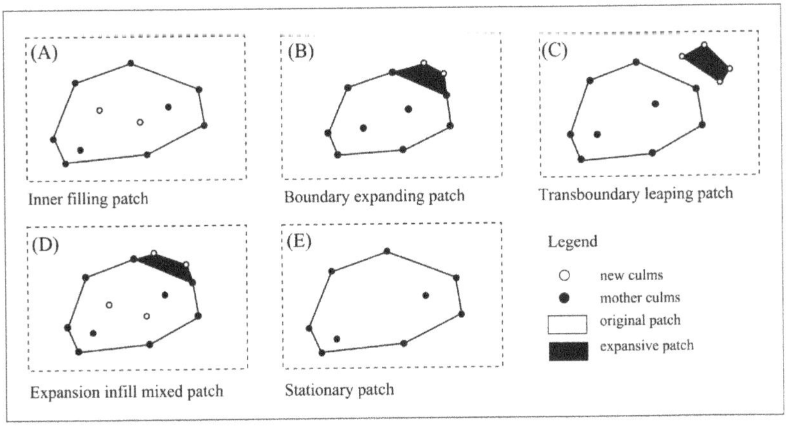

Figure 4. The patch types of bamboo expansion into neighboring forest. (**A**) Inner filling patch (IF), (**B**) Boundary expanding patch (BE), (**C**) Transboundary leaping patch (TL), (**D**) Boundary expansion–inner infilling mixed patch (BM), and (**E**) Stationary patch (ST).

IF: new culms grow inside the patch enclosed by the mother culms, and the patch area remains unchanged and the density of culms increases (Figure 4A). BE: new culms grow at the edge of original patch, resulting in an increase in the area and the density remains unchanged at original patch (Figure 4B). TL: new culms grow outside of the original patch, which form a new patch (or individual plant) separate from original patch (Figure 4C). BM: new culms grow both inside and at the edge of original patch resulting in the patch area and increased culm density (Figure 4D). ST: no new culms appear in or around the original patch (Figure 4E).

2.4.4. Expansion Indexes and Expansion Rate Calculations

We calculated the expansion indexes, expansion distances, and expansion rates of the bamboo area using the functions *EI*, D_{AB}, and *Ea*, respectively. In function (5), *EI* is the expansion index of bamboo patch, with a value range of $[-1, 1]$. When there were no new culms or no mother culms in the patch, the *EI* equals -1 and 1, respectively. N_n is the number of new culms, and N_m is the number of mother culms. In function (6), D_{AB} is distance between culms A and B of bamboo; x_A and y_A are the *x* and *y* coordinate values of culm A in plot; and x_B and y_B are the *x* and *y* coordinate values of culm B in plot. In function (7), *Ea* is the rate of area expansion; S_0 and S_{0+t} are the sums of bamboo patch areas at the times of *0* and *t*.

Expansion indexes:

$$EI = \frac{N_n - N_m}{N_n + N_m} \quad (5)$$

Expansion distance:

$$D_{AB} = \sqrt{(x_A - x_B)^2 + (y_A - y_B)^2} \quad (6)$$

Expansion rate of area:

$$Ea = \left[\left(\frac{S_{0+t}}{S_0}\right)^{\frac{1}{t}} - 1\right] \times 100\% \quad (7)$$

2.5. Statistical Analysis

We used one-way analysis of variance to examine variations in the distribution patterns, patch characteristics, and expansion indexes of bamboo at different times and different locations. We used Pearson correlation coefficient to analyze the correlation among various parameters of bamboo expansion, including year, x-axis, y-axis, total culm number, number of mother culms, number of new culms, area, density of new culms, total density, expansion index, maximum expansion distance, and mean expansion distance. This statistical analysis was performed using SPSS 22.0 statistical and Matlab 2018a software, and graphs were created using Origin 2022 and Matlab 2018a software.

3. Results

3.1. Spatial Distribution of Trees and Bamboo

In the evergreen broadleaf forest at the bamboo–broadleaf forest transition zone (Figure 3 top), there are numerous live trees and a small number of dead ones. In the middle of the transition zone, the evergreen broadleaf forest gradually transformed into a mixed forest of bamboo and broadleaf trees due to the expansion of bamboo. This transformation resulted in an increase in the number of culms and dead trees (Figure 3 mid). As the evergreen broadleaf forest was transformed into a bamboo forest, consequently, the density of the culms, particularly those aging culms, significantly increased, and fewer live trees persisted there (Figure 3 bottom).

3.2. Expansive Patch Parameters

Bamboo exhibited a patchy expansion pattern at the bamboo-broadleaf forest transition zone 2017 to 2021 (Figure 5). Over the course of five years (three on-years and two off-years), the number of bamboo patches increased from 48 to 52, representing a growth of 8.33%. Additionally, the total and average areas of bamboo patches experienced a growth of 53.78% and 41.95%, respectively. Specifically, the values increased from 1145.28 m^2 to 1761.24 m^2 for the total area and from 23.86 m^2 to 33.87 m^2 for the average area (Table 1). Furthermore, the perimeter of the patch increased from 19.90 m to 22.74 m, and the average number of culms in each patch increased from 10.44 ind./m^2 to 18.17 ind./m^2 (Table 2).

Table 1. The changes in patch parameters for bamboo from 2017 to 2021.

Indexes of Patches	2017	2019	2021
Number of patches (ind.)	48	49	52
Total area of patches (m^2)	1145.28	1600.32	1761.24
Area of maximum patches (m^2)	68.82	66.24	132.81
Mean area of patches (m^2)	23.86 ± 11.58 ^{b}	32.71 ± 17.26 ^{a}	33.87 ± 20.80 ^{a}
Expansion rates (%)	15.89	18.21	4.91
Index of maximum patches (%)	6.01	4.14	7.55

Different letters mean significant differences in mean area of patches among the three years at $\alpha = 0.05$.

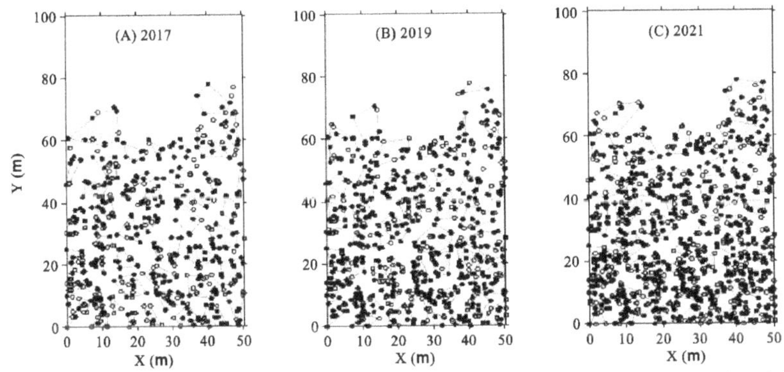

Figure 5. The patterns and processes of bamboo expansion between 2017 and 2021. The solid black dots represent mother culms, while the hollow circles represent newly sprouted culms.

Table 2. The amount, size, and shape for the patches of bamboo in bamboo–broadleaf forest transition zone from 2017 to 2021.

Indexes	2017	2019	2021
Perimeter of patches (m)	19.90 ± 4.50 [b]	22.10 ± 6.05 [ab]	22.74 ± 5.74 [a]
Culm number per patch (ind./patch)	10.44 ± 3.12 [c]	15.76 ± 5.19 [b]	18.17 ± 5.14 [a]
Culm density(ind./m^2)	0.54 ± 0.30 [a]	0.63 ± 0.35 [a]	0.73 ± 0.79 [a]
Mean expansion distance (m)	3.89 ± 0.86 [a]	3.92 ± 1.02 [a]	3.97 ± 0.98 [a]
Shape index (*SI*)	1.21 ± 0.14 [a]	1.14 ± 0.06 [a]	1.21 ± 0.55 [a]
Elongation index (*G*)	4.28 ± 0.51 [a]	4.05 ± 0.22 [a]	4.29 ± 1.96 [a]
Fractal dimension (*FD*)	1.05 ± 0.07 [a]	1.01 ± 0.05 [b]	1.01 ± 0.05 [b]

Different letters indicate significant differences in culm number per patch, culm density, mean expansion distance, shape index, elongation index, and fractal dimension among the three years at $\alpha = 0.05$.

3.3. Changes in Expansion Patch Types

The type of bamboo expansion patches shifted from the boundary expanding patch (BE) to the inner filling patch (IF) and the boundary expansion–inner infilling mixed patch (BM) from 2017 to 2021 (Figures 4 and 5). The percent of IF and BM to the total patch increased from 10% and 39% to 21.15% and 60.03%, respectively; however, the percentage of BE to the total patch decreased from 38.75% to 9.62% (Figure 6A).

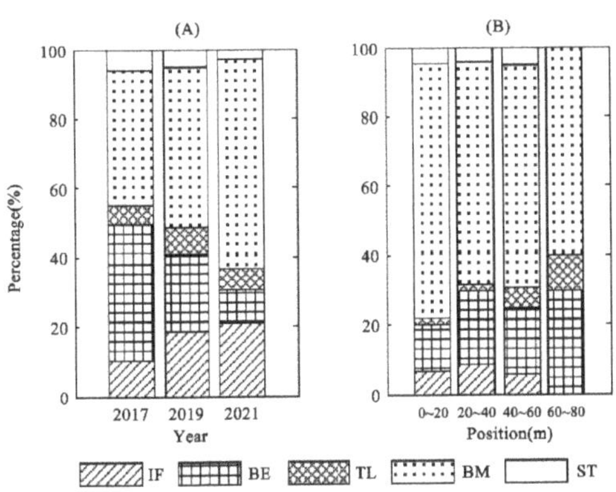

Figure 6. The fraction of bamboo patches according to year (**A**) and transition zone position (**B**) in bamboo–broadleaf forest transition zone. IF—Inner filling patch, BE—Boundary expanding patch, TL—Transboundary leaping patch, BM—Boundary expanding and inner filling mixed patch, and ST—stationary patch.

In terms of the position of the bamboo and broadleaf forest transition zone, within the bamboo forest position (0~20 m), BM emerged as the dominant type of bamboo patch (Figure 6B), comprising 72.95% of the patches, followed by the BE (13.51%), IF (7.12%), ST (4.47%), and TL (1.95%). However, in the bamboo expansion front position (60~80 m), BM decreased to 60.05%, the IF and ST disappeared, and the BE and TL increased to 30.36% and 9.59% (Figure 6B).

3.4. Expansion Indexes and Expansion Rates

The expansion indexes and the expansion rates of bamboo demonstrated a declining trend from 2017 to 2021 (Table 3). In 2017, the expansion indexes for the 0~20 m and 60–80 m ranges of the bamboo–broadleaf forest transition zone were recorded as −0.18 ± 0.36 and 0.17 ± 0.23, respectively. However, by 2021, these indexes had decreased to −0.39 ± 0.20 and −0.28 ± 0.24. The average expansion indexes decreased from −0.25 ± 0.35 to −0.40 ± 0.17, while the average expansion rates increased from 1.55 ± 0.83 m/2a to 1.01 ± 0.36 m/2a.

Table 3. The expansion indexes and rates of bamboo in bamboo–broadleaf forest transition zone from 2017 to 2021.

Type	Position (m)	2017	2019	2021	Average
Expansion index	0~20	−0.18 ± 0.36 Ba	−0.37 ± 0.30 Bb	−0.39 ± 0.20 Bb	−0.32 ± 0.30 B
	20~40	−0.35 ± 0.27 Cb	−0.09 ± 0.31 Aa	−0.43 ± 0.15 Bc	−0.27 ± 0.30 B
	40~60	−0.35 ± 0.37 Cb	−0.12 ± 0.24 Aa	−0.44 ± 0.12 Bc	−0.30 ± 0.29 B
	60~80	0.17 ± 0.23 Aa	−0.12 ± 0.01 Ab	−0.28 ± 0.24 Ac	−0.10 ± 0.26 A
	Average	−0.25 ± 0.35 a	−0.21 ± 0.31 a	−0.40 ± 0.17 b	−0.29 ± 0.29
Expansion rate (m/2a)	0~20	1.44 ± 0.43 Ba	1.16 ± 0.50 Bab	0.91 ± 0.33 Bb	1.15 ± 0.47 B
	20~40	1.47 ± 0.97 Ba	1.64 ± 0.47 ABa	0.92 ± 0.28 Bb	1.37 ± 0.70 AB
	40~60	1.74 ± 0.80 ABa	1.55 ± 0.56 ABb	1.13 ± 0.34 ABc	1.46 ± 0.63 AB
	60~80	2.11 ± 2.98 Aa	2.04 ± 0.47 Aa	1.64 ± 0.34 Ab	1.89 ± 1.27 A
	Average	1.55 ± 0.83 a	1.44 ± 0.55 a	1.01 ± 0.36 b	1.32 ± 0.64

Different capital letters indicate significant differences in expansion indexes and expansion rates between different positions, and different small letters indicate significant differences in expansion indexes and expansion rates between different years at $\alpha = 0.05$.

The expansion indexes of bamboo showed an increasing trend along the direction of bamboo expansion (from bamboo forest to bamboo expansion front), between 2017 and 2021 (Table 3). At a 0~20 m range of the bamboo–broadleaf forest transition zone (bamboo forest), the expansion indexes in 2017, 2019, and 2021 were observed to be −0.18 ± 0.36, −0.37 ± 0.30, and −0.39 ± 0.20, but at a 60~80 m range of the transition zone (bamboo expansion front), these indexes increased to be 0.17 ± 0.23, −0.12 ± 0.01, and −0.28 ± 0.24, respectively. The average expansion indexes increased from −0.32 ± 0.32 to −0.1 ± 0.26, and the average expansion rates increased from 1.15 ± 0.47 m/2a to 1.89 ± 1.27 m/2a as the bamboo expanded from a 0 to 20 m to 60~80 m position.

3.5. The Relationship among the Variables for Bamboo Expansion Parameters

The x-axis is not related to all parameters of bamboo expansion, and the y-axis (from bamboo forest to broadleaf forest) is highly and significantly positively correlated with patch area, maximum expansion distance, and average expansion distance, while it is highly and significantly negatively correlated with the density of new culms and total culms (Table 4). This indicates that bamboo primarily expands along the y-axis direction, and the expansion distance gradually increases as the y-axis values increase, while the number of culms decrease with the increase in the y-axis.

Table 4. The relationship among the variables for bamboo expansion in the bamboo–broadleaf forest transition zone.

Variables	YE	X	Y	TCN	NMCs	NNCs	A	DNCs	TD	EI	MXED
X	0.07										
Y	−0.01	0.04									
TCN	0.58 **	−0.06	0.00								
NMC	0.56 **	−0.07	−0.04	0.92 **							
NNC	0.27 **	0.00	0.07	0.59 **	0.23 **						
A	0.20 *	0.07	0.59 **	0.43 **	0.36 **	0.34 **					
DNC	−0.06	−0.14	−0.43 **	−0.19	−0.36 **	0.28 **	−0.56 **				
TD	0.27 **	−0.09	−0.58 **	0.06	0.25 **	−0.36 **	−0.56 **	0.25 **			
EI	−0.21 **	0.00	0.11	−0.25 **	−0.58 **	0.59 **	−0.02	0.57 **	−0.51 **		
MXED	−0.25 **	0.07	0.31 **	−0.20	−0.40 **	0.32 **	0.19 *	0.13	−0.55 **	0.64 **	
MED	−0.35 **	0.02	0.28 **	−0.39 **	−0.49 **	0.03	0.09	0.03	−0.49 **	0.49 **	0.81 **

YE: year; X: x-axis; Y: y-axis; TCN: total culm number; NMCs: number of mother culms; NNCs: number of new culms; A: area; DNCs: density of new culms; TD: total density; EI: expansion index; MXED: maximum expansion distance; MED: mean expansion distance. "*" or "**" indicate significant or highly significant correlation at $\alpha = 0.05$.

4. Discussion

Bamboo plays an important role in the construction of the forestry economy [43,44]. However, the expansion of bamboo also gives rise to a range of ecological challenges [12,25,45], such as simplified community structure [9], reduced plant diversity [7], and changed terrestrial water and nutrient cycles [11]. Therefore, it is essential for us to comprehend the processes and mechanisms of bamboo expansion in order to provide guidance for the management and control of bamboo expansion. Our results show that the expansion process of bamboo into the surrounding broadleaf forest follows a patch-based expansion process. From 2017 to 2021, in the 5000 m^2 bamboo–broadleaf forest transition zone plots, the number of bamboo patches increased from 48 to 52, and the patch area increased from 1145.28 m^2 to 1761.24 m^2, which aligns with the findings from the macro-scale results (Table 1). For instance, Isagi and Torii (1997) conducted a comparison of aerial photographs and land-use maps of Tanabe Town, Kyoto, Japan, from 1953, 1975, and 1985 [21]. They observed that the number of bamboo patches in the area increased from 24 to 112 and then to 174, with an average area of 0.67 hm^2, 1.56 hm^2, and 1.75 hm^2, respectively. Furthermore, Ding et al. (2006) applied remote sensing and geographic information system technology to study the bamboo forest patches in a 4284 hm^2 area of Tianmushan in Zhejiang, China [22]. Their findings revealed an increase in the number of bamboo forest patches from 3 to 6, with the largest patch area expanding by nearly 54 times. This shows that the expansion of bamboo into the surrounding areas is characterized by patch expansion at both the macroscopic remote sensing scale and the ground community scale.

It is worth mentioning that this research primarily focuses on revealing the patchy expansion pattern at the community scale. According to the spatial relationship between new and mother culms, the expansion patches of bamboo can be divided into five types (Figure 3): inner filling patch (IF), boundary expanding patch (BE), transboundary leaping patch (TL), boundary expansion–inner infilling mixed patch (BM), and stationary patch (ST). Our findings indicate that the type of bamboo expansion patches shifted from BE in the early stage of bamboo expansion (bamboo expansion front and year 2017) to BM and IF in late stage of bamboo expansion (bamboo forest and year 2021), a shift which fully reflects the expansion process and pattern of bamboo, where it first spreads outward and then fills inward (Figure 5). The expansion distance and the correlation between the expansion index provide additional perspectives that support this rule. Specifically, the expansion distance at the expansion front is greater compared to that within the bamboo forest (Table 3). Moreover, there is a significant positive correlation between the expansion distance and the y-axis (representing the transition from bamboo forest to broadleaf forest) (Table 4; Figure 3).

This shift in patch type may be attributed to the expansion costs involved. On the one hand, bamboo is a clonal plant with physiological integration characteristics, and the mature culm provides nutrients and energy (investment cost of mature bamboo) for the new culms through underground rhizomes [28]. However, this process inevitably leads to nutrient and energy losses for the mature culms [29]. Therefore, as bamboo expands outward and occupies a new plot, it fills the interior with new culms, thereby enhancing the absorption capacity of the bamboo patches or cloned fragments for water, nutrients, and other nutrients. The expansion outward occurs only when the accumulated resources within the patches can adequately support the growth and development of the next generation.

On the other hand, when the number of culms reaches a certain threshold (where the available resources within a patch are insufficient to support the growth of all the culms), it triggers intensified intraspecific competition among them [46]. This heightened competition leads to stunted growth or even the death of some culms, resulting in the cloning fragment directly losing all nutrients in the dead culm (incurring cost losses). However, when the bamboo population reaches a critical mass, expanding outward to establish new culms serve multiple purposes. Firstly, it alleviates intraspecific competition [47]. Secondly, the newly established culms absorb resources from their surroundings (new plots), which are then transferred back to the mother culms as nourishment once the new culms mature [48]. This compensates for the resources consumed by the mother bamboos during the establishment of new culms and the growth of bamboo rhizomes. Therefore, when culm numbers accumulate to a certain extent, expanding outward not only avoids cost losses, but also yields benefits for the cloned fragments or bamboo patches. In essence, bamboo expansion does not occur in a strictly sequential manner based on age. Rather, it follows a strategic process of step-by-step expansion, filling, and subsequent expansion and filling again. Some scholars, in their studies on other clonal plants, also believe that the reproduction of cloned plants has a certain strategic control. This strategy manages the position, size and other characteristics of new (daughter) ramets, enabling them to more effective utilization of space, nutrients, and other resources [49–51].

In addition, our findings demonstrate that along the bamboo–broadleaf transition zone, from bamboo forest to broadleaf forest, the average expansion index of bamboo increased from −0.32 to −0.10, and the average expansion rate increased from 1.15 to 1.89 m/2a (Table 4). This rate is lower compared to the expansion rates reported by Okutomi et al. (1996) for bamboo forests in the suburbs of Kyoto, Japan, which found that the expansion rates of bamboo into shrublands, wasteland, and deciduous broadleaf forests from 1961 to 1987 were 3.0 m/a, 2.5 m/a, and 1.8 m/a, respectively [20]. The discrepancy in expansion rates can be attributed to the differences in stand types. The study conducted by Okutomi focused on shrublands, wasteland and deciduous broadleaf forests, and our study area was located in a protected area, with secondary evergreen broadleaf forests. Shrublands and wasteland have a low canopy density and more niche space, which creates conditions for the encroachment of bamboos. On the other hand, evergreen broadleaf trees tend to be more competitive compared to deciduous broadleaf trees due to their lower resource demand and higher utilization efficiency, such as of light and water [52,53]. Consequently, the rate at which bamboo invades shrublands, wastelands, and deciduous broadleaf forests is higher compared to its expansion rate into evergreen broadleaf forests. Zheng et al. (2023) found that the process of bamboo expansion into evergreen broadleaf forests is hindered by competition from broadleaf trees in the Tianmu Mountain Nature Reserve [54].

5. Conclusions

Bamboo expansion is a process in which bamboo proliferates and expands based on the existing spatial pattern. Our research results indicate that this process occurs in the form of patches, and the types of patches change with the expansion time, showing a "first spreading and then filling" pattern. Our study reveals, for the first time, the specific process and pattern of bamboo expansion, providing a reference for the prevention and control of bamboo plant expansion. However, the reasons why bamboo adopts this expansion pattern are speculative based on previous research findings. Further exploration is needed to understand the mechanisms behind bamboo's adoption of the "first spreading and then filling" expansion strategy. Additionally, building upon this study, it is crucial to include research on the growth patterns of other clonal plants, especially rhizomatous clonal plants, which often exhibit expansive or invasive behaviors. This will contribute to a better understanding of the control of expansive plants (invasive plants).

Author Contributions: Conceptualization, Q.Y.; Software, X.Z. (Xiaoxia Zeng), C.G. and Q.Y.; Formal analysis, H.L.; Investigation, X.Z. (Xiaoxia Zeng), H.L., Jian Lu, X.Z. (Xianglong Zhu), Y.H., C.G., Z.R., D.H. and Q.S.; Resources, H.L. and J.L.; Data curation, X.Z. (Xiaoxia Zeng), J.L., X.Z. (Xianglong Zhu) and Y.H.; Writing—original draft, X.Z. (Xiaoxia Zeng); Writing—review & editing, Q.Y.; Funding acquisition, Q.S. and Q.Y. All authors have read and agreed to the published version of the manuscript.

Funding: This research was supported by the National Natural Science Foundation of China (32060319, 32260335, 32360801, and 41807028).

Data Availability Statement: All data generated or analyzed during this study are included in this published article.

Acknowledgments: We are grateful to the Administration of the Qiyunshan Nature Reserve for providing an experimental platform. We also thank other members of our team for their assistance with the field experiments.

Conflicts of Interest: The authors declare no conflicts of interest.

References

1. Zhang, L. Bamboo Expansion Into Adjacent Ecosystems. In *Bamboo Expansion: Processes, Impacts, and Management*; Springer: Berlin/Heidelberg, Germany, 2023; pp. 19–37.
2. Bai, S.; Wang, Y.; Conant, R.T.; Zhou, G.; Xu, Y.; Wang, N.; Fang, F.; Chen, J. Can native clonal moso bamboo encroach on adjacent natural forest without human intervention? *Sci. Rep.* **2016**, *6*, 31504. [CrossRef]
3. Lima, R.A.; Rother, D.C.; Muler, A.E.; Lepsch, I.F.; Rodrigues, R.R. Bamboo overabundance alters forest structure and dynamics in the Atlantic Forest hotspot. *Biol. Conserv.* **2012**, *147*, 32–39. [CrossRef]
4. Buziquia, S.T.; Lopes, P.V.F.; Almeida, A.K.; de Almeida, I.K. Impacts of bamboo spreading: A review. *Biodivers. Conserv.* **2019**, *28*, 3695–3711. [CrossRef]
5. Xu, Q.F.; Liang, C.F.; Chen, J.H.; Li, Y.C.; Qin, H.; Fuhrmann, J.J. Rapid bamboo invasion (expansion) and its effects on biodiversity and soil processes. *Glob. Ecol. Conserv.* **2020**, *21*, e00787. [CrossRef]
6. Ying, W.; Jin, J.; Jiang, H.; Zhang, X.; Lu, X.; Chen, X.; Zhang, J. Satellite-based detection of bamboo expansion over the past 30 years in Mount Tianmushan, China. *Int. J. Remote Sens.* **2016**, *37*, 2908–2922. [CrossRef]
7. Chen, X.; Chen, X.; Huang, S.; Fang, D. Impacts of Moso bamboo (*Phyllostachys pubescens*) invasion on species diversity and aboveground biomass of secondary coniferous and broad-leaved mixed forest. *Front. Plant Sci.* **2022**, *13*, 1001785. [CrossRef]
8. Li, Y.; Feng, P. Analysis of Chinese bamboo resources based on the ninth national forest resources inventory. *World Bamboo Ratt.* **2019**, *17*, 45–48.
9. Zhang, H.; Xue, J. Spatial pattern and competitive relationships of moso bamboo in a native subtropical rainforest community. *Forests* **2018**, *9*, 774. [CrossRef]
10. Ouyang, M.; Tian, D.; Pan, J.; Chen, G.; Su, H.; Yan, Z.; Yang, Q.; Ji, C.; Tang, Z.; Fang, J. Moso bamboo (*Phyllostachys edulis*) invasion increases forest soil pH in subtropical China. *Catena* **2022**, *215*, 106339. [CrossRef]
11. Shinohara, Y.; Otsuki, K. Comparisons of soil-water content between a Moso bamboo (*Phyllostachys pubescens*) forest and an evergreen broadleaved forest in western Japan. *Plant Species Biol.* **2015**, *30*, 96–103. [CrossRef]
12. Song, Q.N.; Ouyang, M.; Yang, Q.P.; Lu, H.; Yang, G.Y.; Chen, F.S.; Shi, J.M. Degradation of litter quality and decline of soil nitrogen mineralization after moso bamboo (*Phyllostachys pubscens*) expansion to neighboring broadleaved forest in subtropical China. *Plant Soil* **2016**, *404*, 113–124. [CrossRef]
13. Cai, X.; Lin, Z.; Penttinen, P.; Li, Y.; Li, Y.; Luo, Y.; Yue, T.; Jiang, P.; Fu, W. Effects of conversion from a natural evergreen broadleaf forest to a Moso bamboo plantation on the soil nutrient pools, microbial biomass and enzyme activities in a subtropical area. *For. Ecol. Manag.* **2018**, *422*, 161–171. [CrossRef]
14. Yang, Q.P.; Yang, G.Y.; Song, Q.N.; Shi, J.M.; Ouyang, M.; Qi, H.Y.; Fang, X.M. Ecological studies on bamboo expansion: Process, consequence and mechanism. *Chin. J. Plant Ecol.* **2015**, *39*, 110–124.
15. Mertens, B.; Hua, L.; Belcher, B.; Ruiz-Pérez, M.; Maoyi, F.; Xiaosheng, Y. Spatial patterns and processes of bamboo expansion in Southern China. *Appl. Geogr.* **2008**, *28*, 16–31. [CrossRef]
16. Suzuki, S.; Nakagoshi, N. Expansion of bamboo forests caused by reduced bamboo-shoot harvest under different natural and artificial conditions. *Ecol. Res.* **2008**, *23*, 641–647. [CrossRef]

17. Song, X.; Peng, C.; Zhou, G.; Gu, H.; Li, Q.; Zhang, C. Dynamic allocation and transfer of non-structural carbohydrates, a possible mechanism for the explosive growth of Moso bamboo (*Phyllostachys heterocycla*). *Sci. Rep.* **2016**, *6*, 25908. [CrossRef]
18. Zou, G.; Wu, B.; Chen, B.; Yang, Y.; Feng, Y.; Huang, J.; Liu, Y.; Murray, P.J.; Liu, W. What Are the Effects of Moso Bamboo Expansion into Japanese Cedar on Arbuscular Mycorrhizal Fungi: Altering the Community Composition Rather than the Diversity. *J. Fungi* **2023**, *9*, 273. [CrossRef]
19. Xiao, H.; Liu, J.; Li, F. Both alpha and beta diversity of nematode declines in response to moso bamboo expansion in south China. *Appl. Soil Ecol.* **2023**, *183*, 104761. [CrossRef]
20. Okutomi, K.; Shinoda, S.; Fukuda, H. Causal analysis of the invasion of broad-leaved forest by bamboo in Japan. *J. Veg. Sci.* **1996**, *7*, 723–728. [CrossRef]
21. Isagi, Y.; Torii, A. Range Expansion and Its Mechanisms in a Naturalized Bamboo Species, *Phyllostachys pubescens*, in Japan. *J. Sustain. Forest.* **1997**, *6*, 127–141. [CrossRef]
22. Ding, L.X.; Wang, Z.L.; Zhou, G.M.; Du, Q.Z. Monitoring *Phyllostachys pubescens* stands expansion in National Nature Reserve of Mount Tianmu by remote sensing. *J. Zhejiang A&F Univ.* **2006**, *23*, 297–300.
23. Xu, Q.F.; Liang, C.F.; Chen, J.H.; Li, Y.C.; Qin, H.; Fuhrmann, J.J. Retracted article: Running bamboo invasion in native and non-native regions worldwide. *Biol. Invasions* **2017**, *19*, 3459. [CrossRef]
24. Lin, Y.T.; Tang, S.L.; Pai, C.W.; Whitman, W.B.; Coleman, D.C.; Chiu, C.Y. Changes in the soil bacterial communities in a cedar plantation invaded by moso bamboo. *Microb. Ecol.* **2014**, *67*, 421–429. [CrossRef]
25. Ouyang, M.; Yang, Q.; Chen, X.; Yang, G.; Fang, X. Effects of the expansion of *Phyllostachys edulis* on species composition, structure and diversity of the secondary evergreen broad-leaved forests. *Biodivers. Sci.* **2016**, *24*, 649–657. [CrossRef]
26. Liu, X.; Fang, P.; Xiong, Y.; Peng, Q.; Yu, Z.; Luan, F.; Song, Q.; Fang, X.; Yang, Q.; Liu, J. Assessment of the influence of bamboo expansion on Si pools and fluxes in a disturbed subtropical evergreen broadleaved forest. *Catena* **2022**, *213*, 106136. [CrossRef]
27. Liu, J.; Yang, Q.P.; Song, Q.N.; Yu, D.K.; Yang, G.Y.; Qi, H.Y.; Shi, J.M. Strategy of fine root expansion of *Phyllostachys pubescens* population into evergreen broad-leaved forest. *Chin. J. Plant Ecol.* **2013**, *37*, 230. [CrossRef]
28. Li, R.; Werger, M.J.A.; Kroon, H.D.; During, H.J.; Zhong, Z.C. Interactions Between Shoot Age Structure, Nutrient Availability and Physiological Integration in the Giant Bamboo *Phyllostachys pubescens*. *Plant Biol.* **2000**, *2*, 437–446. [CrossRef]
29. Yu, F.H.; Dong, M.; Zhang, C.Y. Intraclonal resource sharing and functional specialization of ramets in response to resource heterogeneity in three stoloniferous herbs. *Acta Bot. Sin.* **2002**, *44*, 468.
30. Liu, X.; Siemann, E.; Cui, C.; Liu, Y.; Guo, X.; Zhang, L. Moso bamboo (*Phyllostachys edulis*) invasion effects on litter, soil and microbial PLFA characteristics depend on sites and invaded forests. *Plant Soil* **2019**, *438*, 85–99. [CrossRef]
31. Pan, J.; Liu, Y.; Niu, J.; Fang, H.; Feng, Y.; Bai, T.; Zhang, M.; Deng, W.; Siemann, E.; Zhang, L. Moso bamboo expansion reduced soil N_2O emissions while accelerated fine root litter decomposition: Contrasting non-additive effects. *Plant Soil* **2022**, 1–15. [CrossRef]
32. Xu, D.W.; Liu, J.F.; He, Z.S.; Zheng, S.Q. Community species diversity after *Phyllostachys edulis* expansion to *Cunninghamia lanceolata* forest. *J. For. Environ.* **2019**, *39*, 37–41.
33. Liu, X.M.; Guo, Y.R.; Liu, R.L. *The Collection of Scientific Exploration of Jiangxi Qiyunshan Nature Reserve*; China Forestry Publishing House: Beijing, China, 2010.
34. Yi, T.P.; Shi, J.Y.; Ma, L.S.; Wang, H.T.; Yang, L. *Iconographia Bambusoidearum Sinicarum*; Science Press: Beijing, China, 2008.
35. Dong, M.; Yu, F.H. Clonal Plant. In *Clonal Plant Ecology*; Science Press: Beijing, China, 2011; pp. 5–9.
36. De, Y.D.; Sheng, L.H.; Lei, Y.X.; Jun, L.H.; Min, C.H.; Yun, L.Z.; Hua, T.T. Investigation and analysis of the mammal resources in Qiyunshan Nature Reserve, Jiangxi province. *J. Cent. South Univ. For. Technol.* **2009**, *29*, 45–50.
37. Huang, X.F.; Tu, Y.G.; Zhang, T.; Chen, J.W.; Sun, Z.Y.; Lu, H.J.; Shan, J.H.; Wang, Z.R.; Huang, S.L. Investigation and diversity of birds in winter in Qiyunshan Nature Reserve, Jiangxi Province. *Chin. J. Zool.* **2008**, *43*, 86–94.
38. Chen, Y.L.; Lin, Y.L.; Su, M.S.; Zhang, L.P.; Zhang, Y.; Chen, H.M. Flora of Macrofungal in Qiyun Mountain National Nature Reserve of Jiangxi. *J. Fungal Res.* **2019**, *17*, 26–34.
39. Kazapoe, R.W.; Amuah, E.E.Y.; Dankwa, P.; Ibrahim, K.; Mville, B.N.; Abubakari, S.; Bawa, N. Compositional and source patterns of potentially toxic elements (PTEs) in soils in southwestern Ghana using robust compositional contamination index (RCCI) and k-means cluster analysis. *Environ. Chall.* **2021**, *5*, 100248. [CrossRef]
40. Xu, H.; Croot, P.; Zhang, C. Discovering hidden spatial patterns and their associations with controlling factors for potentially toxic elements in topsoil using hot spot analysis and K-means clustering analysis. *Environ. Int.* **2021**, *151*, 106456. [CrossRef] [PubMed]
41. Zhou, X.; Zhou, Y.; Zhang, X.; Sharma, R.P.; Guan, F.; Fan, S.; Liu, G. Two-level mixed-effects height to crown base model for moso bamboo (*Phyllostachys edulis*) in Eastern China. *Front. Plant Sci.* **2023**, *14*, 1095126. [CrossRef]
42. Chau, A.L.; Li, X.; Yu, W. Convex and concave hulls for classification with support vector machine. *Neurocomputing* **2013**, *122*, 198–209. [CrossRef]
43. Fei, B. Economic value and research significance of Moso bamboo. In *The Moso Bamboo Genome*; Springer: Berlin/Heidelberg, Germany, 2021; pp. 1–12.
44. Liu, W.; Hui, C.; Wang, F.; Wang, M.; Liu, G. Review of the Resources and Utilization of Bamboo in China. In *Bamboo-Current and Future Prospects*; IntechOpen: London, UK, 2018; pp. 133–142.

45. Liu, W.; Liao, L.; Liu, Y.; Wang, Q.; Murray, P.J.; Jiang, X.; Zou, G.; Cai, J.; Zhao, X. Effects of *Phyllostachys pubescens* expansion on underground soil fauna community and soil food web in a Cryptomeria japonica plantation, Lushan Mountain, subtropical China. *J. Soils Sediments* **2021**, *21*, 2212–2227. [CrossRef]
46. Qiu, J.X.; Tang, M.P.; Lou, M.H.; Shen, L.F.; Zou, Y.Q. Analysis of *Phyllostachys edulis* stand spatial structure and competition based on an improved Hegyi model. *Acta Ecol. Sin.* **2016**, *36*, 1058–1065.
47. Sandhu, H.; Shi, P.; Yang, Q. Intraspecific spatial niche differentiation: Evidence from *Phyllostachys edulis*. *Acta Ecol. Sin.* **2013**, *33*, 287–292. [CrossRef]
48. Dong, M. *Clonal Plant Ecology*; Science Press: Beijing, China, 2011.
49. Pitelka, L.F. Application of the-3/2 power law to clonal herbs. *Am. Nat.* **1984**, *123*, 442–449. [CrossRef]
50. Ye, X.H.; Yu, F.H.; Dong, M. A trade-off between guerrilla and phalanx growth forms in *Leymus secalinus* under different nutrient supplies. *Annals of Botany* **2006**, *98*, 187–191. [CrossRef]
51. Chen, X.S.; Xie, Y.H.; Deng, Z.M.; Li, F.; Hou, Z.Y. A change from phalanx to guerrilla growth form is an effective strategy to acclimate to sedimentation in a wetland sedge species *Carex brevicuspis* (Cyperaceae). *Flora* **2011**, *206*, 347–350. [CrossRef]
52. Kitajima, K. Relative importance of photosynthetic traits and allocation patterns as correlates of seedling shade tolerance of 13 tropical trees. *Oecologia* **1994**, *98*, 419–428. [CrossRef]
53. Llambí, L.; Fontaine, M.; Rada, F.; Saugier, B.; Sarmiento, L. Ecophysiology of dominant plant species during old-field succession in a high tropical Andean ecosystem. *Arct. Antarct. Alp. Res.* **2003**, *35*, 447–453. [CrossRef]
54. Zheng, A.; Lv, J. Spatial patterns of bamboo expansion across scales: How does Moso bamboo interact with competing trees? *Landsc. Ecol.* **2023**, *38*, 3925–3943. [CrossRef]

MDPI AG
Grosspeteranlage 5
4052 Basel
Switzerland
Tel.: +41 61 683 77 34

Forests Editorial Office
E-mail: forests@mdpi.com
www.mdpi.com/journal/forests

www.ingramcontent.com/pod-product-compliance
Lightning Source LLC
Chambersburg PA
CBHW061555210926
51850CB00009B/2109
9783725871100